AVANT-PROPOS

A l'heure actuelle, les travaux de recherche et de développement, qui ont pour but de permettre d'évaluer les performances des systèmes potentiels d'évacuation des déchets radioactifs, constituent l'un des principaux aspects de la gestion de déchets radioactifs. En particulier, les recherches et études in situ sont devenues une composante essentielle des programmes nationaux et internationaux consacrés à l'évaluation des dépôts dans les formations géologiques, notamment à l'appréciation de la méthode, à la caractérisation et à la sélection des sites et à l'aménagement des dépôts. Dans l'actuel programme du Comité AEN de la gestion des déchets radioactifs (RWMC), les activités relatives aux recherches et études in situ consacrées aux dépôts dans les formations géologiques jouent une rôle notable. Afin d'axer les travaux sur les sujets les mieux appropriés dans ce domaine, un "Groupe consultatif sur les recherches et études in situ visant l'évacuation dans les formations géologiques" (ISAG) qui a été créé par le RWMC en 1986, a été chargé de formuler des orientations et des avis en la matière. L'ISAG est composé de personnalités qui occupent des positions clés dans les programmes nationaux de gestion des déchets, notamment dans les établissements publics et privés chargés de ces programmes, dans les centres de recherche in situ et dans les organismes réglementaires. Ce Groupe a pour mission de concourir à la coordination des activités de recherche, d'étude et de démonstration menées in situ dans les pays Membres, en fournissant une tribune pour des échanges d'informations et en assurant la planification d'initiatives conjointes au plan international.

L'une des premières tâche de l'ISAG a consisté à passer en revue les besoins actuels dans le domaine des recherches et études in situ. Au premier rang des sujets ainsi définis figure la réponse de l'excavation, autrement dit les effets potentiels qui peuvent être induits par la construction et l'aménagement d'un dépôt de déchets radioactifs dans les formations géologiques. Plus précisément, on a noté que des contraintes résiduelles, un fluage ou une subsidence potentiels et des fissures induites entraînant une augmentation des possibilités d'écoulement de l'eau souterraine, constituent des phénomènes qu'il faut analyser afin d'en déterminer l'incidence sur les performance d'un dépôt du point de vue de la sûreté et sur la conception des barrières ouvragées dont il convient de doter ce dernier. Tous ces aspects ont ultérieurement été examinés au cours d'une réunion de travail organisée par l'AEN au Canada, avec le co-patronage de l'Energie Atomique du Canada, Limitée (EACL).

Le présent compte rendu reproduit les communications présentées à cette réunion de travail, accompagnées d'un résumé et des conclusions établis par le Secrétariat de l'AEN sur la base des principaux résultats dégagés par les présidents de sessions et des débats qui ont eu lieu pendant la réunion. Les opinions exprimées sont celles des auteurs et n'engagent en rien les pays Membres de l'OCDE.

TABLE OF CONTENTS

TABLE DES MATIERES

OPENING OF THE WORKSHOP

SEANCE D'OUVERTURE

Session I - Séance I

OVERVIEW OF RESEARCH AT UNDERGROUND LABORATORIES
ON EXCAVATION RESPONSE AND ASSOCIATED ACTIVITIES

APERCU DES RECHERCHES EFFECTUEES DANS LES LABORATOIRES
SOUTERRAINS CONCERNANT LES EFFETS DE L'EXCAVATION ET LES
ACTIVITIES ASSOCIEES

Chairman - Président: Dr. K.W. DORMUTH (AECL, Canada)

Session II(a) - Séance II(a)

EXCAVATION RESPONSE IN CRYSTALLINE ROCK
- OVERVIEW OF STUDIES

EFFETS DE L'EXCAVATION SUR LES ROCHES CRISTALLINES
- PRESENTATION DES ETUDES

Chairman - Président: Mr. R.A. ROBINSON
(Battelle, United States)

Session II(b) - Séance II(b)

EXCAVATION RESPONSE IN CRYSTALLINE ROCK
- SPECIFIC GEOTECHNICAL STUDIES

EFFETS DE L'EXCAVATION DANS LES ROCHES CRISTALLINES
- ETUDES GEOTECHNIQUES SPECIFIQUES

Chairman - Président: Mr. A. BARBREAU (DPT/CEA, France)

Session III - Séance III

EXCAVATION RESPONSE IN SALT

EFFETS DE L'EXCAVATION SUR LES FORMATIONS SALINES

Chairman - Président: Dr. R.V. MATALUCCI
(SNL, United States)

Session IV - Séance IV

EXCAVATION RESPONSE IN ARGILLACEOUS ROCK

EFFETS DE L'EXCAVATION SUR LES FORMATIONS ARGILEUSES

Chairman - Président: Mr. R.H. HEREMANS
(NIRAS/ONDRAF, Belgium)

Session V - Séance V

DISCUSSION AND CONCLUSIONS

DISCUSSIONS ET CONCLUSIONS

Chairman - Président: Dr. K.W. DORMUTH (AECL, Canada)

Based upon the discussions, a summary and conclusions were prepared by
the NEA Secretariat and the session Chairmen, and are provided in the
Executive Summary and Conclusions on page 11 of these proceedings.

Un bilan et des conclusions ont été établis sur la base des discussions
par le Secrétariat de l'AEN et les présidents de séances. Ils figurent dans le
résumé et les conclusions à la page 17 du compte rendu.

EXECUTIVE SUMMARY AND CONCLUSIONS

BACKGROUND

     Currently, one of the most important areas of radioactive waste
management is research and development directed towards assessment of the
performance of potential radioactive waste disposal systems. In particular,
in-situ research and investigations have become an integral and essential part
of national and international programmes for evaluation of geological
repositories, including assessment of the concept, site characterisation and
selection, and repository development. At the first meeting of the NEA In-Situ
Advisory Group (ISAG), a review was conducted of current needs in the area of
in-situ research and investigations. A primary topic which was identified
concerned excavation response, that is, the effects that may be induced in the
surrounding rock by the construction and development of a geological
repository for radioactive waste. Specifically, it was noted that residual
stresses, creep or subsidence, and induced fractures possibly leading to an
increased potential for groundwater flow, are phenomena which require analysis
to determine their influence on the safety performance of a repository, and on
the design of engineered barriers for the repository.

     The boundary between the "engineered vault" (comprising waste form,
canister, backfill and sealing material) and the host geological environment
(geosphere) is conceptually one of the least well understood components of the
system. Questions relating to the implications of excavation effects and
damage in this zone for the safety performance and the design of repositories
apply in varying degrees to all geological disposal concepts. The answers are
by no means simple, even though much is already known from experience gained
from tunnelling and mining operations and, more recently, from geotechnical
and rock mechanic investigations at in-situ research facilities in Belgium,
Canada, the Federal Republic of Germany, Sweden, Switzerland and the United
States. Relevant aspects of excavation response which require examination,
through fora such as this international workshop, include: a) the importance
of excavation response to the design and safety of a repository; b) the type
of in-situ tests and measurements conducted; and c) the conceptual and
mathematical models that have been developed. The validation of the latter is
particularly important if they are to be used in support of performance
assessment.

WORKSHOP FORMAT

     The objectives of the workshop were to:

a) review the influence of excavation effects on the engineering design
   and the safety performance of repositories;

b) develop conclusions and recommendations on ways of accounting for excavation effects in clay, salt and crystalline geological media; and

c) provide a forum for discussions between scientists conducting in-situ studies and measurements and those developing predictive models for use in safety assessments of repositories.

Thirty papers were presented on several aspects of excavation response in crystalline (granite, schist, tuff), salt and clay geological media. The workshop was a very useful forum for presenting the "state-of-the-art" of investigations (experiments and modelling) into excavation effects. The opening session presented overviews of research into excavation response at underground laboratories sited in clay, crystalline and salt media. The remainder of the workshop examined, in more detail, various aspects of research in these three media. Particular emphasis was placed on theoretical analysis and modelling, in-situ measurement techniques, and methods available to reduce excavation effects. Summaries by Session Chairmen are provided at the end of each session in the proceedings. A final session provided for discussions on future needs in the area of excavation response, including strategies for model validation and accounting for excavation effects in repository design, and on possible related topics to be addressed at a follow-up NEA workshop on sealing of radioactive waste repositories.

CONCLUSIONS AND RECOMMENDATIONS

Five key considerations emerged from the presentations and discussions held at the workshop:

a) the relevance or importance of excavation response and damage to performance assessment;

b) the extent to which fundamental processes, or the physics of excavation response, are reliably understood;

c) future needs for development of instrumentation and measurement methods;

d) appropriate strategies for model development and validation; and

e) implications for repository design and engineering.

1. The Relevance or Importance of Excavation Response and Damage to Performance Assessment

Some differences in priorities emerged amongst the different geological media with respect to the relevance of excavation effects to performance assessment. For clay repositories, it was concluded that excavation effects, if properly managed, would have no adverse influence on long-term safety performance, due to the "self-healing" characteristics of clay. For salt repositories, while some short-term safety assessment concerns may exist, excavation effects were not viewed as an issue for long-term performance assessments. However, it was noted that a need exists for long-term safety analyses to provide a better understanding of the combined effects of excavation, pressure, heat and radiation as they influence long-term creep phenomena. For crystalline respositories (granite, tuff), excavation effects were not considered relevant to short-term performance; however, it was felt that the implications for long-term safety had yet to be fully analysed.

It was concluded that a need exists for more direct integration of excavation response studies with repository performance modelling. It was recommended, therefore, that enhanced efforts be made to have system performance assessment modellers more directly involved in the initial planning of excavation-effects tests, and to provide advice, possibly in an international forum, on what information requirements exist to enable adequate consideration of excavation effects and damage in performance assessment. Emphasis needs to be placed on the development of more quantitative models to demonstrate the insignificance of excavation effects with respect to the long-term performance of a repository, or otherwise to provide fundamental data for the design of any remedial measures (e.g., sealing and grouting) which may need to be implemented. As a consequence, a need exists to identify the most important hydrological, mechanical and thermo-mechanical parameters associated with excavation damage, and methods for the determination of the magnitudes in media where excavation damage becomes significant to waste isolation.

A final issue concerns the degree of transferability of data, measurement methods and models (i.e., the assessment capability) from one site to another that can be achieved. This is an issue that will require review in the site characterisation programmes of several countries.

2.   The Extent to Which Fundamental Processes of Excavation Response are Reliably Understood

With respect to the status of process studies, considerable progress has been made recently at underground research laboratories in increasing knowledge of the characteristics and potential extent of excavation effects on hydrologic, mechanical and thermo-mechanical rock properties. Nevertheless, it was concluded that further work is needed to precisely identify the relevant phenomena. In particular, it was recommended that the focus of future excavation response testing and analysis should be on identifying governing parameters that will potentially affect waste isolation.

It was concluded that excavation effects in salt are understood well enough for the purposes of repository construction. However, the combined effects of various excavation - induced processes as they affect long-term creep phenomena need to be assessed futher. For clay, since the effects are only expected to be short-term, a detailed understanding may not be important or necessary. However, for crystalline rock, more information is required to determine the need for remedial measures (e.g., design of engineered seals) and to assess the long-term safety implications. In this respect, it was concluded that the phenomena of most potential significance are likely to occur in the vicinity of repository access shafts, drifts and emplaced waste containers.

3.   Future Needs for Development of Instrumentation and Measurement Methods

Discussions at the workshop concluded that the current uncertainties in processes and phenomena related to excavation damage and effects arise primarily from a lack of reliable measurements. Although much recent progress has been made, there does not yet exist reliable instrumentation or measurement techniques for measurements of excavation effects near the host rock/excavation interface. Work is needed to evaluate the accuracy and applicability of existing instrumentation for measurement of rock-mass response, and to develop instrumentation of increased accuracy. Particularly for crystalline rock, suitable methods need to be developed for measuring very low permeabilities in the excavation damage zone. Finally, there is a need to develop instruments of sufficient longevity under in-situ conditions so as to allow testing and monitoring for periods extending to several years.

4.   Appropriate Strategies for Model Development and Validation

Some differences in opinion were expressed as to whether there was a relative need to improve testing and data collection, or to improve model development. However, it was generally agreed that there exists a need to encourage more interaction between experimenters and modellers. The input of modellers is essential in order to identify important parameters, and their significance; greater use of in-situ test data is needed to validate the applicable models. It was also felt that there is a need to identify the acceptable "discrepancy bounds" between in-situ test data and model calculations.

Rock-mass response to excavation was considered to be a very useful tool in validating models for long-term performance assessment. Model predictions and comparisons to in-situ data can substantiate the applicability of the models on a repository scale, and under the actual conditions of a repository environment. A recommendation was made that an international workshop should be organised to address the use of experimental data in the validation of relevant geomechanical, thermal-mechanical and fracture-flow models. Such a workshop would review the key work that has been accomplished (successes and failures) and, in particular, would examine such aspects as limitations, deficiencies, error bounds and uncertainties. Emphasis would be placed on joint presentations by experimenters and modellers, and on actual work completed (numerical analyses, experiments).

It was also recommended that an international benchmark exercise be established to review models designed to analyse the geomechanical and thermal-mechanical response of fractured rock or excavation damage zones (EDZs). It was felt that such an exercise could be held as part of the validation workshop outlined above, or as a separate activity.

5.   Implications for Repository Design and Engineering

The excavation damage zone can have a major effect on the design and effectiveness of seals for repository shafts and drifts, particularly in crystalline and clay media. In recent shaft and drift sealing tests at the OECD/NEA Stripa Project, some techniques were demonstrated that may prove suitable for isolating and sealing such excavation-induced, high-permeable zones around shafts and tunnels in granitic rock. Nevertheless, further work is needed to evaluate repository sealing methods under representative in-situ conditions, in addition to the complementary tests under laboratory conditions. Since hydrologic conditions in the EDZ can be difficult to control, careful documentation of hydraulic properties of this zone is essential. Suggestions also were made that more consideration should be given to reducing the extent of excavation damage through the evaluation of controlled-blasting or boring techniques, by constructing in regions of lower stress, and by examining alternative excavation geometries.

RESUME ET CONCLUSIONS

RAPPEL DES FAITS

    A l'heure actuelle, les activités de recherche et développement
relatives à l'évaluation des performances de dispositifs éventuels
d'évacuation (stockage définitif) de déchets radioactifs sont l'un des aspects
les plus importants de la gestion des déchets radioactifs. Les recherches et
études in situ, en particulier tiennent maintenant une place essentielle dans
tous les programmes nationaux et internationaux visant les dépôts en milieu
géologique : évaluation de ce mode d'évacuation, caractérisation et choix des
sites ainsi que mise au point des dépôts. Lors de leur première réunion, les
membres du Groupe consultatif sur les recherches et études in situ pour
l'évacuation dans les formations géologiques (ISAG) de l'AEN ont examiné les
besoins actuels en matière de recherches et des études in situ. Ils ont
notamment défini comme thème prioritaire les effets des travaux d'excavation,
c'est-à-dire les conséquences que pourraient avoir la construction et
l'aménagement d'un dépôt de déchets radioactifs dans une formation géologique.
L'ISAG a en particulier observé qu'il fallait étudier les contraintes
résiduelles, la subsidence ou le fluage et les fissures induites susceptibles
d'accélérer la circulation de l'eau souterraine pour déterminer l'influence de
ces phénomènes sur les performances des dépôts en matière de sûreté et sur la
conception des barrières ouvragées équipant les dépôts.

    La limite entre la "casemate ouvragée" (comprenant la forme des
déchets, le conteneur, les matériaux de remblayage et de scellement) et le
milieu géologique récepteur (géosphère) est, au plan théorique, l'un des
éléments les moins bien connus du système. A des degrés divers, les questions
relatives aux effets que peuvent avoir sur les performances en matière de
sûreté et la conception des dépôts les excavations et les perturbations
provoquées dans cette zone se posent pour tous les dispositifs d'évacuation en
milieu géologique. Les réponses sont loin d'être simples, bien que l'on
possède déjà de solides connaissances grâce à l'expérience acquise dans les
activités de creusement de tunnels et de mines et, plus récemment, grâce aux
recherches visant la géotechnique et la mécanique des roches menées dans des
installations de recherche in situ de la République fédérale d'Allemagne, de
Belgique, du Canada, des Etats-Unis, de Suède et de Suisse. Les aspects des
travaux d'excavation dont l'examen requièrt l'organisation de rencontres
analogues à la présente réunion de travail sont notamment : a) l'importance
des effets des travaux d'excavation dans la conception et la sûreté d'un
dépôt ; b) le type d'essais et de mesures in situ à réaliser ; et c) les
modèles théoriques et mathématiques qui ont été élaborés, leur validation
revêt une importance particulière s'ils doivent être utilisés pour étayer les
évaluations des performances.

ORGANISATION DE LA REUNION DE TRAVAIL

La réunion de travail avait pour objectifs :

a) d'analyser l'influence des effets des travaux d'excavation sur la
   conception technique et les performances des dépôts en matière de
   sûreté ;

b) de formuler des conclusions et des recommandations sur les moyens de
   prendre en compte les effets des travaux d'excavation dans les
   formations argileuses, salines et cristallines ; et

c) de permettre un échange d'idées entre les scientifiques qui
   réalisent les études et les mesures in situ et ceux qui élaborent
   les modèles prévisionnels destinés à être utilisés dans les
   évaluations de la sûreté des dépôts.

Trente communications ont été présentées sur plusieurs aspects des
effets des travaux d'excavation dans les roches cristallines (granites, les
schistes, les tufs), les formations salines et les argiles. La réunion de
travail a contribué très utilement à faire le point des recherches
(expériences et modélisation) concernant les effets des travaux d'excavation.
Au cours de la séance d'ouverture les intervenants ont donné un aperçu des
recherches relatives aux effets des travaux d'excavation menées dans les
laboratoires souterrains aménagés dans des formations argileuses, cristallines
et salines. Le reste de la réunion de travail a été consacré à une étude plus
approfondie de divers aspects des recherches exécutées dans ces trois types de
formations géologiques. Les participants ont particulièrement mis l'accent sur
l'analyse et la modélisation théoriques, les techniques de mesure in situ et
les méthodes disponibles pour atténuer les effets des travaux d'excavation. A
la fin de chaque séance, les Présidents ont procédé à un résumé des débats. La
séance finale a porté sur l'examen des besoins futurs dans le domaine des
effets des travaux d'excavation, notamment les stratégies applicables à la
validation des modèles et la prise en compte des effets des travaux
d'excavation dans la conception des dépôts ainsi que sur les questions
connexes qu'il conviendrait éventuellement d'étudier lors d'une réunion
complémentaire organisée par l'AEN sur le scellement des dépôts de déchets
radioactifs.

CONCLUSIONS ET RECOMMANDATIONS

Cinq éléments fondamentaux se sont dégagés des exposés et des débats
tenus lors de la réunion de travail :

a) l'intérêt ou l'importance, pour l'évaluation des performances, des
   effets et des perturbations provoqués par les travaux d'excavation ;

b) les limites des connaissances raisonnablement sûres au sujet des
   processus fondamentaux ou de la physique des effets des travaux
   d'excavation ;

c) les besoins futurs visant la mise au point d'une instrumentation et
   de méthodes de mesure ;

d) des stratégies appropriées en vue de l'élaboration et de la validation des modèles ; et

e) les conséquences pour la conception et l'ingéniérie des dépôts.

1. Intérêt ou importance, pour l'évaluation des performances, des effets et des perturbations provoqués par les travaux d'excavation

S'agissant de l'intérêt des effets des travaux d'excavation sur l'évaluation des performances, certaines différences dans les priorités sont apparues selon le milieu géologique considéré. Pour les dépôts aménagés dans l'argile les participants à la réunion de travail ont conclu, qu'à condition d'être bien maîtrisés, les travaux d'excavation n'auraient pas d'effets néfastes sur la sûreté à long terme, en raison des propriétés d'auto-colmatage de l'argile. En ce qui concerne les dépôts aménagés dans le sel bien que certains problèmes puissent se poser quant à l'évaluation de la sûreté à court terme, les participants ont jugé qu'il n'y avait pas lieu de se préoccuper des effets des travaux d'excavation dans les évaluations des performances à long terme. Toutefois, ils ont noté qu'il fallait procéder à des analyses de la sûreté à long terme pour mieux comprendre les effets conjugués des travaux d'excavation, de la pression, de la chaleur et du rayonnement dans la mesure où ils influent sur les phénomènes de fluage à long terme. Quant aux dépôts dans des formations cristallines (granite, tuf), les participants ont estimé que les travaux d'excavation étaient sans incidence sur les performances à court terme ; en revanche, ils ont considéré que l'analyse exhaustive des conséquences pour la sûreté à long terme n'était pas encore achevée.

Il est apparu qu'il fallait intégrer plus directement les études visant les effets des travaux d'excavation à la modélisation des performances des dépôts. Les participants ont donc recommandé que des efforts accrus soient déployés pour associer plus directement les concepteurs des modèles d'évaluation relatifs aux performances des systèmes à la planification initiale des essais concernant les effets des travaux d'excavation et pour formuler des avis, éventuellement dans un cadre international, sur les informations requises pour pouvoir étudier sous tous leurs aspects, dans l'évaluation des performances, les effets et les perturbations provoqués par les travaux d'excavation. Il faut privilégier l'élaboration de modèles plus quantitatifs pour démontrer le caractère négligeable des effets des travaux d'excavation du point de vue des performances à long terme d'un dépôt, ou, dans le cas contraire, communiquer des données fondamentales pour permettre la conception de mesures correctives éventuelles (scellement et injection d'un ciment, par exemple) dont la mise en oeuvre pourrait s'imposer. En conséquence, il faut définir les paramètres hydrologiques, mécaniques et thermomécaniques les plus importants liés aux perturbations provoquées par les travaux d'excavation et les méthodes pour la détermination des ordres de grandeur dans les milieux ou ces perturbations commencent à avoir une incidence sur l'isolement de déchets.

Enfin, les participants se sont inquiétés de la mesure dans laquelle les données et les méthodes de mesure et les modèles propres à un site sont transposables à un autre site (capacité d'évaluation, par exemple). C'est une question qu'il conviendra d'aborder dans les programmes de caractérisation des sites de plusieurs pays.

2. Limite des connaissances raisonnablement sûres au sujet des processus fondamentaux intervenant dans les effets des travaux d'excavation

Quant à l'état d'avancement des études de processus, les laboratoires souterrains de recherche ont beaucoup progressé récemment dans la connaissance des caractéristiques et de l'étendue potentielle des effets des travaux d'excavation sur les propriétés hydrologiques, mécaniques et thermomécaniques des roches. Néanmoins, les participants ont conclu qu'il fallait poursuivre les recherches pour cerner précisément les phénomènes pertinents. En particulier, ils ont recommandé de privilégier la caractérisation des paramètres déterminants susceptibles d'influer sur l'isolement des déchets dans les futurs essais et analyses concernant les effets des travaux d'excavation.

Les participants ont conclu que les effets des travaux d'excavation dans les formations salines sont suffisamment compris aux fins de la construction de dépôt. Toutefois, les effets conjugués de divers processus induits par les travaux d'excavation doivent être évalués de plus près dans la mesure où ils influent sur les phénomènes de fluage à long terme. S'agissant de l'argile une compréhension détaillée n'est peut être pas importante ni nécessaire, les effets escomptés devant être de courte durée. Toutefois, en ce qui concerne les formations cristallines, d'avantage d'informations sont requises pour déterminer les besoins en matière de mesures correctives (par exemple, conception de dispositifs de scellement ouvragés) et pour évaluer les incidences sur la sûreté à long terme. A cet égard, la conclusion a été que les phénomènes potentiellement les plus importants devraient se produire à proximité des puits, des galeries d'accès au dépôt et des conteneurs de déchets déposés.

3. Besoins futurs visant la mise au point d'une instrumentation et de méthodes de mesure

Les débats menés lors de la réunion de travail ont abouti à la conclusion que les incertitudes actuelles entachant les processus et les phénomènes liés aux perturbations et aux effets provoqués par les travaux d'excavation tenaient principalement à un manque de mesures fiables. A cet égard, bien que de gros progrès aient été accomplis ces derniers temps, il n'existe toujours pas d'instrumentation ou de techniques fiables pour mesurer les effets des travaux d'excavation à proximité de l'interface formation réceptrice/excavation. Des études sont requises pour évaluer la précision et l'applicabilité de l'instrumentation actuellement employée pour mesurer la réponse de la masse rocheuse et mettre au point une instrumentation encore plus précise. S'agissant notamment des formations cristallines, il faut élaborer des méthodes adaptées pour mesurer les très faibles perméabilités dans la zone perturbée par les travaux d'excavation. Enfin, il est nécessaire de mettre au point des instruments capables de résister suffisamment longtemps dans les conditions in situ pour pouvoir procéder à des essais et à une surveillance se prolongeant pendant plusieurs années.

4. Stratégies appropriées en vue de l'élaboration et de la validation des modèles

Certaines divergences d'opinion se sont manifestées sur le point de savoir s'il valait mieux améliorer l'exécution des essais et la collecte des données ou plutôt progresser dans l'élaboration des modèles. Toutefois, de l'avis général, il faut inciter les expérimentateurs, d'une part, et les concepteurs de modèle à coopérer plus étroitement. L'apport des concepteurs de modèles, d'autre part, est essentiel pour déterminer les paramètres importants et leur rôle ; il faut mieux exploiter les données tirées des essais in situ pour valider les modèles applicables. Les participants ont également été d'avis qu'il fallait définir l'ampleur des écarts que l'on pouvait tolérer entre les résultats expérimentaux obtenus in situ et ceux donnés par les modèles.

Les participants ont jugé que la réponse de la masse rocheuse aux travaux d'excavation était un élément très utile dans la validation des modèles utilisés pour l'évaluation des performances à long terme. Les prévisions données par les modèles et les comparaisons avec les données obtenues in situ peuvent étayer l'applicabilité des modèles à l'échelle d'un dépôt et dans les conditions y régnant effectivement. Il a été recommandé qu'une réunion de travail internationale soit organisée pour étudier l'utilisation des données expérimentales dans la validation des modèles pertinents simulant les phénomènes géochimiques et thermomécaniques ainsi que l'écoulement dans les fissures. On s'attacherait pendant cette réunion à passer en revue les travaux fondamentaux qui ont été réalisés (avec ou sans succès) et à examiner certains aspects particuliers comme les limitations, les lacunes, les marges d'erreur et les incertitudes. On mettrait l'accent sur la présentation conjointe de communications par des expérimentateurs et des concepteurs de modèles et sur l'examen de travaux effectivement exécutés (analyse numérique, expérience).

Il a également été recommandé de définir un calcul repère international afin de faire le point des modèles conçus pour analyser les réponses géomécaniques et thermomécaniques des roches fissurées ou des zones perturbées par les travaux d'excavation. Les participants ont estimé que ce projet pourrait être réalisé dans le cadre de la réunion de travail relative à la validation esquissée plus haut, ou en tant qu'activité distincte.

5.   Conséquences pour la conception et l'ingéniérie des dépôts

La zone perturbée par les travaux d'excavation peut avoir un effet de première importance sur la conception et l'efficacité des dispositifs de scellement des puits et des galeries d'un dépôt, surtout dans les formations cristallines et argileuses. Les essais de scellement de puits et de galeries, récemment réalisés dans le cas du Projet OCDE/AEN de Stripa, apportent la preuve que certaines techniques sont appropriées à l'isolation et au scellement de ces zones rendues hautement perméables par les travaux d'excavation qui entourent les puits et les galeries creusés dans les formations granitiques. Néanmoins, il faudrait poursuivre les études pour évaluer des méthodes de scellement des dépôts dans des conditions in situ représentatives, en plus de l'exécution d'essais complémentaires en laboratoire. Comme les conditions hydrologiques dans les zones perturbées par les travaux d'excavation peuvent s'avérer difficiles à maîtriser, une caractérisation minutieuse des propriétés hydrauliques de ces zones est cruciale. Certains participants ont suggéré de s'attacher davantage à réduire l'ampleur des perturbations provoquées par les travaux d'excavation grâce à un choix judicieux de techniques d'explosion contrôlée ou de forage, au choix de régions soumises à des contraintes plus faibles pour la construction des dépôts et à l'étude de diverses configurations pour les excavations.

OPENING OF THE WORKSHOP

SEANCE D'OUVERTURE

OBJECTIVES OF THE WORKSHOP / LES OBJECTIFS DE LA REUNION DE TRAVAIL

L.G. Chamney
Radiation Protection and Waste Management Division
OECD Nuclear Energy Agency
Paris, France

Currently, one of the most important areas of radioactive waste management is research and development directed towards assessment of the performance of potential radioactive waste disposal systems. In particular, in-situ research and investigations have become an integral and essential part of national and international programmes for evaluation of geological repositories, including assessment of the concept, site characterisation and selection, and repository development. In the current programme of the NEA Radioactive Waste Management Committee (RWMC), activities related to in-situ research and investigations for geological repositories play a significant role. In order to focus work on the most appropriate topics in this field, the Advisory Group on In-Situ Research and Investigations for Geological Disposal (ISAG) was established by the RWMC in 1986 to provide relevant guidance and advice. ISAG is composed of participants with key positions in national waste management programmes, including the governmental and industry organisations responsible for the programmes, in-situ research facilities and regulatory agencies. The objective of the Group is to assist in the co-ordination of in-situ research, investigation and demonstration activities in Member countries by providing a forum for exchanging information and planning joint initiatives at an international level.

One of the first tasks of the In-Situ Advisory Group was to conduct a review of current needs in the area of in-situ research and investigations; this led to two topics being highlighted as meriting discussion at the international level. The first topic concerned excavation response, that is, the effects that may be induced by the construction and development of a geological repository for radioactive waste. Specifically, it was noted that residual stresses, creep or subsidence, and induced fractures possibly leading to an increased potential for groundwater flow, are phenomena which require analysis to determine their influence on the safety performance of a repository, and on the design of engineered barriers for the repository. The second topic which was identified was a review of the materials and methods available for the backfilling and sealing of repositories located in different geological media. This review was seen as consequent to an examination of excavation response.

With respect to the multi-barrier concept for geological disposal of radioactive waste, the boundary between the "engineered vault" (comprising waste form, canister, buffer, backfill and sealing material) and the host geological environment (geosphere) is conceptually one of the least well understood components of the system. Questions relating to the implications of excavation effects and damage in this zone for the safety performance and the design of repositories apply in varying degrees to all geological disposal concepts. The answers are by no means simple, even though much is already known from experience gained from tunnelling and mining operations.

Recently, there has been considerable effort devoted to research into excavation responses. Geotechnical and rock mechanics investigations are part of in-situ research programmes. A "mine-by" experiment has been conducted at the Climax Mine in Nevada, and one is currently being planned at the Canadian Underground Research Laboratory (URL) to assess the ability to model rock-mass response to excavation and to assess rock-mass damage caused by different excavation techniques. Other studies, for example at the Konrad Mine in the Federal Republic of Germany, examine rock stability following excavation using the room-and-pillar method. In salt and clay formations, creep is particularly important, especially that induced following the introduction of heat. This has been the subject of detailed studies at several in-situ research facilities, including Mol (Belgium), WIPP (United States) and Asse (Federal Republic of Germany).

Several areas can be highlighted that merit specific attention in a workshop, including: a) the importance of excavation response to the design and safety of a repository ; b) the type of in-situ tests and measurements conducted; and c) the conceptual and mathematical models that have been developed. The validation of the latter is particularly important if they are to be used in support of performance assessment.

ISAG therefore recommended to the RWMC that a workshop be organised to examine relevant aspects of excavation response. A small Consultant Group developed the initial objectives and scope of the workshop, and a Technical Programme Committee was established to oversee the planning*. Atomic Energy of Canada Limited (AECL) offered to co-sponsor and host the workshop.

The objectives of the workshop are to:

a) review the influence of excavation effects on the engineering design and the safety performance of repositories;

b) develop conclusions and recommendations on ways of accounting for excavation effects in clay, salt and crystalline geological media; and

---

*   Consultant Group (February 1987): A. Bonne (Belgium), W. Brewitz (Federal Republic of Germany), G. Simmons (Canada), R. André-Jehan (France), P. Bourke (United Kingdom), J. Weidman (Switzerland), P. Zuidema (Switzerland), S. Carlyle (OECD/NEA).

     Technical Programme Committee: A. Bonne (Belgium), W. Brewitz (Federal Republic of Germany), G. Simmons (Canada), C. Voss (United States), L. Chamney (OECD/NEA).

c) provide a forum for discussions between scientists conducting in-situ studies and measurements and those developing predictive models for use in safety assessments of repositories.

Particular emphasis will be placed on theoretical analysis and modelling, in-situ measurement techniques, excavation techniques, and methods available to reduce excavation effects. In meeting the workshop objectives, it is intended that a thorough exchange of views occurs among scientists working on different geological host formations.

The following programme has been developed for the workshop:

Session I   : Overview of research at underground laboratories on excavation response and associated activities

Session II   : Excavation response in crystalline rock:
               a) Overview of studies
               b) Specific geotechnical studies

Session III: Excavation response in salt

Session IV   : Excavation response in argillaceous rock

Session V   : Discussion and conclusions

               –   Rapporteurs (the Session Chairman) will provide a summary of
                    the main points arising from Sessions I-IV, as the basis for
                    discussion and for developing conclusions and
                    recommendations on:

                    a)   future needs in the area of excavation response,
                          including strategies for model validation and
                          accounting for excavation effects in repository design;
                          and

                    b)   possible topics to be addressed at a follow-up NEA
                          workshop on sealing of radioactive waste repositories.

A field excursion to the Underground Research Laboratory (URL) at Lac du Bonnet, where studies on excavation response are in progress or planned, and to the Whiteshell Nuclear Research Establishment (WNRE) at Pinawa, will be held in conjunction with the workshop.

SESSION I

OVERVIEW OF RESEARCH AT UNDERGROUND LABORATORIES
ON EXCAVATION RESPONSE AND ASSOCIATED ACTIVITIES

SEANCE I

APERCU DES RECHERCHES EFFECTUEES DANS LES LABORATOIRES
SOUTERRAINS CONCERNANT LES EFFETS DE L'EXCAVATION ET LES
ACTIVITIES ASSOCIEES

Chairman - Président

K.W. DORMUTH
(AECL, Canada)

ACCOUNTING FOR EXCAVATION RESPONSES IN DEVELOPING AN
UNDERGROUND REPOSITORY IN ARGILLACEOUS FORMATIONS

A.A. Bonne
SCK/CEN
Mol (Belgium)

ABSTRACT

Various concepts have been developed for deep geological radwaste
repositories in argillaceous formations and as a consequence several mining
techniques are envisaged. Immediate excavation effects are expected around
underground openings : convergence, fracturation and pore water pressure
dissipation. It is shown that their influence upon the long term performances
of the system is not significant, providing that their extent and intensity is
kept within limits. Experience is being gained in assessing the overall
immdiate excavation response in argillaceous rock by full scale mine-by tests.

PRISE EN COMPTE DES EFFETS DES TRAVAUX D'EXCAVATION DANS LA MISE
AU POINT D'UN DEPOT SOUTERRAIN DANS DES FORMATIONS ARGILEUSES

RESUME

Divers méthodes ont été mises au point pour l'enfouissement profond de
déchets radioactifs dans des formations argileuses et en conséquence,
plusieurs techniques minières sont envisagées. Les effets immédiats de
l'excavation attendus autour des ouvrages souterrains concernent : la
convergence, la fissuration et la dissipation de la pression interstitielle.
Il est démontré que leur influence sur les performances à long terme du
système n'est pas significative, pour autant que leur extension et leur
intensité soient limitées. Des essais pilotes d'excavation contrôlés (mine-by
test) en vrai grandeur en cours d'exécution contribuent à enrichir les
connaissances concernant les effets des travaux d'excavation sur les
formations argileuses.

1. INTRODUCTION

The common characteristic of the argillaceous formations is that they contain a dominant constituent fraction of clay minerals or related mineral species. They occur with varying structure, texture, paragenesis and in very different geological environments. Lithologic, granulometric, fabric and genetic criteria are the keys to name such rocks ooze, mud, silt, clay, mudstone, claystone, marl, argillite, siltstone, shale, slate, tillite, minette and so on.

The argillaceous formations envisaged in different countries for disposing of radioactive waste are generally stratiform and laterally continuous over large areas. They generally show an almost constant thickness and a gentle dip. In some cases they may contain interposed lithologies of a different nature from the bulk formation, which thus influence the overall behaviour. All this brings about that different schemes for radioactive waste repositories have been considered. Clays, mudstones/siltstones, marls, tillites and shales are the most investigated lithologies for the time being.

2. REPOSITORY DESIGN CONSIDERATIONS

The factors determining the design of a repository are related to the waste characteristics and arisings, to site and geological characteristics, to regulatory guidances, to technological capabilities and to costs /1/. The geological characteristics (essentially those of the host rock) and the technological means are issues of primary concern in the scope of this workshop.

2.1. Strength and Initial Deformation Characteristics

Characteristic stress versus strain evolutions obtained from tests on argillaceous materials are given in fig. 1. Curve I represents the response of argillaceous materials at higher confining pressures and curve II shows the evolution at low confining pressure. The first case (higher confining conditions) indicates a stiffer behaviour of the material (absence of a concave shaped part in the curve). With increasing deviation stress, discontinuities in the sample begin to propagate and an evolution trend goes on until failure. Such test/material evolutions may be strongly influenced by the presence of oriented, penetrative discontinuities, like slaty cleavage or bedding planes, by pore water pressure and by compressive strength conditions.

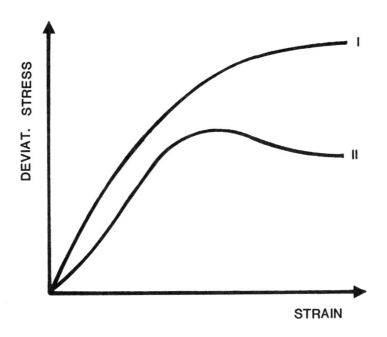

Fig. 1. Typical stress-strain relation for argillaceous rocks

2.2. Design factors

Practice and experience of excavation in deeper clays and weaker argillaceous rocks is very limited, but some experience in shallow weak materials (tunnelling) and deeper stiff rocks (e.g. in the coal fields) has been gained. However specific factors have been proposed to express the ability of a rock to support forces. The "competence factor", F (F = ratio of unconfined uniaxial compressive strength to geostatic pressure /2/) and the "stability factor" N (N = ratio of the product overburden density p times the depth D to the undrained shear strength /3/) of a rock have been proposed as indices for determining when a supporting structure is required or not for excavated openings. Table I shows critical values for F and N and their relation to rock support requirements.

Table I.   Support requirements related to the competence factor (F) and the stability factor (N) of rocks

F	N	support requirement
F<2		support of openings required to withstand the total overburden pressure
2<F<4		support of openings required to withstand a substantial part of the overburden
4<F<10	N>3.5	support required to withstand a small fraction of the overburden (not if creep and tensile strength are favourable)
F>10	N<3.5	no support required except to support loose blocks

Within the large variety of rocktypes belonging to the class of argillaceous rocks, these representative for candidate repository host formations show an extended range of values for F and N. The argillaceous formations under consideration range from medium soft clays to medium strength rocks and the construction techniques appropriate for each type of geomedium will vary accordingly. Primary support (rock bolts, primary lining), drilling and blasting are appropriate for the rocks in the upper ranges of competence. These techniques are rather well known from mining and tunnelling works in non-argillaceous rocks. For the excavation and supporting techniques in weaker argillaceous rocks (clays, mudstones, marls, etc.) less expertise is available.

2.3. Repository concepts

In the context of this workshop we do not consider the seabed disposal option where the target formations are deep ocean clayey sediments. Indeed the waste emplacement techniques proposed until now for this option, viz. the penetrator and deep boring emplacement methods, are not real excavation methods /4/.

The concepts on land consider mined repositories for shallower host rocks of low competence (clay, marls, mudstones, tillites) /1,5/ and deeper more competent geomedia (shales) /6/ or deep borehole disposal for thick and deeper low competent host rocks /7/. For the mined repositories generally one single disposal horizon is considered. In the disposal horizon one foresees a network of galleries or tunnels. The waste packages are to be emplaced either in drilled holes in the floor of the galleries/tunnels or in the galleries or tunnels themselves. For mined repositories in less competent argillaceous host rocks various lining methods can be envisaged. The various concepts are synoptically represented in fig. 2.

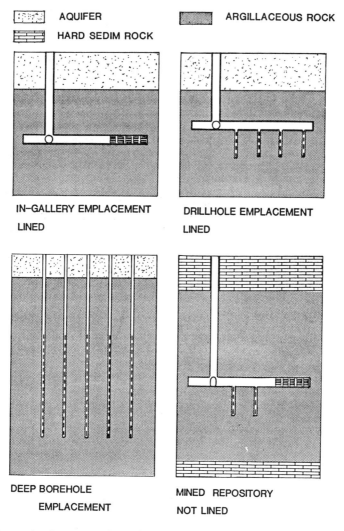

Fig. 2. Schematic lay-out of various repository concepts for argillaceous rocks

The lay-out of the subterranean part of a disposal facility is built up of the following structural parts :

- shafts (may be inclined in the case of shales in a hilly geomorphology) ;
- tunnels or rooms (access, connections, utility, disposal, workshops, tunnelmachine departure rooms, lateral developments from shafts, etc.) ;
- junctions ;
- drillholes.

Shafts or inclined adits will have to be designed on the basis of the very site specific conditions and therefore are not discussed here. Those parts of the shafts or inclined adits which are to be constructed in the host rock itself, require similar considerations as the tunnels or rooms to be excavated in it.
In some concepts waste packages are emplaced in disposal drillholes to be made in the floor of the galleries or rooms. The considerations about competence and lining necessities are also applicable to the disposal drillholes.

2.4. Mining methods

It is obvious that with the variability of competence and plasticity of the argillaceous rocks and the various geological settings also a large variety of mining methods could be applied ; also for each part of the repository, depending on its required dimensions and shape, different mining techniques are applicable. Of the mining activities those related to rock breaking (fragmentation) and supporting are of direct influence on the excavation response. Rock removal operations may have an indirect impact in the sense that they may influence the time span between excavation and supporting. The primary excavation responses will be altered by the subsequent operational conditions during the emplacement of the waste packages (ventilation, drainage of seeping groundwater, oxygen percolation, temperature variation). Room/tunnel/gallery/drillhole filling, extrados backfilling and sealing are the last mining activities which determine the course of the primary excavation responses. On a later stage the near-field phenomena, caused by the presence of the emplaced waste packages will on their turn interfere with the effects produced by the earlier activities. We will concentrate hereafter on the primary effects.

2.4.1. Rock breaking (fragmentation)

The rock breaking method will determine the quality of the excavation and the immediate rock responses. The choice of the method normally depends on the competence of the rock, on the front advance rate required and on the cost.

At present several mechanical methods suitable for the excavation of argillaceous formations /7/ are applying :

- partial face excavators (roadheaders, bucket excavators and impactors) ;
- full face machines (e.g. drum diggers), or
- drill and blast.

Hand excavation methods would only be used locally, e.g. for trimming junctions.

The partial face excavators have the advantage of their versatility, e.g. the same tool is able to perform continuous tunnelling even with changing shape and/or dimensions, all with a minimum overbreak. An important drawback of these machines is their local attack of the front ; the primary relieve is thus punctual and not homogeneously spread over the front. Drill and blast techniques would be required for tunnels/rooms in stronger argillaceous rock but they have a number of serious disadvantages, which make these technique less appropriate. Controlled perimeter blasting is difficult to be performed because slight changes in the rock characteristics (even locally) may influence the blasted volumes. Fracturing in the adjacent rock increases locally the permeability of the rock wall and rough wall surfaces may require smoothing. The full face tunnelling machines require an initial excavation to set the boring machine in position along the tunnel axis ; this may be 30 metres or more in length. This opening needs to be excavated by other means. Full face borers are only efficient for tunnel lengths of more than 300 metres. From the point of view of impact on the front and adjacent rock mass the full face borers attack in an axisymmetrical way over the full front, with a constant and smooth perimeter. The full face excavation method is the method which induces the least instantaneous fracturing processes (see below).

2.4.2. Wall support

The design of an effective wall support system includes three main elements :

- an understanding of the stress configuration around the excavation ;
- a selection of the most appropriate method and materials ;
- an appropriate emplacement technique.

Definitive wall support methods for radwaste repositories would be wall lining or rock reinforcement. The wallrock lining refers to a passive support inside the perimeter of the excavated volume with the purpose to physically restrain the rock and when appropriate tighten the structure. A large variety of lining techniques and methods has been developed, some of them even applying treatment of the rock itself (freezing, injection of cement, etc.). Precast segmental lining, cast in-situ lining, flexible precast lining and sprayed concrete are already everyday practise in excavation works. Precast segmental lining methods require and extrados grouting in order to homogenise at best the stress configuration around the excavation. Rock reinforcement is a rather active method by placing reinforcing items within the rock mass itself. Rock reinforcement makes fractured rock (see section 3) behave as a monolithic mass and some items of the system play the role of tension members of a the composite system. Lining and rock reinforcement are currently also combined in underground works, e.g. in junctions.

Generally speaking, continuous lining methods are rather to be applied in the case of soft, plastic host rocks, while reinforcement methods may be sufficient to retain high competence rocks in the fractured zones and to maintain in these zones an arch of rock in compression around the excavation.

3. EXCAVATION EFFECTS

It is obvious that besides the intrinsic properties of the host rock, the nature, the intensity and the evolution of the primary excavation effects will depend also on the excavation and wall support methods applied, their progress rate and the time lag between excavation and lining. It is not possible to detail here all effects for the large span of rock characteristics involved and for the many applicable mining techniques.

3.1. Convergence

A general feature is that by driving a tunnel or excavating a room an important change occurs in the stress configuration on the local level around the excavation. The resulting strain is delayed over a certain time period and may stabilise or lead to failure. This leads to a **convergence** of the walls which is accompanied by a loss of mechanical resistance of the material (softening) along with a plastic deformation. The description of the convergence depends on the assumed type of response, most frequently assumed to be elastic, elasto-plastic or elastic-brittle. For a detailed overview of the convergence phenomena the reader is referred to specialised literature /8/.

3.2. Fracturing

As we already mentioned, the **fracturing** of the rock is one of the phenomena that may occur in the near field of the excavation. The immediate surrounding walls of an excavated argillaceous rock may show fractures of different origin : 1) existing natural fractures in the rock mass (joints, bedding planes, cleavages), 2) stress concentration effects around the excavation and 3) fractures due to the excavation methods. The fracture spacing varies from centimetre scale to decametre scale, the aperture from invisible for the naked eye up to several centimetres in the case of the open fractures. The fractures may be filled or void. The strength of the bulk depends on physical characteristics of the fracture filling (planarity, aperture, roughness, strength of the infilling minerals, etc.). The natural fracturing in the rock will vary considerably with lithologic rock type, depth and location and it may be enhanced by the excavation. In the formation selection procedure one normally excludes rocks with a natural fracture pattern. The stress configuration around an excavation varies with the strength of the rock, the depth of the excavation, the ratio of the initial pre-mining vertical and horizontal stress and with the shape (e.g. circular or horseshoe shaped cross-section) and dimensions of the excavation. For instance with a high pre-mining vertical to horizontal stress ratio, the stress in the immediate roof of the excavation is tensile. This will tend to cause roof failure, e.g. by separation of the bedding planes in shale.

3.3. Hydrologic response

A third important effect of the excavation concerns a local hydrologic perturbation. By the excavation of rooms and tunnels and their operation afterwards the (fluid) pressure inside the rooms is reduced to about atmospheric conditions by ventilation/climatisation and pumping. It means that locally a dissipation in the interstitial pressure will be established and a drain of the interstitial fluids from the host rock towards the openings will occur. The permeability of the host rock (eventually altered by convergence and fracturation in the excavation near field), the wall support technique applied and the local hydraulic gradient will determine the drain rate towards the openings. Mathematical formulations have been developed for predicting the water inflow in openings and tunnels /19/. Inflow rates of 350 cubic meters of water per minute have been calculated for a repository in shale /7/.

An other effect, which is not directly an excavation effect, is the penetration of air into the wall rock as a consequence of the ventilation/climatisation of the opening which provokes a physico-chemical alteration of the host-rock and as a consequence entails in changes in the geomechanical characteristics.

4. LONG-TERM PERFORMANCE CONSIDERATIONS

For a mined repository system the source term to be considered in the long-term performance assessments is to be emplaced in excavated openings and it is evident that both the waste itself and the excavation affect the same part of the system, viz. the near-field. The most relevant scenario to be considered in the performance assessments is the natural degradation scenario. The leaching mechanism of radionuclides enclosed in a waste matrix is very complex and much information is missing to model accurately all the processes involved.

For assessing the importance of the near field phenomena and of the perturbations in the immediate environment of the excavated rooms we here refer to results that have been obtained in the case of the Boom clay at Mol. In the case of the normal evolution scenario one assumes that due to natural degradation radionuclides are released from the waste package, the diffuse into the host rock and finally a very small fraction of the released radionuclides escape from the host formation into the over and underlying aquifers (see fig. 3). For these conditions it has been shown that different source types yield very comparable migration fluxes into the clay.

Fluxes were calculated for two very different extreme source conditions : a pulse source and a long lasting flux source of constant strength. In the case of a pulse source the immediate and complete dissolution of all radionuclides present in the waste was assumed. In the case of the flux source the radionuclides are assumed to dissolve completely at a constant dissolution rate over 100,000 years. The analytical MICOF-code /9/ was used to perform these calculations. As boundary condition a zero- concentration at infinite distance from the source was applied. The fluxes of radionuclides reaching the upper aquifer are calculated by making use of best estimate parameter values for ^{135}Cs, ^{99}Tc and ^{129}I. This last radionuclide has been considered here because it is not adsorbed or retained by the clay and because of its long half-life. The parameter values applied in the calculations for ^{129}I are given in Table II. The calculated fluxes for the two source conditions are given in figures 4, 5 and 7. In the case of ^{135}Cs (fig. 5), with a retention factor of 400, the difference between the maximum fluxes is about 1 % and a shift of about 50,000 years can be observed between the two curves. For ^{99}Tc (fig. 6), with a retention factor of 107, the maximum differences are about 1 %. More important differences occur for ^{129}I, with a retention factor of 1 ; the difference between the maximum fluxes is about

30 % ; the occurrence time of the maximum flux shifts from 36,000 years for the pulse source to 120,000 for the flux source. It can thus be concluded that for the case of a very performing geological barrier (in casu a 50 metres thick barrier with excellent retention characteristics) and provided the near-field disturbances are limited there is no need for a complicated source term model.

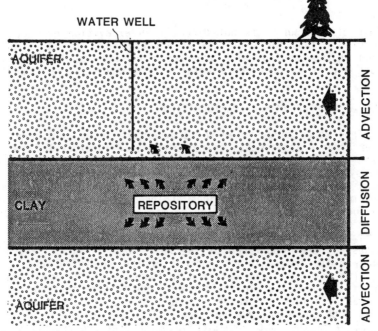

Fig. 3. Synoptic schema of the normal evolution scenario for a repository in the Boom clay at Mol

Table II. Parameter values for ^{129}I-flux calculations

Parameter	Value
half-life (y)	16 E+7
retention factor R	1
source strength S(Bq.m^{-2})	E+9

The excavation response can be seen as a physical change in the near field rock volume. The intrinsic physical parameters influencing the radionuclide migration, which are susceptible to be altered by the excavation response, are the hydraulic conductivity, the effective porosity and the dispersion/diffusion constant. In order to evaluate the effect of the excavation response again a number of exercises for the case of the Boom clay at the Mol site can be made. The worst case assumptions we can make is to deny any retardation function to the zone altered by the excavation in the

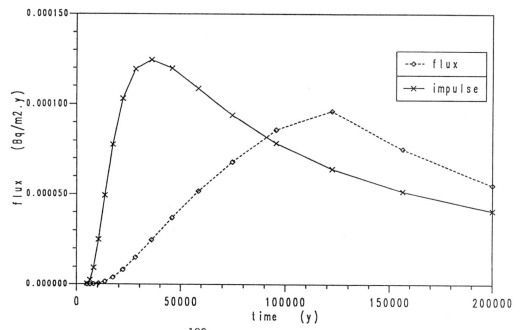

Fig. 4. Calculated flux of ^{129}I at the interface Boom clay overburden (natural degradation scenario / Mol site) for pulse and concentration source conditions

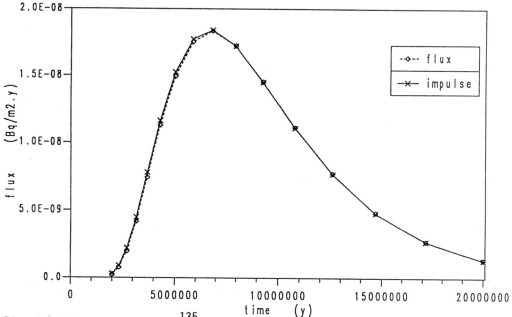

Fig. 5. Calculated flux of ^{135}Cs at the interface Boom clay overburden (natural degradation scenario / Mol site) for pulse and concentration source conditions

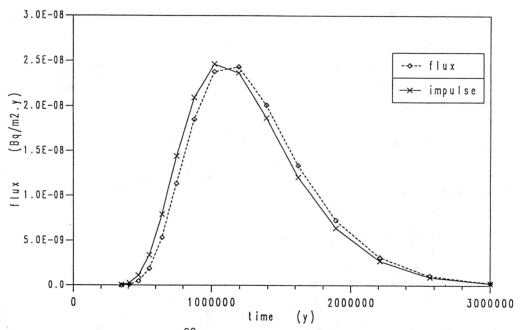

Fig. 6. Calculated flux of ^{99}Tc at the interface Boom clay overburden
(natural degradation scenario / Mol site) for pulse and concentration
source conditions

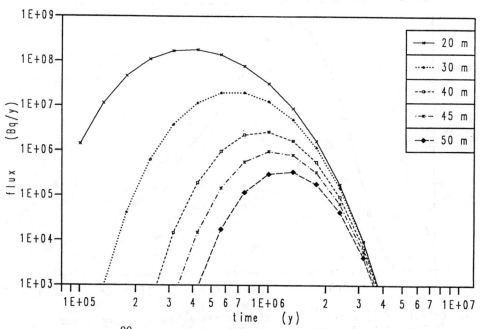

Fig. 7. Flux for ^{99}Tc assuming various clay barrier thicknesses. Normal
evolution scenario, reference case Boom clay at Mol

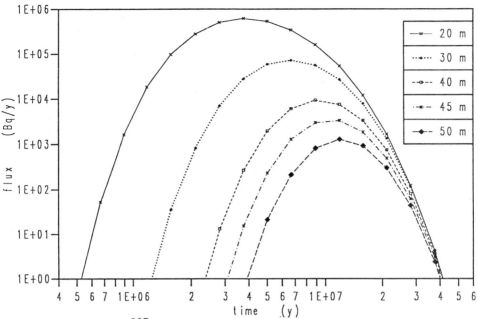

Fig. 8. Flux for ^{237}Np, assuming various clay barrier thicknesses. Normal evolution scenario, reference case Boom clay at Mol

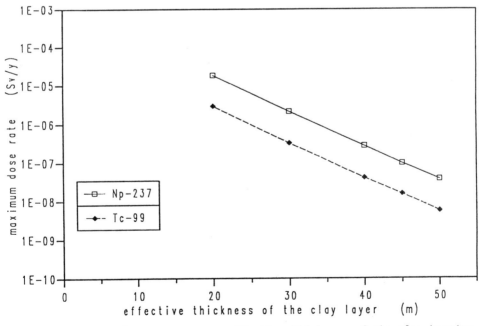

Fig. 9. Maximum dose rate versus effective thickness of the clay barrier, normal evolution scenario, reference case boom clay at Mol

host rock. In other words, the thickness of the argillaceous barrier is reduced from 50 to 40 m or even less, and again the flux intensity and transit time of the maximum flux is calculated. Figures 7 and 8 show the results obtained for ^{99}Tc and ^{237}Np, which appeared from the performance assessments studies as the most important radionuclides for the long-term. Transformation of these fluxes into dose to man (see fig. 9) shows that halving the thickness of the clay barrier yields a two orders of magnitude increase of the dose. For the very specific case we mentioned here, the dose increment by 50 % reduction of the barrier thickness is still within the 99 % upper percentile obtained by stochastic analysis for the same natural degradation scenario (see fig. 10).

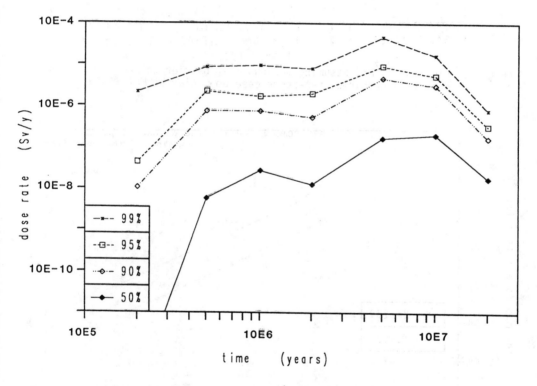

Fig. 10. Statistical distribution of calculated individual dose rates for the normal scenario at Mol (well pathway)

5. EXCAVATION EFFECT ALLOWANCES

The principle of the multiple barrier concept requires that each barrier should be optimised, taken into account its proper contribution to the overall performance of the system. The optimization principle for the natural barriers is incorporated in the site selection criteria and specifications, but necessitates also that any disturbance should not bring about inadmissible weakening of the barrier performances.

For the argillaceous rocks, some excavation methods (e.g. reinforcement by anchors in the junction structures) may disturb the argillaceous barrier over important distances. As a consequence there is a need to define allowable excavation disturbances, which however cannot be defined independently from the requirements ensuring excavation stability for the repository lifetime up to the decommissioning. Also the site and formation specific circumstances have to be taken into account. Once allowable perturbances have been defined these will have their importance for :

- selection of rock breaking methods (dimensions, type, progress rate of front, time gap between excavation and lining, etc.) ;
- selection of the wall support methods (types, convergence limits, etc.) ;
- backfill and sealing methods ;
- site characterisation requirements ;
- monitoring during operational period.

C.M. Koplik et al. /10/ proposed already in 1979 a methodology for defining excavation performance criteria. However no explicit excavation performance criteria for argillaceous formations have been formulated since. Now that general concepts and repository schemes have been designed for the argillaceous environment, time has come to define the allowable perturbances induced by the excavation technique before the optimization of the repository design is launched and detailed concepts are established.

6. SPECIFIC IN-SITU TESTS RELATED TO EXCAVATION RESPONSES

Although there exists a large variability in the geotechnical properties of argillaceous rocks a standard scheme for their excavation is the rock breaking and wall supporting. All other conditions being the same the overall excavation response will depend on the techniques used in both steps. Individual specific parameters and particular response effects (e.g. strength, loading, convergence) may be determined on samples or by small scale in-situ tests (e.g. dilatometer test), but the overall excavation response can only be assessed in-situ under real excavation circumstances. Therefore a method of in-situ reconnaissance or pilot testing has been set-up, which is currently known as a Mine-by test /11/.

The mine-by test is a large scale in-situ test which is performed to assess the extent of the disturbed rock around an underground opening, the effect of the excavation (support, fracturation, interstitial pressure dissipation) and the deformation characteristics of the ground mass (modulus of deformations and creep parameters). The response is monitored by using instrumentation that is installed from pilot adits (shafts, rooms or tunnels) before excavation starts. Auxiliary in-situ and laboratory testing (mapping, pressiometer tests, strength tests and creep tests on cores, deformation of lining, loading of support etc.) are normally performed during or after the excavation in order to complement the theoretical approaches.

The mine-by test can also be used for :

- direct demonstration of rock mass behaviour in case appropriate numerical modelling is not amenable or available ;
- prototype testing in optimisation purposes.

A limited number of well instrumented mine-by tests in argillaceous rock types have been performed already. A listing of such tests is given in Table III.

Table III. Mine-by tests in argillaceous rocks

project/country	rock type	depth (m)	refer.
Seelisberg Tunnel (CH)	marl	200-900	/12/
Kisenyama Pow. Plant (J)	chert/slate	250	/13/
Dinorwic Pumped Stor (UK)	slate	450	/14/
Drakenberg Pump. St. (SA)	siltst./mudst.	200	/15/
Ohnaruto Tunnel (J)	siltst./mudst.	15	/16/
Hades URF (B)	clay	245	/17/
Hades URF (B)	clay	223	/18/

7. CONCLUSIONS

A large variety of mining methods can be envisaged for the construction of geologic repositories in argillaceous rocks. The direct effects related to the excavation of openings and occurring in the immediate vicinity of the excavation are : a convergence of the walls, the fracturation of the surrounding rock and the local pore water pressure dissipation around the opening. The intensity, extent and evolution of these effects around the opening will largely depend on the properties of the argillaceous rock, the methods of rock fragmentation and of wall lining.

The near-field effects due to the presence of radioactive wastes will occur in a rock volume that will have been submitted to alterations caused by the excavation. When considering the long term performances of the overall system, one has to assess the consequences of the excavation response within the context of all ulterior near-field effects.

On the condition that the performance assessment exercises for the case of the Boom clay might be generalized and provided that in the site selection procedure the selection criteria take into account a sufficient safety factor for geological barrier thickness and that the excavation effects are kept within precalculated allowable limits, it is to be expected that these effects will not have a significant influence upon the long-term performances of the system.

REFERENCES CITED

/1/ Bonne, A. and Manfroy, P. : "Factors involved in updating the concept of a repository of HLW, ILW and alpha-bearing wastes in a deep clay formation on land", Proc. Symp. on Siting, Design and Construction of Underground Repositories of Radioactive Wastes, IAEA, Vienna, 1986, 507-519.
/2/ Muir Wood, A.M. : "Tunnels for roads and motorways" Quarterly Journal of Engineering Geology, 5, 1972, 111-126.
/3/ Hudson, J.A. and Boden, J.B. : "Geotechnical and tunnelling aspects of radioactive waste disposal", Proc. Int. Tunnel. Conf., Brighton (1981), 1982.
/4/ An. : "Seabed Disposal of High Level Radioactive Waste", OECD, Paris, 1984.
/5/ Bechai, M; and Heystee, R.J. : "Radioactive waste Management in Shallow Tunnels in Glacial Till or Clayey Soil ; Geotechnical and Hydrogeological considerations", Proc. Symp. on Siting, Design and Construction of Underground Repositories of Radioactive Wastes, IAEA, Vienna, 1986, 151-168.
/6/ Chapman, N. and Tassoni, E. : "Feasibility studies for a radioactive waste repository in a deep clay formation", EUR 10061 EN, 1985.
/7/ An. (report prepared by Parsons Brinckerhoff Quade & Douglas, Inc.) : "Technical support for GEIS : radioactive waste isolation in geologic formations, Volume 12, Repository preconceptual design studies : Shale", Y/OWI/TM-36/12, (1987).
/8/ Gesta P. : "Tunnel Support and Lining", in Tunnels et Ouvrages Souterraines, 73, (1986), pp. 13-38.
/9/ Put, M.J. : "A Unidirectional Analytical Model for the Calculation of the Migration of Radionuclides in a Porous Geological Medium", Rad Waste Management and the Nuclear Fuel Cycle, 6, 3-4, 1985, 361-390.
/10/ Koplik, C.M. et al. : "Information Base for Waste Repository Design /Vol. 6/ Excavation Technology", (1979), NUREG/CR-0495.
/11/ Roberds, W. et al. : "In situ Test Programs Related to Design and Construction of High-Level Waste (HLW) Deep Geologic Repositories", (1983), NUREG/CR-3065, Vol. 2.
/12/ Letsch, U. : "Seelisberg Tunnel : Huttegg Ventilation Chamber", in Proc. Int. Symp. on Field Measurements in Rock Mechanics, Zurich, 1977, P. 577.
/13/ Yoshida, M and Yshimura, K. : "Deformation of Rock Mass and Stress in concrete lining around the machine hall of Kisenyame Underground power plant", in Proc. Sec. Congress of Int. Soc. for Rock Mechanics, Belgrade, (1970), p. 593.
/14/ Douglas, T.H. et al. : "Dinorwic Power Station Rock Support of the Underground Caverns", in Proc. Fourth Int. Congress of Rock Mechanics, Montreux, (1979), p. 361.
/15/ Sharp, J.C. et al. : "The Use of Trial Enlargement for the Drakensburg Pumped Storage Scheme", in Proc. of the Fourth Congr. of the Int. Soc. for Rock Mechanics, Montreux, (1979), p. 617.
/16/ Hata, S.C. et al. : "Field Measurements and Consideration on Deformability of the Izumi Layers, Rock Mechanics suppl. 8, 1979, p. 349.
/17/ Rousset, G. et al. : "Mechanical Behavior of Galleries in Deep Clay / Study of an Experimental Case (experimental drift at Mol, Belgium)", Pasco.
/18/ Neerdael, B and De Bruyn, D. : "Excavation Response Studies in the Mol Facility", this workshop.
/19/ Goodman, R.E. et al. : "Groundwater inflows during Tunnel Driving", Bull. Int. Assoc. of Engin. Geologists, 2, 1965, p. 39.

ACCOUNTING FOR EXCAVATION RESPONSES IN DEVELOPING
AN UNDERGROUND REPOSITORY IN HARD CRYSTALLINE ROCK

Gary R. Simmons
Atomic Energy of Canada Limited
Whiteshell Nuclear Research Establishment
Pinawa, Manitoba ROE 1LO, Canada

ABSTRACT

Several countries are studying the concept of disposing of nuclear fuel
waste in a repository excavated deep in hard crystalline rock. Surrounding
the repository and the emplaced waste, there will be a volume of rock that
will be affected by the excavation, operation and sealing operations involved
in waste disposal. This rock undergoes both immediate and time-dependent
changes that may alter the rock mass properties locally. These changes, and
their implications for repository operation and sealing, are not well
understood. This paper discusses excavation responses, and their possible
significance for waste disposal. It also identifies some research programs in
progress, or planned, to study them.

PRISE EN COMPTE DES EFFETS DES TRAVAUX D'EXCAVATION DANS LA MISE
AU POINT D'UN DEPOT SOUTERRAIN AMENAGE DANS UNE FORMATION
CRISTALLINE DURE

RESUME

Plusieurs pays étudient actuellement la possibilité de stocker
définitivement les déchets nucléaires dans un dépôt creusé en profondeur dans
une formation cristalline dure. Autour du dépôt et des déchets mis en place,
on trouvera un volume rocheux qui aura été perturbé par les travaux liés à
l'excavation, à l'exploitation et au scellement qui accompagnent le stockage
définitif des déchets. Cette formation subit des changements immédiats et
progressifs qui peuvent modifier localement ses propriétés. Ces changements et
leurs incidences sur l'exploitation et le scellement du dépôt ne sont pas bien
compris. L'auteur analyse les effets des travaux d'excavation et leurs
conséquences possibles pour le stockage définitif des déchets. Il évoque
également quelques programmes de recherche, en cours ou prévus, pour étudier
ces effets.

INTRODUCTION

This workshop has been organized by the OECD/NEA to consider the implications of the disturbance, created in the rock mass due to excavation, on the design and safety of repositories for geological disposal of nuclear fuel waste. This paper provides an introduction to the issue of excavation disturbance in hard crystalline rock. Rather than cover the subject comprehensively, my objective is to stimulate discussion. Other presentations will provide an overview of the current state of knowledge of excavation disturbance.

The disposal concept being studied by many countries is the emplacement and sealing of corrosion-resistant containers of nuclear fuel waste deep in stable, hard crystalline rock. In these concepts there are multiple barriers to the release of radionuclides, including some or all of the following: the waste form, the container, the buffer, the backfill, the geosphere and the biosphere (see Figure 1). The waste containers are emplaced in a repository which is a series of shafts, tunnels and rooms that give the operator access to the desired location within the rock body at the selected site.

The most variable component in a disposal system is the geosphere. It is being studied extensively to increase confidence in our ability to effectively isolate high-level radioactive waste. One area of study is the volume of rock immediately surrounding all repository excavations. This rock will be excavation-disturbed; that is, it will be altered in an immediate, and possibly in a time-dependent way, by the excavation process and by the stress redistribution resulting from the removal of constraint on the system.

The design and performance assessment of nuclear fuel waste repositories must consider the excavation-disturbed zone (EDZ) and evaluate its relevance. The objective of a repository is to isolate high-level radioactive waste in a way that imposes an "acceptable risk" on man and the environment. "Acceptable risk" varies amongst countries, and is defined for the operating and the post-closure phases of a repository by national regulations. Operating safety includes consideration of potential industrial, radiological and environmental impacts that create risk for the current and immediate future generations. Post-closure safety includes consideration of the potential radiological and environmental impacts, caused by the migration of radionuclides away from the repository, that may create risk for generations much further in the future. In order to assess the relevance of an EDZ, we must understand the mechanisms causing it and the magnitude and extent of the disturbance.

CAUSES OF EXCAVATION DISTURBANCES

Excavation disturbances around openings in hard crystalline rock are primarily caused by the excavation method and by the stress redistribution in the rock mass as, and after, the opening is created. The disturbances of interest are the creation, extension or reactivation of fractures that result in either a modified structural condition or modified hydraulic conductivity in the rock mass around the excavations.

Excavation Method

The energy imparted on the rock mass by the act of excavating an opening causes damage in the rock. The extent and nature of this damage is dependent on the total amount of the energy and the degree to which the energy is focussed. Continuous excavation methods normally involve the boring of openings in hard crystalline rock. These methods break the rock by applying high and very localized compressive force (nearing a point load) to the rock causing a localized shear failure. The local nature of the loading is believed to have minimal effect on the rock mass remaining around the excavation.

The drill and blast excavation method utilizes explosives to rapidly release large amounts of energy into the rock mass. This process causes the rock to fail and fragment over an area that is controlled by the properties of the explosive and the rock mass, and by the dimensions and spacing of the explosives and the drill holes in which they are loaded. Based on Swedish experience in granitic rock, a generalized relationship between explosive energy, or charge density, and the peak particle velocity imparted to the rock has been developed [1]. Figure 2 shows the relationship for several charge densities, and also identifies the peak particle velocities that are estimated to cause disturbance in the rock mass. These data do not account for the effects of excavation size, in situ stresses and rock mass conditions which do affect the extent and uniformity of the damage zone.

The energy released to the rock mass by the drill and blast excavation method may be controlled by adjusting the size and spacing of the blast holes, and the charge density and dimensions of the explosive. The objective is to limit the energy released by the explosive in each blast hole to the level required to provide adequate fragmentation of the rock. This will result in the minimum disturbance in the rock mass outside the excavation perimeter. There are several techniques for designing blast rounds to achieve this objective in tunnelling including the Swedish blast design method [2] and the cratering theory [3]. Specific examples of the application of controlled blasting methods are the work done in the Edgar Mine in Colorado for the United States Department of Energy (US/DOE) [3, 4] and work done by Atomic Energy of Canada Limited (AECL) in its Underground Research Laboratory (URL) [5, 6].

There have been no tests that compare the EDZs caused by the continuous and drill and blast excavation methods in hard crystalline rock. Such tests would require that both methods be used to excavate openings of similar dimension in essentially the same volume of rock. Comparison of the EDZs would allow the damage due to blasting to be isolated, and analytical methods for estimating the disturbance caused by excavation to be developed and validated. One test was planned for the Grimsel Rock Laboratory in Switzerland [7]. The bored excavation was completed but the drilled and blasted excavation was cancelled.

Stress Redistribution

The creation of an opening in a rock mass under triaxial stress causes a response in the rock mass. Because constraint on the remaining rock mass is locally removed, the rock mass will move toward the excavation free face. To maintain the static equilibrium, the stress field will alter. In the volume of rock immediately surrounding the opening the response may result in displacements on pre-existing fractures, the creation of new fractures or the reactivation of healed fractures on various scales.

The spacial extent of the disturbance resulting from stress redistribution is controlled by several factors including the shape of the excavation, its orientation relative to the stress field, the frequency and spacing of fractures, the properties of the rock and the fractures, and the magnitude, orientation and anisotropy of the stress field. We must understand the underlying physics of the mechanisms causing this disturbance in order to develop models with which to assess its significance.

Kelsall et al. [8] have published an analysis of a circular excavation in basalt under isostatic stress conditions, and have included the effect on the hydraulic conductivity of various sets of pre-existing fractures adjacent to the opening. They suggest that the redistribution of stresses around an excavated opening might affect the conductivity in three ways:

"- by fracturing of originally intact rock due to excessive compressive or tensile stresses,
- by opening or closing of pre-existing fractures due to changes in the normal stresses acting across the fractures or to shearing along the fractures, and
- by loosening of the crystal structure in response to reduced confining stresses".

Only the first two mechanisms are considered by the authors to be relevant in hard crystalline rocks.

For the case considered by Kelsall et al., the initiation of new fractures in originally intact basalt was considered highly unlikely because the assumed stresses were low relative to the strength of the rock. In granitic rock bodies of the Canadian Shield, such as that containing AECL's URL (Figure 3), the stresses are higher and deviatoric. Creation of new fractures around excavations has been seen in the URL at several locations including the 240 Level and the shaft below the 240 Level [9]. The major principal stresses that have been determined at the URL are shown in Figure 4 as a function of depth.

In Kelsall et al.'s analyses of fracture set hydraulic conductivity changes they utilized the aperture/permeability relationships based on the cubic law of fracture flow, while acknowledging that there is much continuing research on the validity of the cubic law in natural fracture systems. For their comparative analyses they felt that it provided a reasonable result. Two cases were considered in the analyses, an elastic case and an elasto-plastic case (Figure 5). The effect of the non-elastic zone around the excavation for the elasto-plastic case is quite pronounced in this example.

Because stress redistribution is associated with changes in the porosity of the rock, it has implications for repository safety. Computer codes for analysing the hydraulic significance of porosity changes are under development. Some of the mechanisms necessary to simulate the stress redistribution contribution to the EDZ are not well understood, particularily the mechanical and hydraulic response of a rock mass containing newly created and existing fractures. Our understanding of the physics of fracture creation and propagation, and of naturally occurring fractures, must be improved, and computer models must be developed to simulate these responses. Techniques must also be developed and proven for measuring these effects in situ.

POSSIBLE EFFECTS ON A REPOSITORY

Repository Concepts

The concepts being developed internationally for nuclear fuel waste repositories cover a spectrum of configurations and environments. The specific details are governed by the characteristics of the media in which they will be sited, the form of the waste being handled and regulations of the siting country. There are two general concepts for repositories in hard crystalline rocks. In the first, the repository systems are designed to contain and isolate the waste containers individually or in small groups. The movement of water to, and radionuclides from, the individual containers is inhibited. In the second, the repository systems are designed to isolate the entire waste inventory from continuing contact with the groundwater system. In both types of repository, the dominant mechanism for groundwater and radionuclide transport near the waste containers is generally diffusion. Examples of both repository types are in conceptual development.

The first repository type is the basis for most of the concepts and designs being developed. One example is the conceptual design study of a used-fuel disposal centre now under way in Canada for Atomic Energy of Canada Limited [10]. This study is determining the technical feasibility of immobilizing and disposing of used CANDU[1] fuel bundles deep in stable rock bodies in the Canadian Shield (see Figure 6). The repository is a series of haulageways and rooms excavated at a depth of 1000 m in plutonic rock. Access and service shafts excavated at the extreme ends of the repository provide personnel and material access and handle services and ventilation. The in-floor borehole emplacement alternative was used in the study, but the in-wall borehole emplacement and in-room emplacement alternatives are equally viable (Figure 7). The in-room emplacement concept is an example of emplacing waste containers in small groups.

This type of repository requires the application of engineered barriers to retard the movement of groundwater to the nuclear fuel waste, and the transport of radionuclides from the nuclear fuel waste to the accessible environment and man. When installed, these barriers are intended to provide a low permeability, and possibly a chemically active, pathway, for groundwater and solute movement. Several disposal system components comprise the engineered barriers, including the waste form (low solubility), the container (corrosion-resistant, limited life), the buffer material packed around the container (low permeability, chemically active) and the backfill, shaft and borehole seals (low permeablility). While these components play different roles in the various national concepts, they all contribute to the isolation of the nuclear fuel waste in the repository.

The presence of an EDZ surrounding the excavations would be potentially detrimental to this type of repository. Possible operational effects would include increased instability in the crown and walls of the excavations, requiring more ground support and increased vigilance to control loose ground. The impact of an EDZ on waste isolation would include the possible presence of a zone of enhanced radial and/or tangential hydraulic conductivity surrounding all excavations, including the waste emplacement boreholes if they are part of the concept. This zone would provide a preferential pathway for radionuclide

---

1)CANadian Deuterium Uranium

migration to the major groundwater flow systems. It would also be present in the rock around emplacement rooms, haulageways and shafts and would require special treatment at the location of seals.

The second type of repository is conceived to isolate the entire waste inventory from the saturated groundwater system. One example of this type is the WP-cave repository concept developed in Sweden [11]. In this concept the volume of rock mass in which the waste would be emplaced would be entirely surrounded by a low conductivity hydraulic shield of sealing material installed in an annular excavation. Inside the hydraulically-shielded volume of rock mass the emplacement areas would be excavated, filled with nuclear fuel waste and sealed. Outside the hydraulic shield, a series of excavations and boreholes would create a hydraulic cage that would divert moving groundwater around the hydraulically-shielded emplaced waste. Because of the hydraulic cage and shield the hydraulic gradient across the emplacement volume would be low and the dominant mechanism for groundwater and radionuclide transport would be diffusion. Because this concept utilizes a zone of enhanced hydraulic conductivity outside the shield, an EDZ may be beneficial.

Other examples of the second type of repository are those planned to be above the local groundwater table, such as the US/DOE concept for the repository in the Yucca Mountain area of Nevada. This repository will be a few hundred metres above the water table in a partially saturated rock mass. Surface waters that enter the rock mass will move downward toward the water table, and a portion of these waters will pass through the repository. The sealing systems are being developed to allow this water to move through the repository toward the water table without undue delay. The effect of an excavation-disturbed zone on the safety of this repository concept must be assessed.

Effects of the Excavation-Disturbed Zone

The effects of an EDZ on a nuclear fuel waste repository occur in the operating and possibly in the post-closure phase. The significance to waste isolation is minimal in the operating phase. The main concerns are the potential for rock loosening on the crown and the walls of the excavations, and the possible increases in groundwater inflow to the excavation caused by the excavation disturbance. These situations are already encountered in civil, mining and geotechnical engineering projects and are primarily an operating cost factor. An unnecessarily extensive EDZ could be a significant operating difficulty and cost factor, because some repository excavations will be open and operated for 60 to 100 years.

Waste isolation in the post-closure phase of a repository may be significantly affected by the presence of an excavation-disturbed zone. Prior to repository excavation, the rock mass will contain well defined pathways for groundwater flow. These will be characterized during the development of the repository and will likely be an important part of the safety assessment. An EDZ created during construction of a repository may substantially alter the flow system, and therefore may invalidate the model used for initial licensing. The EDZ would then have to be realistically incorporated into the revised model, and that model revalidated. The likelihood of a significant effect is not known because the techniques for measuring and modelling the properties of the excavation damage zone are not developed sufficiently.

The conditions in the rock mass adjacent to an emplaced container are important in isolating the waste packages and in achieving retrieval with a high degree of confidence. The retrieval of a waste package may be achieved safely and cost effectively by re-entering through existing excavations when the container, the sealing materials (if used) and surrounding rock are stable and do not move with time. An EDZ could conceivably contribute to movement, or spalling, of the surrounding rock, and could result in movement or jamming of the containers in the emplacement boreholes, particularily in concepts that do not use buffer or packing materials between the container and the borehole wall.

An EDZ may cause a more hydraulically conductive zone to form around the excavations, and would affect the design and construction of the shaft, tunnel, and borehole seals. In a study of shaft seals by Golder Associates [12] for AECL, the groundwater transit time upward from the bottom of a shaft in a hypothetical groundwater flow system was calculated for the various components of the shaft seals, using porous media modelling methods. For a 1-m-thick EDZ surrounding a 6-m-diameter shaft, the transit time from the repository horizon to a major groundwater flow channel 640 m above was as low as 5100 years. In this calculation the hydraulic conductivity of the EDZ was assumed to be 10^{-9} m/s, as compared to 10^{-9} m/s for the backfill and 10^{-12} m/s for the buffer and the concrete in the sealed shaft. In the summary of the study the authors identify the EDZ as the most critical zone in the shaft sealing system. The authors also noted that the hydraulic conductivity of the EDZ might be too low for treatment with existing conductivity reduction methods.

Design and construction methods for shafts, tunnels, emplacement rooms and boreholes must be selected to minimize the EDZ. The methods for installing seals and controlling the hydraulic conductivity of the EDZ must be developed and demonstrated, and must then become design and performance criteria in developing, operating and sealing a repository.

SOME RESEARCH AND DEVELOPMENT PROGRAMS STUDYING THE EDZ

With the effects of the EDZ on repository performance uncertain, it remains an issue that must be dealt with before designing repository systems. To put it into proper perspective, the sensitivity of repository safety to the properties of an EDZ must be analysed. To do this the EDZ must be represented in the safety assessment models in a realistic way. To develop these models, research into the controlling mechanisms and properties of the EDZ is required. Major research and development issues include

- understanding the fundamental mechanisms controlling the formation of the excavation-disturbed zone,
- developing reliable methods for measuring the extent and properties of the excavation-disturbed zone, and
- developing computer models for estimating the extent and properties of the excavation-disturbed zone.

When progress is made on these issues, it will be possible to reasonably assess the significance of the excavation-disturbed zones. With this capability the EDZ could be dealt with as one factor in the "design as you go" approach to repository development, seal application and safety assessment. As it may never be possible to design or analyse away the EDZ, methods must be developed to control and improve its properties. Such research is included in the OECD/NEA International Stripa Project Phase 3, and in the programs of the US/DOE and AECL.

Some research and development relevant to these issues has already been completed or is in progress. Examples are discussed below.

Stripa Mine, Sweden

In the Swedish-American co-operative program at the Stripa Mine, one test included measurements of the effect of the EDZ. The Macropermeability Experiment [13] (Figure 8), was performed to improve techniques for characterizing the permeability of large volumes of low-permeability rock. Fifteen radial boreholes were drilled from the test room and instrumented to measure pressure gradients. This technique created a large pressure sink to perturb a relatively large volume of the fracture flow system. The test drift was sealed off and equipped with a constant-temperature air-circulation system that was controlled to evaporate all water seeping into the room. The water inflow rate was determined from measurements of the mass flow rate and the relative humidity of the entering and exiting air streams.

The head measurements done during the experiment indicated that the rock mass hydraulic conductivity away from the opening was about 10^{-10} m/s. However, near the excavation there was a "skin zone" with a significantly lower hydraulic conductivity caused by the excavation disturbance. Kelsall et al. [8] conducted an analysis of this experiment that included several simplifying assumptions, and determined that a zone of reduced hydraulic conductivity should have developed around the opening.

Colorado School of Mines CSM/ONWI Facility

One aspect of the program in the Colorado School of Mines/Office of Nuclear Waste Isolation (CSM/ONWI) facility [14] was the characterization of a rock mass in the vicinity of an excavation. The test room was constructed using 10 blast rounds, 7 designed using the Swedish blasting method [4] and 3 designed using a tunneling application of the Livingston cratering theory [3]. A major effort then involved characterizing the rock mass surrounding the excavation to identify and quantify the disturbed zone. The arrangement is shown in Figure 9. The EDZ was assessed in radial boreholes, 76-mm in diameter, drilled into the rock affected by 7 excavation blast rounds. The assessment included monitoring variations in modulus (E), permeability (K), cross-hole ultrasonic velocities (Vp), rock quality designation (RQD) of the borehole core and in the stress field. The excavation-disturbed zone documented by Hustrulid and Ubbes for blast round number 4 are shown in Table I.

The stress measurement techniques were not successful because of the fractured nature of the rock mass.

The instrumentation used for the modulus measurements, the permeability measurements and the crosshole ultrasonic measurements performed to expectations during these tests. These results indicate the variability in the determination of the extent of the EDZ using the selected techniques.

TABLE I: Comparison of the Extent of Excavation Disturbance
in Blast Round Number 4 Determined by Four Techniques

Borehole Position	Depth of Disturbed Zone (m) Determined by Four Techniques			
	E	K	Vp	RQD
RUW	0.6	0.6	---	0.0
RUE	0.6	0.9	0.3	0.9
ROU	0.3	0.0	0.3	0.9
RHE	0.9	0.9	0.9	0.9
RDD	0.9	0.9	0.3	0.9
RDE	0.9	0.9	0.6	0.9
ROW	0.6	1.2	---	0.9

Grimsel Rock Laboratory, Switzerland

A test has been conducted in the Grimsel Rock Laboratory to define and
quantify the rock mass damage caused by mechanical tunnel excavation. A
schematic illistration of the test is shown in Figure 10. The test area is
entirely within the Grimsel granodiorite.

The original plan for the test included a precharacterization of the
volume of rock, excavation of a drill and blast test section and a mechanically
excavated test section, followed by post-test characterization of the disturbed
zones. The EDZ caused by each type of excavation could then be directly
compared. The mechanically excavated test section was completed in 1983, but
the drill and blast test section has been indefinitely postponed.

Prior to excavation of the test section, a series of boreholes parallel
and perpendicular to the axis of the test section were drilled to allow
characterization of the rock prior to excavation. The remainder of the
characterization boreholes shown in Figure 10 were drilled for post-test
characterization. The techniques applied to assess the excavation disturbed
zone included the following:

- displacement measurements using the sliding micrometer and distometer
  during excavation,
- deformation modulus using a borehole dilatometer and flat jacks in slots,
- cross-hole velocity measurements, and
- fracture permeability measurements in boreholes using straddle packers
  and between slots cut in the tunnel floor.

The data were analysed and compared with computer simulations.

Egger [15] concluded that there was no structural instability due to
excavation distubance; there was rock damage in the shear zones near the
excavation; and there was no measurable change in the rock mass strength or
hydraulic conductivity near the excavation opening that could be attributed to
excavation disturbance. However, he also cautioned that the excavation
disturbance in the shear zones may influence the potential for radionuclide
migration and should not be disregarded.

The Underground Research Laboratory, Canada

There are studies of excavation disturbance in progress at AECL's URL, and others are being planned. This program will be described later in the workshop by Mr. P.A. Lang and aspects of it will be presented by P.A. Lang, P.M. Thompson et al., R.A. Everitt et al., C.D. Martin and R. Koopmans and R.W. Hughes in papers in these proceedings.

SUMMARY

The EDZ has the potential of affecting the operational and the post-closure safety of repositories, depending on several characteristics specific to the site and the concept of the repository. There will be a disturbance around each opening, due to the excavation method and to the redistribution of stresses. The physics of these processes must be better understood. Equipment and techniques must be developed to allow field measurements of the extent and properties of the disturbed zone. Computer codes should be developed to simulate the creation and properties of the EDZ as a tool for use in analysing field measurements and assessing the safety implications. Some research and development has been done, and more is planned, to provide the capabilities necessary to assess the significance of the EDZ and to account for it in the design, development and sealing of repositories.

In parallel, research and development programs must be continued on methods for designing and emplacing seals in the presence of an EDZ, and for controlling the properties of the EDZ where this is important to repository system performance.

REFERENCES

1. Holmberg, R. and P.A. Persson : "Design of Tunnel Perimeter Blasthole Patterns to Prevent Rock Damage", Trans. Instn. Mining Met. Sect. A, 89, 37-40, 1980.

2. Langefors, U. and B. Kihlstrom: "The Modern Technique of Rock Blasting", Almquist and Wiksell, Stockholm, Sweden, 1963.

3. Sperry, P.E., G.P. Chitombo, and W.A. Hustrulid: "Hard Rock Excavation at the CSM/OCRD Test Site Using Crater Theory and Current United States Controlled Smooth Wall Blasting Practices", OCRD Technical Report BMI/OCRD-4(4), 1984. #

4. Holmberg, R.: "Hard Rock Excavation at the CSM/OCRD Test Site Using the Swedish Blast Design Techniques", OCRD Technical Report BMI/OCRD-4(3), 1983. #

5. Kuzyk, G.W., P.A. Lang and G. Le Bel: "Blast Design and Quality Control at the Second Level of Atomic Energy of Canada Limited's Underground Research Laboratory", Proc. Large Rock Cavern Symposium, Helsinki, Finland, 1986, pp. 147-158.

6. Favreau, R.F., G.W. Kuzyk, P.J. Babulic, R.A. Morin, and N.J. Tienkamp: "The Use of Blast Simulations to Improve Blast Quality", Proc. First International Symposium on Rock Fragmentation, Keystone, Colorado, 1987, pp. 424-435.

7. NAGRA: "Grimsel Test Site - Overview and Test Programs", National Cooperative for the Storage of Radioactive Waste Technical Report 85-46, 1985. *

8. Kelsall, P.C., J.B. Case and C.R. Chabannes: "Evaluation of Excavation-induced Changes in Rock Permeability", Int. J. Rock Mech. Min. Sci. & Geomech. Abstr., Vol. 21, No.3, 1984, pp. 123-135.

9. Martin, C.D.: "Shaft Excavation Response in a Highly Stressed Rock Mass", Proc. Workshop on Excavation Response in Deep Radioactive Waste Repositories - Implications for Engineering Design and Safety Performance, OECD, Paris, in preparation.

10. Atomic Energy of Canada Limited CANDU Operations in association with J.S. Redpath Mining Consultants Limited, Golder Associates and The Ralph M. Parsons Company: "Used-Fuel Disposal Centre - A Reference Concept", Atomic Energy of Canada Limited Technical Record, in preparation.

11. Sagefors, I.: "Mining Technique for High Level Nuclear Waste Repository", Trans. Inst. Mining Met. Sect. A, 96, 1987 July.

12. Golder Associates in association with Dynatec Mining Limited: "Conceptual Design of Shaft Seal For a Nuclear Fuel Waste Disposal Vault", Atomic Energy of Canada Limited Technical Report, in preparation.

13. Gale, J.E., P.A. Witherspoon, C.R. Wilson, and A. Rouleau: "Hydrogeological Characterization of the Stripa Site", Proc. Geological Disposal of Radioactive Wastes - In Situ Experiments in Granite, OECD, Paris, 1983, pp 79-98.

14. Hustrulid, W. and W.F. Ubbes: "Results and Conclusions from Rock Mechanics/ Hydrological Investigations: CMS/OWNI Test Site", Proc. Geological Disposal of Radioactive Wastes - In Situ Experiments in Granite, OECD, Paris, 1983, pp 57-75.

15. Egger, P.: "Field Study of Rock Damage Around a Gallery", Proc. Second International Symposium - Field Measurements in Geomechanics, Kobe, Japan, 1987, pp. 541-550.

16. Baumgartner, P. and G.R. Simmons: "The Canadian Nuclear Fuel Waste Disposal Concept and Related Research Programs", Presented at the Acres Annual Geotechnical Seminar on Storage and Treatment of Wastes, Niagara Falls , Ontario, 1986 April 18 & 19. Also issued as Atomic Energy of Canada Limited Report AECL-9050.

---

available from National Technical Information Service, U.S. Department of Commerce, 5285 Port Roual Road, Springfield VA 22161.

* available from NAGRA, Parkstrasse 23, 5401 Baden, Switzerland.

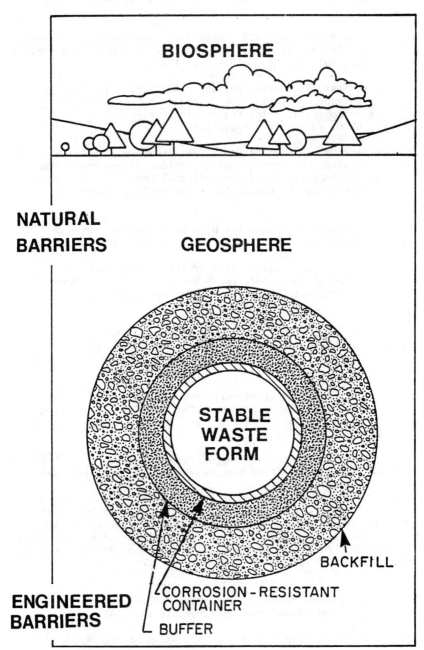

Figure 1:   The Multiple Barrier Disposal Concept.

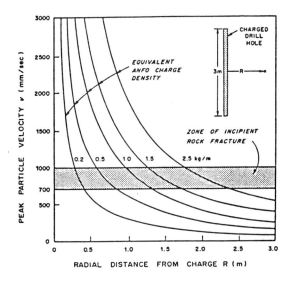

Figure 2:   Method of Estimating the Thickness of Blast Damage Zone In Relation to Explosive Charge Density. [1]

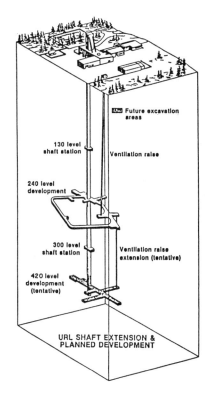

Figure 3:   The Underground Research Laboratory Arrangement.

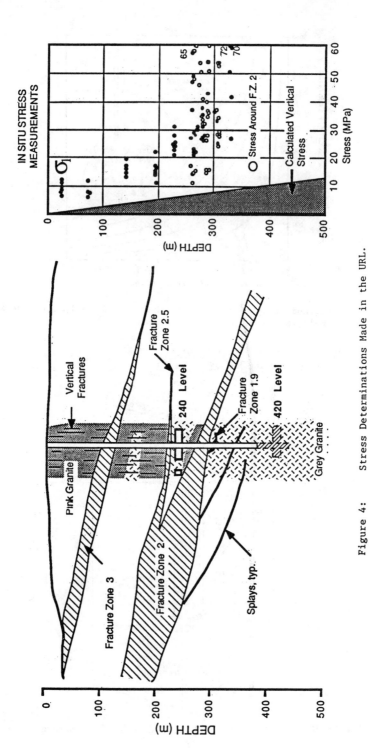

Figure 4:   Stress Determinations Made in the URL.

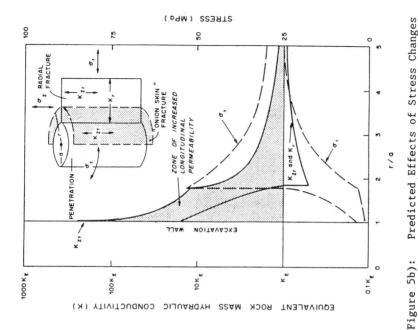

STRESS (MPa)

Figure 5b): Predicted Effects of Stress Changes on Rock Mass Hydraulic Conductivity (elasto-plastic stress analysis for fractured basalt-isotropic initial stress conditions). [8]

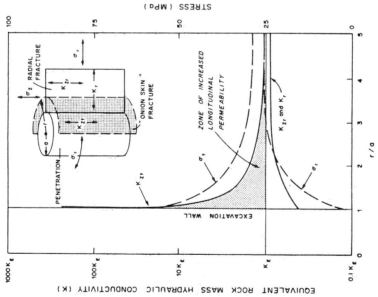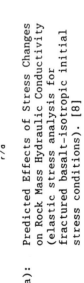

STRESS (MPa)

Figure 5a): Predicted Effects of Stress Changes on Rock Mass Hydraulic Conductivity (elastic stress analysis for fractured basalt-isotropic initial stress conditions). [8]

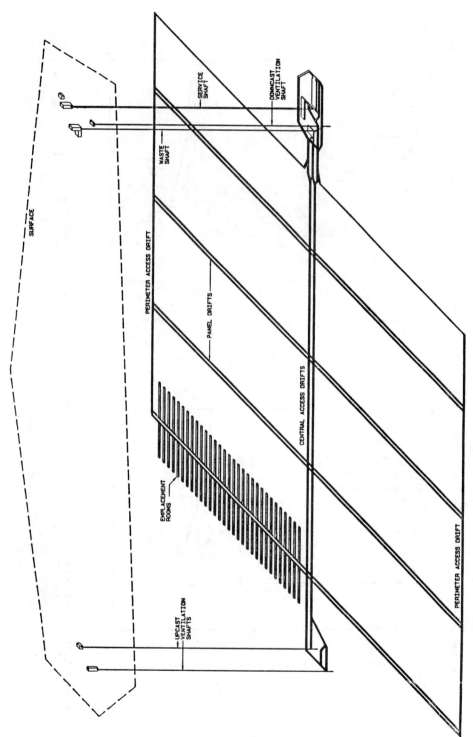

Figure 6: Used Fuel Disposal Centre – A Reference Concept Layout. [10]

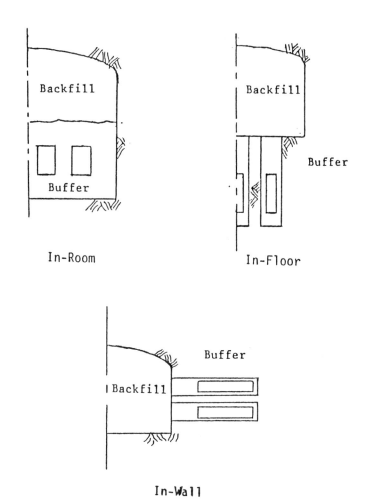

In-Room

In-Floor

In-Wall

Figure 7:   Nuclear Fuel Waste Container Emplacement Alternatives. [16]

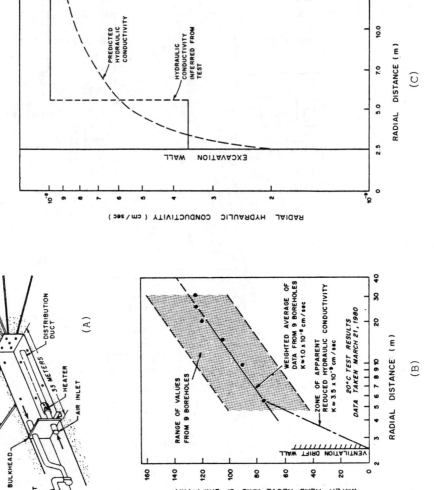

Figure 8: Stripa Macropermeability Experiment. [8]

- 64 -

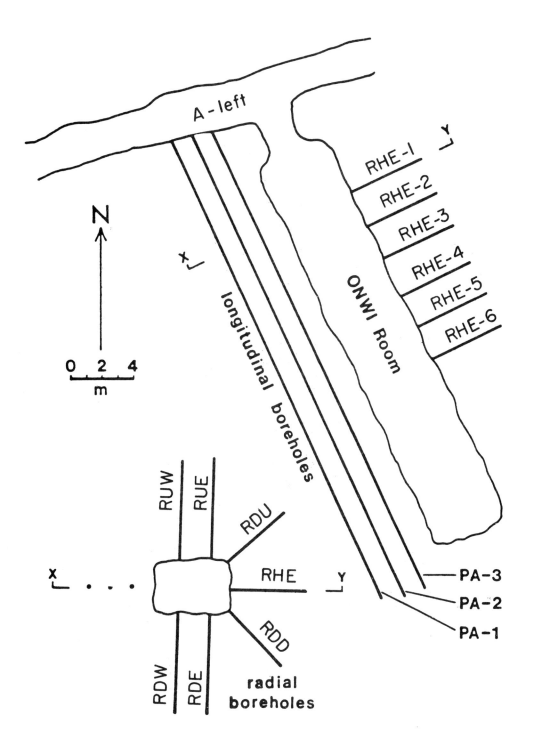

Figure 9:   CMS/OWNI Test Room Arrangement. [14]

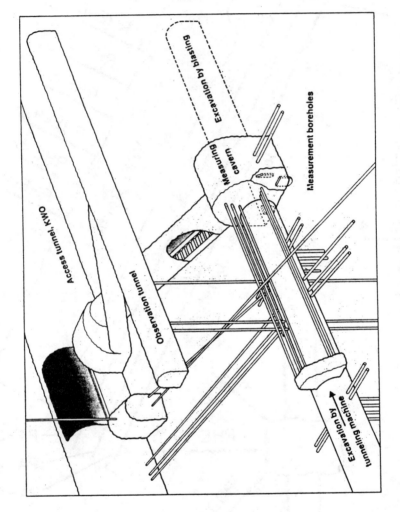

Figure 10: Schematic of the Rock Damage Study Test at the Grimsel Rock Laboratory. [7]

INVESTIGATIONS OF EXCAVATION EFFECTS AT THE ASSE SALT MINE

Dr. W. Brewitz
Prof. Dr., K. Kühn
Dipl.-Geol. M.W. Schmidt
Gesellschaft für Strahlen- und Umweltforschung mbH München
Institut für Tieflagerung
Braunschweig, Federal Republic of Germany

ABSTRACT

The Asse mine, initially a potash mine later converted into a rock salt mine, was operated between the years 1906 and 1964. After production terminated the mine was selected as an underground laboratory for the disposal of low- and intermediate-level radioactive waste. Since most of the old mine workings in the Halite formations had not been backfilled, extensive rock mechanical investigations were carried out right from the beginning of the waste disposal operations.

The results have shown large rock stress alternations and rock mass deformations which have taken place due to the creeping behaviour of rock salt. FE-calculations confirmed distinct spatial as well as temporal changes of the natural stress field. Greatest deformations correlate with the central area of the mine excavations.

RECHERCHE SUR LES EFFETS DES TRAVAUX D'EXCAVATION
DANS LA MINE DE SEL DE ASSE

RESUME

La mine d'Asse était au départ une mine de potasse qui a été ultérieurement transformée en mine de sel gemme et a été exploitée de 1906 à 1964. Lorsque la production a été arrêtée, on a utilisé la mine comme laboratoire souterrain de recherche pour le stockage des déchets de faible et moyenne activité. La majorité des anciennes chambres creusées dans les formations de halite n'ayant pas été comblées, on a pu entreprendre d'importantes études de la mécanique des roches dès le début des activités de stockage.

Il ressort de ces travaux que le fluage du sel gemme a entraîné dans la masse rocheuse d'importantes modifications des contraintes et des déformations de grande ampleur. Des calculs fondés sur le principe des éléments finis ont confirmé une évolution dans le temps et dans l'espace du champ des contraintes. On a pu corréler les déformations les plus importantes avec la principale zone d'excavation minière.

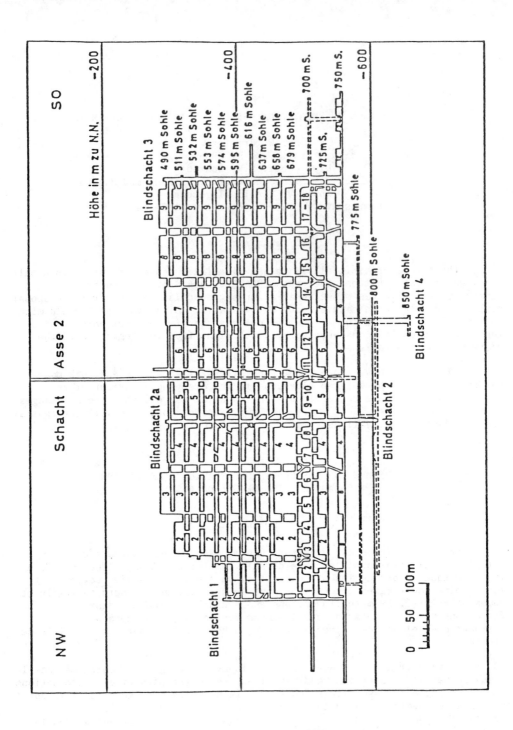

Fig. 1: Profile of the old workings in the Asse salt mine.

Introduction

The Asse Mine II was developed as a potash mine in the years 1906 till 1908. It is located about 25 kilometers southeast of Brunswick in Lower Saxony. Asse is the name of a hill site stretching from northwest to southeast over a distance of 8 kilometers. Its highest elevation measures 234 meters above sea level. The ridge is a result of the salt anticline which is covered by 220 m of Buntsandstein formations at its top.

The mining took place between 490 and 750 m levels. After the potash production had ceased in 1925 rock salt was mined until 1964. While the potash workings were backfilled most of the 140 stopes in rock salt measuring up to 60 x 40 x 15 m were left open. When the mine terminated production entirely in 1964 there existed a total cavity volume of about 3,5 million m^3 [Fig. 1]. This fact and the general suitability of the mine were the reasons for being selected as testing site for the disposal of LLW and ILW. Since 1965 the mine is used as a pilot repository as well as research and development site for radioactive waste disposal. Today after developing the Asse mine down to 925-m-level large scale in situ experiments are being prepared or have already been performed in order to develop and to demonstrate suitable techniques for the disposal of HLW/ILW. Some tests aim at the determination of sensitive parameters for the safety analysis for a future HLW-repository in a virgin salt dome, possibly at Gorleben.

In particular the following in situ experiments have to be stated:

- Retrievable HLW-test-disposal in boreholes (GSF, ECN)

  For this test 30 canisters containing a HLW simulate consisting of a glass matrix doped with Sr-90 and Cs-137 have been produced. The test field on the 800-m-level consists of two parallel test galleries, separated by a pillar of 10 m thickness. Eight disposal boreholes will soon be ready for the experiment to begin.

- Retrievable ILW-test-disposal in boreholes (KFA, GSF, BGR)

  The test field on the 800-m-level has been completed with various geotechnical measuring devices. The initial monitoring of the rock stresses has been started. The actual test with the insertion of casks with dissolver sludge and casks with spent HTGR fuel elements will begin in 1989.

- Thermal simulation of direct disposal of spent fuel elements (KfK, DBE, BGR, GSF)

  Backfilling in situ experiments have already been performed in order to develop suitable techniques for embedding 65 Mg casks in drifts with a cross-sectional area of 15 m^2. The development of the test field on the 800-m-level is in progress.

- Dam construction in rock salt (DBE, BGR, GSF)

Above the actual test field two galleries for the instrumentation of measuring boreholes and the performance of geoelectric as well as gravimetric measurements have been developed on the 865-m-level. The excavation of two galleries for the construction of two pilot dams has been started. Initial laboratory experiments and FE-calculations supported the conceptual design of these dams with regard to their short- and long-term stability and impermeability.

General Aspects of Rock Mechanics in Salt Mines

Due to the past mining activities and the existing unbackfilled openings above 750-m-level large scale alterations took place in particular inside the salt structure. Such rock stresses caused long-term deformation processes in and around the existing mine cavities. According to the viscous flow of rock salt a quasi-stationary creeping is being measured at many locations in the Asse mine.

For evaluation of the rock mechanical stability of the entire Asse mine the well established methods of civil engineering are suitable to a very limited extent only. Instead today's state of art provides for a geotechnical safety analysis for subsurface cavities which consists of the following equivalent and consecutive steps.

- Evaluation of practical experiences in similar subsurface cavities in respect of deformation behaviour and mechanical stability.

- Numerical assessment of rock stress changes and the corresponding deformations for construction, operation and post-operation phase of the mine by Finite Element methods.

- Monitoring of the mine deformation and comparison with the assessed convergence rates and rock stresses for verification of applied data base and numerical codes.

In the course of such a stability analysis the surrounding rock mass is being treated as both as building material as well as mechanical support.

Subject to the degree of confidence in the geotechnical data base such a rock mechanical analysis has to be extended over a sufficient period of time in particular if significant changes of the boundary conditions such as extension of cavity volume and/or temperature changes are expected.

Specific parameters of interest are

- rock mechanical deformations in the mine as well as on surface
- stress and carrying capacity of the rock mass affected by mining
- rheological behavior of the rock formations
- fracturing of rock mass
- permeability of rock mass

Respective investigations for determination of these parameters are to be performed where possible, i. e. in shafts, drifts, galleries, caverns and boreholes.

Main task of rock mechanical investigations and stability analysis are

- proof of mechanical stability of underground openings

- dimensioning of large diameter drifts and caverns

- prevention of rock falls and fracturing

- control of the mechanical effects of an artificial temperature increase in the rock mass

- development of technical supports (backfilling) and barriers (dams)

These basic principles were the basis for the geotechnical evaluation of the old workings of the Asse mine as well as for the planning of the newly developed test sites below the 750-m-level. These subjects have been worked up by the Institut für Tieflagerung in collaboration with a number of institutes of universities and in particular with the Federal Institute for Geosciences and Mineral Resources of the Federal Republic of Germany.

In general the non-elastic properties of rock salt and its creeping behaviour permit mining in large stopes virtually without any safety measures such as rock bolts etc. Even today about 50 years after excavation a large number of these stopes is accessible in the Asse mine. Sections of greater rock pressure can be recognised by the deformation of rock pillars, but the stopes as such are still in good shape. A visible kind of deformation is the limited spalling of rock faces especially in the vicinity of old workings. By convergence these rooms have lost and are still loosing part of their sizes. This process seems to follow approximately a closure function that is linear with time. In comparison with other rock types this process will come to an end only when an equilibrium of acting stresses is reached.

Although this deformation doesn't cause any problems either to salt mines nor to radioactive waste repositories in salt there is the risk that the water bearing country rocks might be affected. This could result in a water intrusion into the mine workings with severe consequences to the mine's safety. At a number of potash and rock salt mines such a scenario has been experienced. For this reason rock mechanical investigations are a crucial point in any site investigation for a radioactive waste repository in salt.

Rock Mechanical Investigations In and Around Old Mine Workings

In the old mining area of the Asse Mine an extensive mine survey is being carried out in order to identify and to locate rock mass deformations where possible. At specific locations in the mine the following phenomena have been observed:

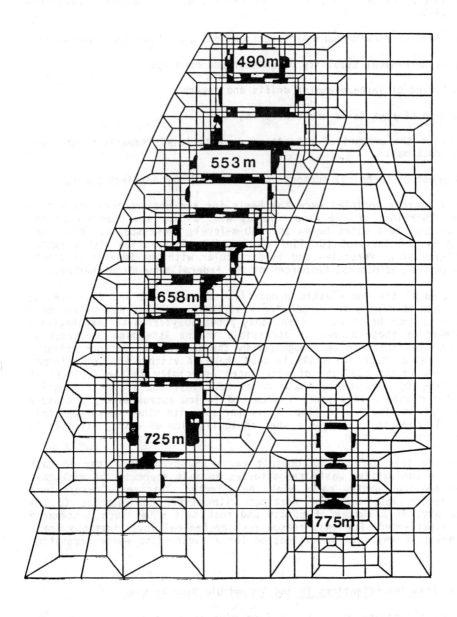

Fig. 2: Zones of tensile stress around the mine workings of Asse II
(tension cut off).

- uplift of footwall in mining chambers and drifts
- rock fall from hanging walls underneath large rooms (edge-shaped plates up to 2,5 m in size and 30 to 100 cm in thickness)
- spalling of pillars between stopes

For exact quantification convergence measurements are being carried out in regular intervals. Altogether 13 measuring stations have been installed between 490- and 700-m-levels located in the center of the old mining area as well as at its periphery. From the long-term measurements it is determined that convergence increases down to 637-m-level. The convergence rate is faster in the central section of the mining area than in other mine workings. The vertical and lateral deformations of rock pillars are being monitored at ten pillars by extensometers. Seven of them have been selected in the central part of the mine close to the southern flank of the salt anticline where rock stresses are increased.

The results achieved so far have shown that the pillar deformation follows at all measuring points a constant rate while its size increases with depth. The data confirm the model calculations performed by LANGER, BGR [Fig. 2] [6]. It was stated by BGR that "by caving of the pillar/stope system, the rock mass reaches an equilibrium state which is characterized by stress redistribution as a consequence of creep deformation. In this a stationary stress can be established for the pillar system which does not exceed the rock pressure. Rock mass failures as a result of the material tensile strength being exceeded are limited to the immediate vicinity of the workings and the pillar system. This failure phenomenon, limited to mine chambers will also continue in the future, however, it does not represent a hazard for the overall load bearing system of the mine. Due to the creep capacity of the salt mass, a sudden collapse of the pillar system needs not to be expected. Over the long term, however, it can not be excluded that a gradual reduction of the mine workings together with advanced deformation and slow destruction of the pillar system could occur".

For verification of the results of FE-calculations all data derived from high precision deformation measurements were evaluated more specific. The following factors were taken into consideration for the calculation of the convergence of stopes, drifts and galleries:

- distance (r) of the measuring point from the central stoping area
- time (t) of development
- rock mass temperature (T)
- depth and size of individual opening
- depth and size of total stoping area
- rock mass density (γ)

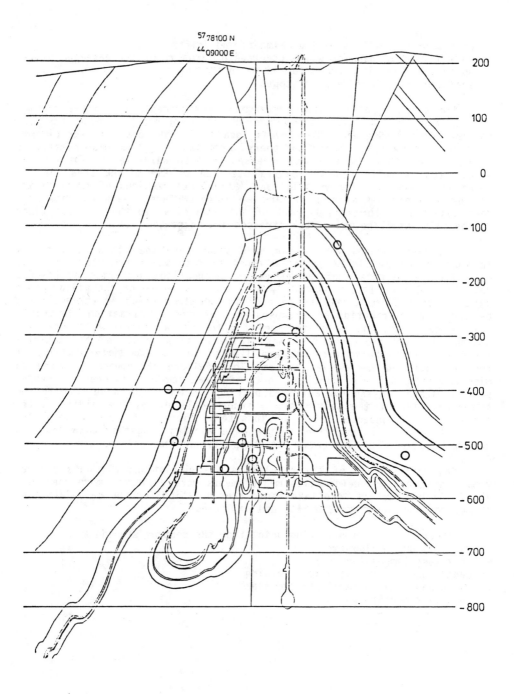

Fig. 3: Locations of the registered seismic events in the Asse Mine
(June 1984 - August 1986).

From extensometer and convergence measurements at various stations and the levelling survey at surface results were achieved in respect of the time dependent deformation of the mine. The different phases of the creeping of rock salt were determined by the evaluation of the deformation rate. The following characteristics can be stated on the basis of regression analyses [11]:

- the settling rate of the surface above the mining area reaches not more than 1 mm/year at the maximum

- the convergence of the mine sections in the younger halite formation is influenced by the acting stress field in the adjacent salt rock formations and the overburden in the southern flank

- lateral displacements are directed from the core of the salt anticline towards the main stoping area at the southern flank

- the intensity of those displacements is changing in accordance with the distance to the main stoping area

- separately surveyed supporting elements such as pillars and roof pillars show uniform deformation rates and are in a non-critical condition of the secondary creeping.

These results of the deformation measurements have confirmed so far the prognosis of the FE-calculations of BGR, that at present no serious consequences regarding the mine's stability have to be considered. The uniformity of the deformation process shows in addition that changes of the rock mechanical situation have not to be expected in the near future.

Microseismic Monitoring of Rock Falls

The above stated results have been achieved by deformation measurements at very specific locations in the mine. In order to get a more overall picture of the deformation processes a microseismic array has been set up in such a way that sudden rock mass relaxations as a result of increased stresses can not only be registered but also be located. This indirect method for identification of deformation processes is in particular a well established monitoring system in mines where there is a significant risk of methane gas eruptions.

The array consists of 11 geophones which are placed on the 490-m-, 750-m- and 800-m-levels. Of those geophones seven register the vertical and four the horizontal components of the deformations. The measurements have reached high accuracy. Today it is possible not only to separate small scale from large scale relaxations but also to differ between relaxation processes and rock falls. From June 1984 till August 1986 240 seismic events were registered of which only 18 have not been located [Fig. 3]. The seismic energy of those events ranges from 10 to 1000 Joule. The registered signal amplitudes can be classified as low, what admits the interpretation that increased rock stress concentrations do not exist in the Asse mine.

The locations of the registered seismic events reveal an area of increased activity in the central mine section and in the east. The explanation for that might be the existance of more brittle salt formations like polyhalite and kieserite near the stopes as well as tectonic effects. On the other hand the rock mass below the 800-m-level was identified as an area with very few events which is virtually inactive in regard of its microseismicity.

Rock falls are known from many locations within the former stoping area. By these seismic investigations rock falls have also been identified in stopes which today are no more accessible. In this way these measurements supplement directly the deformation measurements. They have confirmed the concentration of rock falls in the vicinity of the central stoping area where creeping rates in pillars and convergences are highest.

Geological and Geotechnical Exploration Below 750-m-Level

In the beginning of the 80's a geological exploration programme was started together with the development of the deeper mine sections. This programme was supplemented by 5 deep drillings from the surface. From the evaluation of drilling cores and borehole logs a better knowledge of the external structure of the salt anticline and an advanced geotechnical classification of the salt as well as the covering rock strata were achieved. Summarizing the results of these geotechnical and geological investigations of the Asse structure, a subdivision of eight homogeneous strata sections in the salt and country rocks can be defined.

From these three sections have been identified in the salt series

-   rock salt of the Staßfurt series
-   carnallite seam
-   rock salt of the Leine series

and five sections in the country rock

-   cap rock
-   Buntsandstein
-   Muschelkalk
-   Keuper
-   jurassic and cretaceous formations.

On the basis of specific physical and rock mechanical data from drill cores an advanced FE-model of the Asse is presently being developed by H. K. NIPP, BGR [2].

Preliminary results show that below the old mine workings and with increasing depth and distance a significant change towards the primary rock stress distribution can be expected.

To prove these predictions the following in situ stress measuring techniques have been applied:

- overcoring
- rock stress monitoring
- flat jack
- hydraulic fracturing.

In particular hydraulic fracturing was experienced as most suitable for the determination of stress fields. This was proved by comparing the in situ measurements with the results of an additional accompanying FE-calculation. From the computation of the stress distribution in the near field of a measuring borehole it was derived that the induced fractures must cut at least 50 cm into the rock mass for measuring the uneffected stress field. The orientation of these fractures can be located by seismic methods. Fig. 4 shows the distribution of radial and tangential stresses in the vicinity of a drift on the 775-m-level. The influence of stress redistribution in the near field of the drift can be neglected in a radial distance of 10 m and more from the wallface.

In order to understand the immediate excavation effect on new drifts, geotechnical measuring devices were installed in boreholes from adjacent galleries prior to drift excavation. The readings were taken when the tunneling of the drift approached and passed the instrumentation.

Fig. 5a shows in principle the position of a vertical four point extensometer installed from the pilot tunnel which is located above the drift to be excavated. The displacement history of these anchor points of 1,5 m, 4,5 m, 10,5 m depth from the floor of the pilot tunnel can be seen from Fig. 5b. The sudden increase of displacements beginning after 22 days is the result of the passing of the continuous miner on 865-m-level, 45 m below the pilot tunnel. About 20 days after passing the station a distinct decrease of the deformation rates to lower values can be observed. Similar results were achieved from horizontal extensometers which were directed towards the ILW test gallery on 800-m-level (Fig. 5c). The same procedure can be adopted for measurements of stress changes.

Figure 6a shows in principle the position of stress monitoring boreholes aside and atop of drifts to be excavated. In these boreholes either hydraulically or pneumatically operated stress gauge systems can be installed. The hydraulic stress gauge system is a flat jack system developed by Glötzl whereas the pneumatic type AWID was developed by GSF. The orientation of the individual pads provides for stress measurement in different directions. The example (Fig. 6b) shows increased stresses in the side walls and reduced stresses in the roof after development of a drift on 865-m-level.

Stress measurements

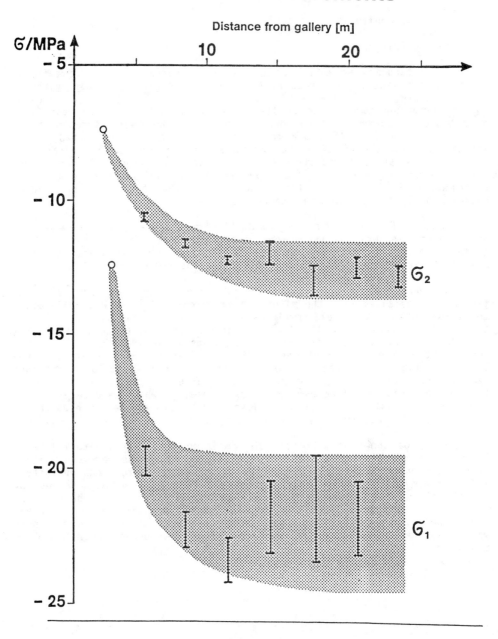

Fig. 4: Stress distribution in the surrounding of a gallery (775-m-level) determined by hydraulic fracturing.

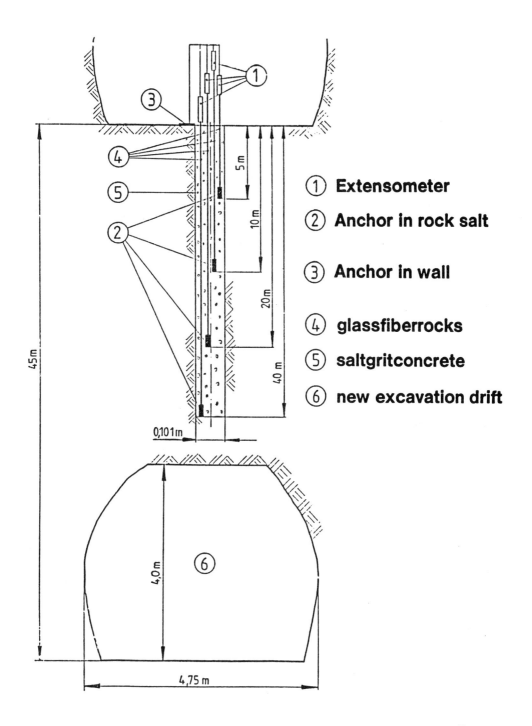

① Extensometer

② Anchor in rock salt

③ Anchor in wall

④ glassfiberrocks

⑤ saltgritconcrete

⑥ new excavation drift

Fig. 5a: Position of the vertical four point extensometer between pilot tunnel and new excavation drift.

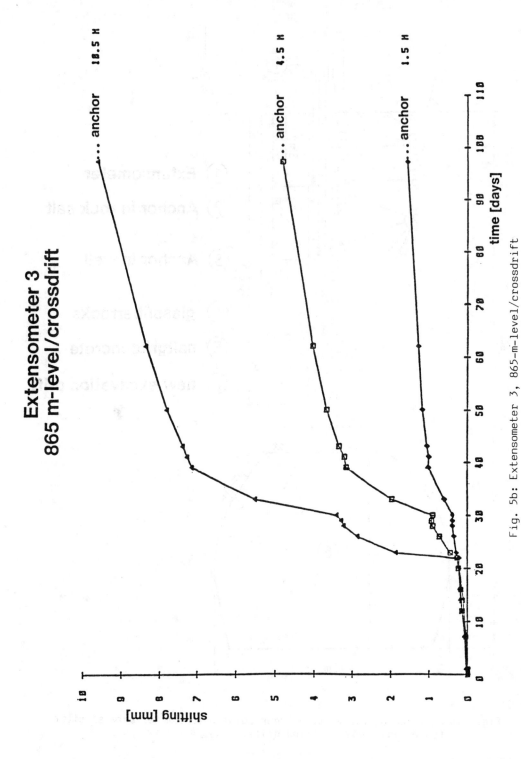

Fig. 5b: Extensometer 3, 865-m-level/crossdrift

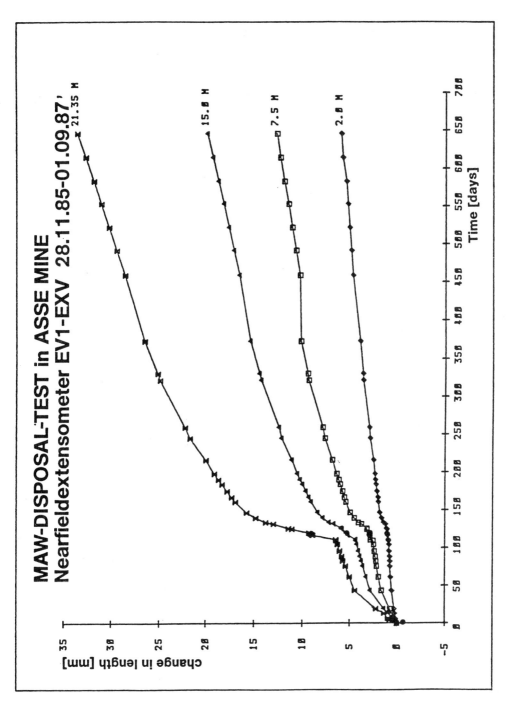

Fig. 5c: Nearfield extensometer, MLW-disposal test site

Stress-Monitor-Stations
Flat Jack "System Glötzl"

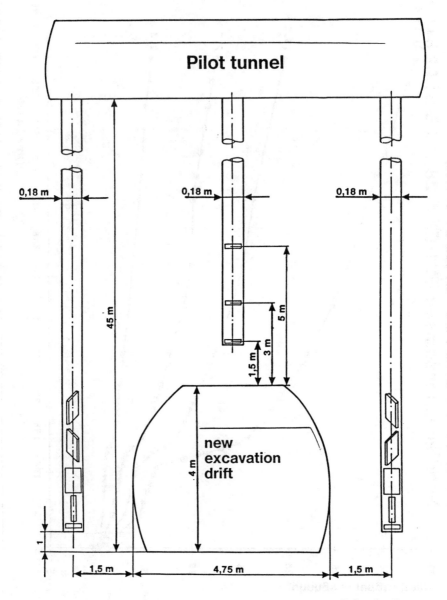

Fig. 6a: Principle position of stress monitoring boreholes in the vicinity of the new excavated drift.

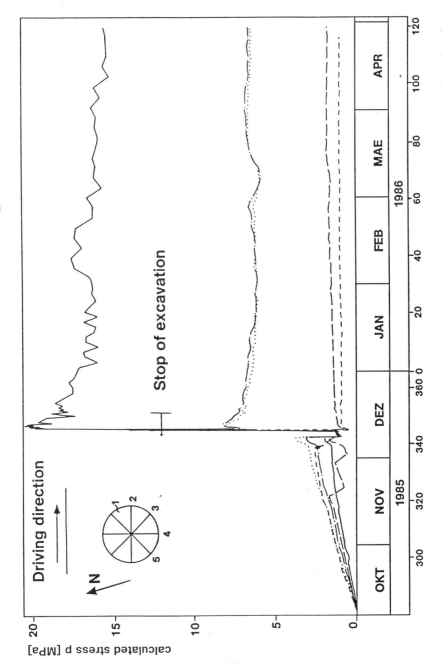

Fig. 6b: Results of the stress monitoring station in the side wall of a drift on the 865-m-level

Deformation Measurement in a Cavern at 979-m-Level

A prototype cavern with a volume of 10.000 m^3 was excavated during the years 1976/77 for field testing and demonstration of a specific disposal method for conditioned intermediate radioactive waste in 200-1-drums. The cavern is situated in the central part of the Asse salt anticline within an unworked area of the Staßfurt Halite formation. The top of this cavity is situated at a depth of 959 m, at that time 185 m below the deepest existing workings.

The cavity has a shape of a prolate ellipsoid, main axis ratio 1.34:1, with a cone superimposed in the top area. The transition areas are smoothed between the ellipsoid and conical part of the cavern as well as between the top of the cone and the central access shaft of elliptical cross section. The total height of the cavity is 37 m with a maximum diameter of 24.1 m at a depth of 979,2 m below surface. The cavity has been excavated by conventional mining. The excavation work was accompanied by comprehensive geotechnical in situ measurements.

First convergence measurements at the central level at 979,2 m depth were carried out on February 16, 1977. The corresponding observations at a lower level at 989,2 m were initiated about two months later on April 12, only a few days before the end of the excavation phase. The measurements at both levels finally had to be discontinued in February 1978 after an observation period of 371 and 316 days respectively. Thereafter the cavern was no more accessible for further manual measurements.

The time dependent horizontal convergence measured at both levels in different directions is very uniform irrespective of these directions. Within the observation period a mean absolute diameter convergence of 134 $\pm$ 1.8 mm was measured at the upper level corresponding to a mean relative convergence of 0.5 % with reference to the measuring points 1,5 m deep in the side walls. The result at the lower level is 77 $\pm$ 0.8 mm or 0.4 % respectively with reference to the same wall depth.

Assuming that opposite reference points converge constantly with time in radial directions the specific absolute radial displacements can be determined for different directions and different wall depths.

For the initial 371 days of observation the absolute deformation of the surrounding rock mass was determined by combined convergence and extensometer measurements. The time dependent radial displacement amounted to 56 mm at 3.4 m wall depth and to about 12 mm at 40.3 m respectively. After that only relative displacements were determined from the extensometer readings. During the following 4 years the relative displacements increased continuously. The specific rates referring to the individual anchor points increase with increasing wall depth. The total average radial strain of the rock mass up to a depth of 40 m was calculated to 0,4 % after 5 years of observation. These data history indicates typical steady state creep at this part of the mine.

Final Remarks

From the described rock mechanical characteristics it becomes clear that the Asse anticline consists of two different areas:

- the old mining area above 750-m-level with a very relaxed stress field
- the newly developed mine section between 750 and 965 m with a more undisturbed stress field.

The large scale in situ demonstration tests mentioned before will be performed on the 800-m-level or deeper. With respect to the less complex and from this point of view more favourable rock mechanical situation in this part of the mine the results being expected can with a higher degree of confidence be applied to other sites, e. g. to Gorleben. In addition it has to be stated that the preliminary planning of a HLW-repository in a virgin salt dome provides for the development of disposal galleries at about 800 m depth. These rock mechanical aspects as well as the geology/mineralogy of the rock strata are additional criteria for the transferability of data.

Bibliographic References

[1] Dreyer, W.: "Underground Storage of Oil and Gas in Salt Deposits and Other Non-hard Rocks", Stuttgart, 1982.

[2] Dürr, K.; Graefe, V.; Liedtke, L. & Meister, P.: "Evaluation of Salt Rock Pillar Stability Utilizing Numerical Calculations, Mine Survey and In Situ Rock Mechanics Measurements", Proc. 6th. Int. Salt Symposium; Toronto, 1983.

[3] Glötzl, R.; Kappei, G.; Meyer, Th. & Schmidt, M. W.: "New Approach to the Long-term Determination of Stress Fields to Reduce Overloading Risks Underground", Proc. Large Rock Caverns, Helsinki, 1986.

[4] Hente, B.; Quijano, A. & Dürr, K.: "Microseismic Observations in a Salt Mine with Reference to the Mine Survey Results", Proc. 4th Conf. on Acoustic Emission, Microseismic Activity in Geologic Structures and Materials, University Park, Pennsylvania, 1985.

[5] Kühn, K.: "A Review of In Situ Investigations in Salt", Proc. of a Joint CEC/NEA Workshop of In Situ Experiments in Underground Laboratories for Radioactive Waste Disposal, Brüssel, 1984.

[6] Langer, M.: "Main Activities of Engineering Geologists in the Field of Radioactive Waste", Bull. of the Int. Assoc. of Engineering Geology, 34 (1986).

[7] Lux, K. H. & Rokahr, R. B.: "Laboratory Investigations and Theoretical Statements as a Basis for the Dimensioning of Caverns in Rock Salt Formations", Proc. 1st Conf. on the Mechanical Behavior of Salt, Clausthal-Zellerfeld, 1984.

[8] Quast, P. & Schmidt, M. W.: "Disposal of MLW, LLW in Leached Caverns", Proc. 6th Int. Salt Symp., Toronto, 1983.

[9] Schmidt, M. W.; Quast, P. & Hawickenbrauck, E.: "Engineering, Geological and Safety Technological Aspects for the Final Disposal of In Situ Consolidated Radioactive Waste in Hardrock and Salt Formations", Proc. 5th Int. Congr. Intern. Assoc. of Engineering Geology, Buenos Aires, 1986.

[10] Staupendahl, G. & Schmidt, M. W.: "Field Investigation of the Long-term Deformation Behaviour of a 1000 m^3 Cavity at the Asse Salt Mine", Proc. 1st Conf. on the Mechanical Behavior of Salt, Trans Tech Publ., Clausthal-Zellerfeld, 1985.

[11] Dürr, K.: "Deformationsmessungen zur Beurteilung der Standsicherheit in einem ehemaligen Salzbergwerk", Proc. Vth Int. Symp. for Mine Survey, Varna, 1982.

Session I - Séance I

CHAIRMAN'S SUMMARY - RAPPORT DE SYNTHESE DU PRESIDENT

K.W. DORMUTH (AECL, Canada)

* *

The three overview papers provided overviews of research performed to investigate excavation response in clay, crystalline rock and salt. The comments that follow are based on the opinions of the authors and the chairman, and are specific to these three media.

Relevance of Excavation Effects to Repository Design

In crystalline rock, the effects of excavation on the rock structure, which include the development of damaged zones around openings, are essentially permanent. These effects are clearly important to the engineering and effectiveness of seals for repository shafts and drifts. Excavation effects are not expected to prove important to the short-term performance of a repository, nor are they expected to present problems for construction and operation. Their implications to long-term safety have yet to be fully analysed.

In clay, the effects of excavation are very important to the design of the excavation and support of the walls. These effects include convergence of walls, fracturing of surrounding rock and dissipation of pore-water pressure. Because of the self-healing characteristics of clay, they are not expected to be important to the long-term safety of the facility, providing some care is taken to keep their range and intensity within reasonable limits during construction. Near-field analyses must take account of the alteration caused by excavation damage.

In salt, there are important short-term effects, possibly for 500 years. In the long-term, excavation effects will not be significant because of the self-healing characteristics of the salt. However, the combined effects of excavation, heat and radiation as they influence long-term creep are important to long-term safety analysis. At the Asse mine, large rock-mass deformations take place due to the creeping of the salt, and there are distinct changes in the stress field in space and time. The greatest deformations occur in the central area of the excavated mine.

Current Understanding of Excavation Response

It is not certain whether our knowledge of excavation response in crystalline rock is sufficient to allow design of suitable repository seals and assessment of long-term safety implications. Equipment and techniques, not currently available, are required to better determine the extent and properties of the disturbed zone. Mathematical models are required to analyse the behaviour of the disturbed zone and to assess its safety implications.

Excavation effects are not well understood in clay. However, since the effects are short-term, such a detailed understanding may not be important or necessary. In salt, excavation effects are understood well enough for the purpose of repository construction. However, the long-term combined effects of excavation, pressure, heat and radiation as they affect the creep of the rock are not well understood.

Requirements for Instrumentation Development

Work is needed to evaluate the accuracy and applicability of existing instrumentation for measurement of rock-mass response, and to develop instrumentation of increased accuracy. Instrumentation development is also required to produce instruments of sufficient longevity under in-situ conditions to allow testing and monitoring for periods of many years.

Sealing-Related Comments

The excavation response zone has a major effect on the design and effectiveness of repository seals. Therefore, it is vital that repository seals be tested under representative in-situ conditions so that the influence of the excavation response zone can be properly evaluated. In the case of crystalline rock, significant work may be required to define the nature of the excavation damage zone sufficiently well for reliable shaft and drift seal design.

General Comments

There has not been sufficient involvement of repository performance modelling done to assess the importance of excavation response to repository performance. Even though the characteristics of the damage zone cannot be very accurately quantified in many cases, scoping studies based on repository performance requirements are needed to clarify issues, define important characteristics and parameters, and help define the need for measurement and remedial action. Closer co-ordination is needed among the scientists, engineers and system modellers to properly define the issues and develop experimental strategies.

It should be realised that the response of the rock mass to excavation can be a very useful tool in validating models for long-term performance assessment. The prediction and subsequent comparison with measurement of the relatively short-term effects can be used to substantiate the applicability of the models on the actual scale and under the actual conditions of the repository environment.

SESSION II(a)

EXCAVATION RESPONSE IN CRYSTALLINE ROCK
- OVERVIEW OF STUDIES

SEANCE II(a)

EFFETS DE L'EXCAVATION SUR LES ROCHES CRISTALLINES
- PRESENTATION DES ETUDES

Chairman - Président

R.A. ROBINSON
(Battelle, United States)

EXCAVATION RESPONSES IN DEVELOPING
UNDERGROUND REPOSITORIES IN FRACTURED
HARD ROCK IN SWITZERLAND

R. W. Lieb(1), P. Zuidema(1) and M. Gysel(2)
Nagra(1) and Motor-Columbus Consulting Engineers Inc.(2)
CH-5401 Baden, Switzerland

ABSTRACT

Keywords: Hydraulic short circuits, hydraulic cage and near-field
hydraulics, excavation-induced stress states and their influence on engineered
barriers and permeability, chemical effects, smoothness of the excavation
surface and skin effect.

At the Nagra Grimsel test site (granite), where excavation effects
caused by tunnel boring were investigated, no significant influence of rock
decompression on hydraulic permeability could be measured. In addition, Nagra
has conducted preparatory studies on decompressed zones in the fractured
gneiss of the Piz Pian Grand and in the marls of the Oberbauenstock/Wellenberg
which are potential host rocks for an L/ILW repository.

EFFETS DES TRAVAUX D'EXCAVATION DANS LES DEPOTS
SOUTERRAINS EN COURS DE MISE EN PLACE DANS UNE FORMATION
DURE EN SUISSE

RESUME

Mots-clés : court-circuits hydrauliques, cage hydraulique et phénomènes
hydrauliques dans le champ proche, état de contrainte induit par les travaux
d'excavation et leur influence sur les barrières ouvragés et la perméabilité,
effets chimiques, régularité de la surface de l'excavation et effet de peau.

Au laboratoire souterrain de Grimsel, où la Société coopérative
nationale pour l'entreposage des déchets radioactifs (CEDRA) a étudié les
effets induits par le creusement d'un tunnel, le relâchement des contraintes
dans la roche n'a pas entraîné de modifications mesurables de la perméabilité
hydraulique. En outre, la CEDRA a entrepris des études préparatoires sur les
zones de relâchement des contraintes dans le gneiss fissuré de Piz Pian Grand
et dans les marnes de l'Oberbauenstock/Wellenberg, formations susceptibles de
servir de roche réceptrice pour un dépôt de déchets de faible et moyenne
activité.

INTRODUCTION

The performance of underground repositories for the final disposal of nuclear wastes strongly depends on the governing hydraulic conditions in the rock formations surrounding the repository. Low water flow rates through the rock formations as a whole are required to provide long enough transport times for radionuclides from the repository to the biosphere. It is generally recognized that the excavation and construction of the repository itself may significantly influence these hydraulic conditions.

An excavated open tunnel will affect the hydraulic flow field in the host rock formation by means of its drainage effect. Even a backfilled and perfectly sealed tunnel may still change the virgin hydraulic conditions in the rock since the excavation operation, with the induced stress changes in the rock around the opening, may alter the hydraulic permeability of the rock around the tunnel.

Excavation effects are being investigated by Nagra, the Swiss National Cooperative for the Storage of Radioactive Waste by evaluation of data from existing underground structures, theoretical analysis, modeling and specific in-situ tests.

In 1983/84 the Grimsel Test Site (GTS) was constructed in the granites and granodiorites of the crystalline Aar-massif in central Switzerland. The first major investigation program at the GTS was the test on excavation effects [1].

The Grimsel test results on excavation effects are directly relevant for two repository options considered in Nagra's disposal program, namely the HLW repository in crystalline bedrock in Northern Switzerland and the possible L/ILW site in the gneiss at Piz Pian Grand in the Alpine region. Other types of hard, fractured rock (marl) are investigated by Nagra for the potential L/ILW sites at the Oberbauenstock or Wellenberg. Although being a sediment, the marl at these two sites is best conceptualized as a hard, fractured medium.

THE IMPORTANCE OF EXCAVATION EFFECTS

Excavation effects may influence the performance of a repository system in many ways. Some examples from the Swiss program are as follows:

Hydraulic Short-Circuits

The altered zone around tunnels or shafts may increase the longitudinal permeability of the host rock along these structures and may thus create hydraulic short-circuits from the repository to the biosphere. The whole layout of a repository design may be affected by such considerations.

In the case of the Wellenberg L/ILW site, for example [2], Nagra at first had planned a direct access tunnel from the surface to the repository. Hydraulic calculations, however, showed the necessity for an indirect adit reaching the repository from the hillside. This arrangement (Fig. 1) substantially increases the hydraulic flow time and avoids flow along the tunnel from the repository to the portal. Lateral seepage from the repository into the access gallery is prevented by an adequate distance. Furthermore, excavation effects along the loop behind the repository will increase the drainage effect of that portion of the adit, thereby lowering the hydraulic potential and reducing the flux through the repository.

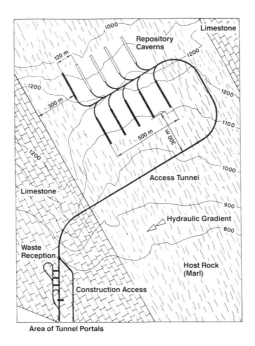

Area of Tunnel Portals

Figure 1.   Preliminary layout of the potential Wellenberg L/ILW
            repository.

Hydraulic Cage and Near-Field Hydraulics

        The groundwater flow <u>around</u> a repository in fractured hard rock is of
interest not only for the radionuclide transport in the rock but also as a
boundary for the near-field barrier system, with respect to radionuclide release and barrier degradation. The groundwater flow <u>through</u> the repository determines the relative importance of the different radionuclide transport mechanisms. In addition, it substantially affects the long-term integrity of the
near-field barriers and the rate of nuclide release.
        The groundwater flow through a repository has been calculated by
Höglund [3] and the results show that the presence of a zone with increased
permeability around the caverns has great impact on the hydraulic flow field
inside and around the repository. Such a zone can be advantageous regarding the
release rates and the integrity of the barriers by forming a hydraulic cage
around the repository, i.e., by diverting the flux around the repository instead of through it.
        Flow conditions in and around the repository are governed by the geome-
try of the system, geological features of the host rock and the magnitude of
the various permeability values relative to each other (Fig. 2). Quantitative
determination of the excavation effects is important in this connection.

A significant reduction of overall near-field releases can be achieved by appropriate positioning of the waste containers with respect to the fracture zones with increased permeabilities.

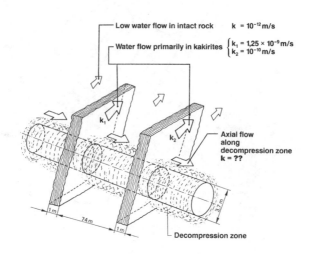

Figure 2. Near-field hydrological model from "Projekt Gewähr 1985" [4].

Excavation-Induced Stress States
(influence on engineered barriers and on near-field host rock permeability)

The engineered barriers, such as tunnel linings, bentonite backfill and canisters may be mechanically affected because of excavation effects. Excavation of a repository tunnel induces stress and strain changes in the host rock. Stresses in the vicinity of the tunnel may exceed rock strength leading to fracturing. The rock pressures acting on the tunnel lining and backfill material may vary in time depending on the long-term behavior of the rock mass affected by the excavation.

In some types of fractured hard rocks (e.g., marl), rheological effects may play a dominant role because of the long time-spans involved. Rock creep rates, which might be neglected for ordinary tunnel structures with a life span of 100 years, might, in the case of a final repository, result in important stresses or fissuring of the technical barriers at later times. Rock mechanical methods used in analyzing such rheological effects are available. However, the measuring of relevant rheological parameters of low to very low creeping rock is still considered to be difficult. Rock mechanical stress changes resulting from the excavation may cause time-dependent changes of the permeability in the near-field area (decompressed zone) due to closing of fissures and for certain rock types, due to rock swelling [5].

Other uncertainties are related to the continuum/discontinuum aspect of the rock around the tunnel. The real "block and joint structure" may have a large influence not only on the stressing of the engineered barriers but also on the near-field hydraulics. Fairhurst and Hart [6] describe a numeric block model which may be used to study permeability changes in joint systems near a tunnel under varying stress conditions in the rock. Fairhurst [7] advances one step further when he describes a hybrid analysis model, in which the near-field discontinuum model is imbedded in a continuum far-field model (Fig. 3).

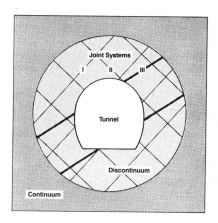

Figure 3. Hybrid model for rock mechanical hydraulic analysis [7].

Chemical Effects

Measures taken for support or consolidation of the host rock may alter the chemical near-field environment. Such measures could, for example, be rock bolting, prestressed anchors or low and high pressure grouting using, for instance, cement-bentonite-water mixes, which are used to counter rock mechanical and hydraulic excavation effects. Chemical changes on rock surfaces in fractures may affect the sorption capacity for radionuclides. Smaller permeability may have to be paid for by less predictable sorption behavior when using grouting procedures. Special durable mixes will have to be developed in order to maximize long-term stability.

Smoothness of the Excavation Surface

Regular shape and smoothness of the excavated rock surface is an important prerequisite for homogeneous backfilling/sealing of a repository tunnel. Edges and holes in the rock surface, therefore, have to be avoided or smoothened out after excavation. In general, mechanical excavation with tunnel boring machines or road headers results in smooth surfaces. Appropriate blasting techniques often deliver good results as well. Experience shows, however, that the mechanical excavation normally induces the least possible fissuration and, therefore, the smallest possible permeability change.

Skin Effect

From water head measurements in ventilation tests at the Stripa mine during the SAC-Phase (Swedish-American Cooperative Program), Wilson et al. [8] derived the indication of a "skin" of lower permeability between the drift wall and the zone more inside the granite formation. Based on the thickness of the skin layer (about 2.5 m), the average (radial) skin permeability was analyzed to be about one-third of the average permeability of the rock mass. Wilson et al. [8] remark: "Compressive rock stresses around the drift, chemical precipitates left in the rock by the evaporating water, and two-phase flow from gases coming out of solution in the ground water are some of the processes which may be responsible for the presence of the skin."

On the other hand, it is noted that a decompressed zone around a tunnel is subject to a volume increase and that shear stresses may increase. Therefore, it is to be expected that existing fissures are enlarged or that new fissures are generated [5]. Thus the decompressed zone would tend to have an increased hydraulic permeability except in cases where the excavation-induced deformations are purely elastic.

Skin effects may be responsible for some strange effects measured or observed during experiments in underground galleries and caverns.

SWISS INVESTIGATIONS ON EXCAVATION EFFECTS

Excavation effects can be assessed by theoretical analysis, by laboratory model tests, by in-situ experiments or by a combination of all of them. Amongst others, the following quantitative investigations of such effects were carried out by Nagra:

Mean Permeabilities of Potential Host Rocks

An important task is the assessment of the mean permeability of the undisturbed rock formation outside the decompressed zone. Since the Nagra repository projects are sited in fractured hard rock, it was attempted to evaluate experience with approximately 25 Swiss pressure tunnels which cross through a broad range of fractured hard rock formations in the Alps. Mean permeabilities of geological formations extending over a few to many kilometers were derived by Gysel [9] from water losses measured in these pressure tunnels. The typical measuring arrangement is shown in Fig. 5.

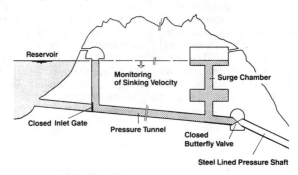

Figure 4. Arrangement of water loss measurements in pressure tunnels.

Gysel then tried to classify fractured hard rocks in the Swiss Alpine region according to their mean formation permeability as follows:

Table I: Permeability of Fractured Hard Rock [9]

Geological formation	Permeability (m/s)
Crystalline schists and gneisses	(*) 10^{-8} to 10^{-9}
Bündnerschiefer (clayey/calcareous schists) and Flysch	10^{-7} to 5×10^{-8}(*)
Granite and porphyric granite	5×10^{-7} to 10^{-7}
Dolomite and limestone (not karstified)	5×10^{-7} to 10^{-7}(*)
Tertiary sandstone (folded Molasse)	5×10^{-6} to 10^{-6}

*Range extends by a factor of five for a few measurements.

Decompressed Zone in Granitic Gneiss

The theoretical approach adopted by Nagra so far for the evaluation of hydraulic excavation effects starts with analyzing the deformations due to the excavation. The deformations of the decompressed zone around the tunnel are then allocated fully to increases in fracture width which are subsequently converted to permeability increases using cubic laws.

In the case of the Piz Pian Grand site, the elastic-plastic analysis of the repository access tunnel of 6.0 m diameter, approximately 1,000 m below ground in granitic gneiss resulted in an altered zone (plastic or decompressed zone) having a thickness of approximately 11 m (Fig. 5).

This result is based on the conservative assumption that no rock support is acting before the deformations induced by the stress changes are fully developed. It was further assumed that the volume increase of the decompressed zone manifests itself in two orthogonal sets of fracture systems which are both parallel to the tunnel axis. A third system perpendicular to the tunnel axis cannot develop due to the planar strain condition in each tunnel cross section. From the assumed spacing of the water-bearing fractures (1 m in the example), the permeability increase relative to the mean permeability of the host rock was computed by using laboratory test results by Louis [10] on the permeability of orthogonally fractured rock.

Permeability increases due to shear movement on fractures were disregarded in this case. Their mechanism has been investigated in laboratory tests by Barton [11].

As indicated in Fig. 5, the decompressed zone was divided into a number of circular rings. Maximum and minimum permeability values delimit the shaded range. Maximum values are based on the assumption that the pre-existing rock fissues are opened up in the decompressed zone. Therefore, the permeability caused by the fissure width increase has to be superimposed to the pre-existing mean rock permeability. Minimum values assume that the new fissures are not connected to the pre-existing joints. In this case, no superposition of pre-existing and newly opened up permeability is assumed.

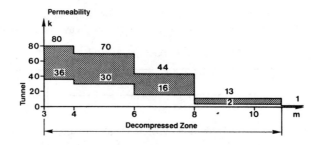

Figure 5. Piz Pian Grand L/ILW project site. Normalized permeability distribution of the decompressed zone around the access tunnel; mean permeability of undisturbed rock = 1.0.

Studies of Swelling Potential and Field Tests on Marl

The marl formation at the potential L/ILW sites of Oberbauenstock and Wellenberg contains some tectonized calcite-containing fracture systems. The marls exhibit a moderate potential for rock swelling at stress changes. Nagra investigated whether the swelling potential under certain conditions tends to close existing fissures in the rock, notably in the excavation-induced decompressed zone [5].

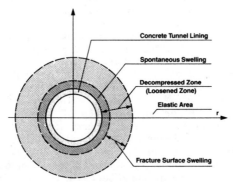

Figure 6. Decompressed zone with definition of swelling zones in marl.

It was found that the spontaneous swelling, which takes place after the tunnel excavation has taken place, occurs only in an inner part of the "plastic" zone because of the relatively high overburden of 500 to 1,000 m. The high stress level of the primary stress state, therefore, tends to minimize the swelling zones around the tunnel [12]. On the other hand, Gysel [5] defined a "fracture surface swelling" which may occur on the surfaces of open fractures, fracture fissures, etc., because of circulating water. This phenomenon can also occur in areas where the continuum theory predicts no swelling. A self-healing process is therefore expected because of the combined spontaneous and fracture surface swelling depending, however, on the intensity of the swell parameters of the marl.

Test on Excavation Effects at the Grimsel Test Site (GTS)

The major part of the excavation test at the GTS [1] was executed in 1983, i.e., at a relatively early stage of the Swiss repository program. Hence, the investigation program for this test was based on rock mechanical and hydraulic considerations with limited input from performance assessment. Measurements of the greatest possible accuracy were made in and around a horizontal tunnel of 30 m in length and 3.50 m in diameter before, during and after it was excavated by a full-face tunnel boring machine.

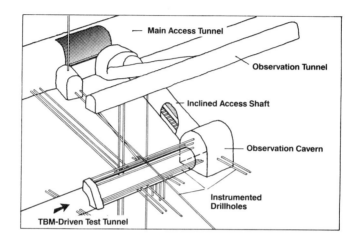

Figure 7. Grimsel Test Site: Tests on excavation effects.

The test section lies 450 m below ground in a parallel textured granodiorite formation with a compressive strength of 120 MPa and a rock matrix permeability in the order of less than 10^{-11} m/s. It is intersected by three steeply inclined schistose fracture zones. The total length of the 38 fully cored boreholes (dia 101 mm) drilled parellel and perpendicular to the tunnel axis amounts to nearly 600 m.

The main parameters measured were displacements and rock deformability in boreholes using sliding micrometers and dilatometers; deformation (convergence) of the tunnel profile; p- and s-wave velocities, rock stresses and permeability by hydraulic single hole injection tests. The instrumentation worked well and provided consistent data. However, changes of the above-mentioned parameters and notably changes of the permeability found after excavation can largely be explained by elastic deformation of the pierced rock mass and the change in hydraulic boundary conditions due to the tunnel (hole in the former full rock plate). A relatively substantial increase in permeability was found in the mentioned fracture zones, but no significant increase could be measured in areas of sound rock, not even in the immediate vicinity of the tunnel wall.

More recently, P. Baertschi (Nagra) has conceived a vacuum bell (dia 15 cm), which can be attached to the rock surface and tightly sealed in a circular notch sawed into the rock. It allows performance, under accurately controlled conditions, of long-term vacuum or ventilated tests directly at the tunnel surface. Application to date focussed on solute transport, capillary action and extraction of pore water for chemical analysis. The equipment appears to be well suited for further investigation of skin effects.

Some typical data from the excavation test at the GTS has been summarized by Egger [1]. More details will be published in due course in a Nagra Technical Report (NTB). The test did not show the formation of a tube of altered rock with increased permeability along the circular drift excavated by the tunnel boring machine. However, the GTS case reflects a stress distribution around the tunnel which, due to the comparatively small overburden of 450 m and the high strength of the granodiorite, did not induce plastic deformation or fracturing of the rock.

CONCLUSIONS

A number of potential excavation effects in fractured hard rock are known in principle. Uncertainty still exists with regard to the interaction and the relative importance for system performance of the individual effects. This depends largely on the type and quality of the host rock and on system design.

Both the hydraulic effects as well as the long-term rock mechanical behavior are considered to be crucial. Further work should be directed primarily at identifying and understanding the excavation effects. Thereafter, it must be established which parameters are governing these effects and how they are best measured.

Based on such fundamental work, it should then be possible to determine for a particular case (host rock and repository system) which excavation effects are relevant and important, and to conceive an investigation program to quantify them.

REFERENCES

[1]   Egger, P.: "Nagra, Grimsel Test Site, Tests for Excavation Effects (Machine-Excavated Tunnel), Preliminary Summary of Test Results and their Interpretation", Federal Institute of Technology, Lausanne, Internal Nagra Report, Baden, September 1986.

[2]   Nagra, Hydrodynamische Modellierung Wellenberg, Motor-Columbus Consulting Engineers Inc., Baden, Nagra NTB 87-18, Baden, in press.

[3]   Höglund, L.O.: "Working Report, Near-Field Hydrology", Kemakta Consultants/Stockholm, Internal Nagra Report, Baden, August 1986.

[4]   Nagra: "Projekt Gewähr 1985, Nuclear Waste Management in Switzerland: Feasibility Studies and Safety Analysis", Nagra NGB 85-09, Baden, June 1985

[5]   Gysel, M.: "Auflockerungszonen um Stollen und Kavernen im Valanginienmergel", Motor-Columbus Consulting Engineers Inc., Baden, Nagra NTB 85-30, Baden, January 1985.

[6]   Fairhurst, Ch. and Hart, Roger D.: "Verification and Validation of Coupled Mechanical/Water Flow Effects in Rock Masses. Some Possibilities and Limitations", Proc. GEOVAL-87 Symposium, Stockholm, April 7-9 1987.

[7]   Fairhurst, Ch.: "Summary of Session C1, Case Histories", Proc. of the
       International Symposium on Large Rock Caverns, Helsinki, 25-28
       August 1986, Vol. 3, 1709-1712.

[8]   Wilson, C.R. et al.: "Large-Scale Hydraulic Conductivity Measurements in
       Fractured Granite", Int. J. Rock Mech. Min. Sci. & Geomech. Abstr., 20
       No. 6 (1983), 269-276.

[9]   Gysel,   M.:   "Bestimmung   der   Felsdurchlässigkeit   aufgrund   von
       Stollen-Abpressversuchen", Wasser, Energie, Luft, Vol. 76, No. 7/8,
       1983.

[10]   Louis, C.: "Strömungsvorgänge in klüftigen Medien und ihre Wirkung auf
       die Standsicherheit von Bauwerken und Böschungen im Fels", Veröffentli-
       chungen des Instituts für Bodenmechanik und Felsmechanik der Universität
       Karlsruhe, Heft 30, 1967.

[11]   Barton, N. et al.: "The Modelling and Measurement of Superconducting
       Rock Joints", Proc. of the 26th US Symposium on Rock Mechanics/Rapid
       City, SD, 26-28 June 1985.

[12]   Gysel, M.: "Design of Tunnels in Swelling Rock", Rock Mechanics and Rock
       Engineering, Vol. 20, No. 4, 1987.

CALCULATIONS OF FLOW ALONG FRACTURES
USING STRUCTURAL DATA

L. Liedtke and A. Pahl
Federal Institute for Geosciences and Natural Resources
Hannover, Federal Republic of Germany

ABSTRACT

The Federal Institute for Geosciences and Natural Resources (BGR) is
conducting flow tests in fracture systems in the Grimsel Rock Laboratory,
within a cooperation program with NAGRA (Switzerland) and GSF (Munich, FRG).
For this purpose, it is absolutely necessary to combine structural mapping and
hydraulic tests.

The use of statistics for a computerised evaluation of flow systems is
discussed using examples of fracture systems. In addition, the influence of
boreholes on flow in the fractures was investigated. Model calculations of
the flow through the fractured rock that has been mapped in detail yielded
estimates of the flow paths in the rock between ground level and drifts at
great depth.

CALCULS DE L'ECOULEMENT LE LONG DES FISSURES
A PARTIR DE DONNEES SUR LA STRUCTURE DE LA FORMATION

RESUME

Dans le cadre d'une collaboration avec la Société coopérative nationale
pour l'entreposage des déchets radioactifs (CEDRA, Suisse), et la Société pour
la recherche sur les rayonnements et l'environnement (GSF, Munich, RFA),
l'Institut fédéral des Sciences de la terre et des matières brutes (BGR)
réalise des essais d'écoulement dans des systèmes fissurés au laboratoire
souterrain de Grimsel. Pour ce faire, il est absolument nécessaire de
conjuguer le levé de cartes structurales et la réalisation d'essais
hydrauliques.

Partant d'exemples de systèmes de fissures, les auteurs examinent dans
quelle mesure les statistiques peuvent être utilisées pour une évaluation
informatisée de systèmes d'écoulement. En outre, les auteurs ont étudié
l'influence sur l'écoulement dans les fissures des trous de forage exécutés.
Il a été procédé à une simulation mathématique de l'écoulement à travers la
roche fissurée dont on a établi un levé détaillé ; on a ainsi obtenu des
estimations des cheminements d'écoulement dans la roche entre le niveau du sol
et les galeries aménagées à grande profondeur.

1. INTRODUCTION

The flow behavior and the rate of propagation of dissolved pollutants must be known in order to assess the propagation of toxic substances and other pollutants in repositories in mines in the case of a leaking waste container. In-situ hydraulic experiments have been made to determine the flow paths in fractures and faults in a virtually impermeable rock matrix in the Grimsel Rock Laboratory in Switzerland. The Grimsel Rock Laboratory is situated in a transition zone between the Grimsel Granodiorite and the Central Aare Granite, the test site itself is in the Central Aare Granite. The test site and surroundings are described in detail in [1 & 2].

2. TRACER TESTS

Ten core boreholes were drilled to a depth of up to 135 m in the rock of the test site. The location and orientation of these boreholes were chosen according to the geological structures and hydraulic aspects. Numerous injection tests and geological drill cores showed the possible flow paths, which together with the results of simple numerical models provided information on the amounts of water that could be injected and the level of the water in the observation boreholes. If the rock were not jointed, but uniformly permeable like gravel or sand, then these experiments would also provide information on permeability and flow paths. But because the water flows almost only in fractures and fracture zones, tracer tests must be conducted to determine the flow rate.

On the basis of these permeability tests and geological drill cores, several open fractures could be located. Two intersecting fractures are shown in Figure 1, together with the tunnels of the rock laboratory, the test site, and the boreholes for the flow test.

These two intersecting fractures do not have a geometrically regular form and consist of several more or less parallel fractures. To determine the flow rate, all of the boreholes at the test site are sealed off by pneumatic packers except for the two intersecting fracture systems (Fig. 1). The two fracture systems are divided into four sections in this figure. Water was injected into section 4 at a uniform rate of 10 - 15 L/min and a constant rate of flow of water from section 1 was the prerequisite for assuming steady-state conditions. Freshwater and saline water were injected alternatingly. The electrical conductivity was 200 µS/cm. The brine was injected for 5 h and the freshwater for 19 h. This cycle was repeated three times. The electrical conductivity of the injected water was measured at the point of injection (the source), the point of ejection (the sink) and several points within the rock. The electrical conductivity of the water as a function of time at the sink is shown in Figure 2 [1]. The rate of injection was 15 L/min for this test and travel times of 80 and 88 min for the brine was determined.

3. NUMERICAL MODEL AND THE COMPUTER PROGRAM

3.1 General

The numerical program package DURST (DURchströmung und SToff transport) was used to calculate the flow field and transport of substances. This is a finite-element program system for calculating flow and tranport in fractured rock and can be divided into three main parts: the flow model, the transport model, and the graphics package.

Both steady-state and non-steady-state groundwater flow in porous media, as well as flow in fractured media, can be calculated with the flow model. The general Darcy flow law for inhomogenous, anisotropic aquifers and the non-linear flow law for joint and tube flow are used to describe the flow. The program uses isoparametric hexagonal elements to represent the continuum, isoparametric rectangular elements for the fractures, and straight line elements for the cylinders. By superposing the nodes, the elements can degenerate to trianglular, prismatic, or tetragonal elements.

The propagation of an ideal dissolved substance with flowing groundwater can be simulated with the transport model. This can be done for both porous media and water in fracture systems. A velocity field from a flow model is needed to calculate the transport paths. In the simplest case, the calculation can be limited to purely advective transport. Diffusive transport can be additionally modelled on the basis of the Fick's First Law; either isotropic molecular diffusion or orthotropic diffusion or dispersion as a function of velocity can be used. Moreover, the transport model can simulate the non-conservative behavior of radioactive substance according to the first-order rate law if the value of the half-life is given.

The transport model is based on the following considerations:
The groundwater serves as tranport medium for a substance with a concentration

$$C = \frac{m_c}{B} = \rho\, c, \tag{1}$$

where
$$\rho = \frac{m}{B}, \tag{2}$$

m = mass of the solution with a volume B,
m_c = mass of the substance in a volume B,
ρ = density of the solution.

The movement (steady-state or non-steady-state) of the groundwater, i.e. solution, is described by velocity tensors $v = (v_x, v_y, v_z)$ within a cartesian coordinate system. It is assumed that the density of the solution ρ has no influence on the velocity field $\underline{v}$, which is the total volume through which the water flows and B is a volume selected for monitoring the transport. The following mass balance can then be set up for the dissolved substance:

$$\int_B \frac{\partial C}{\partial t}\, dV = \int_B r\, dV - \int_{\partial B} \underline{J} \cdot \underline{n}\, dA, \tag{3}$$

where r is the rate of production (r > 0) or decomposition (r < 0) of the substance and $\underline{J} = (J_x, J_y, J_z)$, which represent the velocity of the substance:

$$\underline{J}(C) = \underline{v}\, C - \underline{D} \cdot \text{grad } C, \tag{4}$$

where $\underline{v}$ C is the advective flow rate, $\underline{D}$ is the symmetrical diffusion tensor and $\underline{D}$ · grad C is the dispersion flow rate. $\underline{J}$ · $\underline{n}$ is the velocity of the sub-stance emerging from the monitored volume B through the surface ∂B, where n is the vectors normal to the surface ∂B. The right side of Equation (3) describes the increase in the mass of the substance with time in volume B, the left side gives the corresponding local concentration changes.

The dispersion flow rate is symmetric

$$W = -\underline{D} \cdot grad\ C \qquad\qquad (5)$$

and can be written out as follows:

$$\begin{bmatrix} W_x \\ W_y \\ W_z \end{bmatrix} = - \begin{bmatrix} D_{xx} & D_{xy} & D_{xz} \\ D_{yx} & D_{yy} & D_{yz} \\ D_{zx} & D_{zy} & D_{zz} \end{bmatrix} \cdot \begin{bmatrix} \partial C/\partial_x \\ \partial C/\partial_y \\ \partial C/\partial_z \end{bmatrix} .$$

In the case of anisotropy, the dispersion coefficients can be reduced to D (in m²/sec). In the case of orthotropy, coefficients are needed for the direction of flow and perpendicular to the flow.

3.2 Geometrical Model

The planes 1 and 4 in Figure 1 are rotated about their line of inter-section so that they are superposed to form a square plate. This plate model is divided into 1296 elements 1.33 m to a side defined by 4 nodes each. The injection point (source) and exit point (sink) are represented by two elements about 32 m apart on a diagonal of the plate model. The changes in the water levels in the observation boreholes are shown in Figure 3 for an injection rate of 15 L/min. These levels were calculated using the flow model and are shown only for the nodes of the macro-elements. Each macro-element consists of 9 elements. The velocity tensor calculated from these levels is needed to determine the parameter variations for the breakthrough curve.

4. VARIATION OF THE PARAMETERS

As mentioned above, tracer tests were carried out to determine the flow rate of the water in the fractures. The time between the injection is started and the first appearance of a measurable change of concentration of the tracer at the sink is termed the breakthrough time. The breakthrough time is a func-tion of flow rate in the fracture, dispersion D (in m²/sec), etc. Chemical reactions, biological transformations, radioactive decay rates, and, if pre-sent, anisotropic dispersion are neglected owing to the short test period and the type of tracer (NaCl).

When steady-state flow is assumed, the numerical iteration shows the transport time interval to be a function of the Courant number (Eq. 7) and the Gitter-Peclet number (Eq. 8) [3 & 4]:

$$C_0 = |\ \Delta t \cdot u\ /\ \Delta x_i\ | \leqslant 1, \qquad\qquad (7)$$

where Δt is the time interval,
  u is the contribution of the velocity vector u,
  $\underline{u}$ = (u_x, u_y, u_z), and

Δx_i is the size of the elements; and

$$Pe = \left| \frac{u \cdot \Delta x_i}{D} \right| \leqslant 2, \tag{8}$$

where D is the dispersion coefficient (in m²/sec).

   If these boundary conditions are neglected, large amplitude oscillations occur at the source and sink. A value is assumed for the dispersion coefficient that will hinder this undesired effect, together with an increase in
the number of elements in the affected parts of the model.

   The dispersion coefficient was varied in four steps from $2 \cdot 10^{-5}$ to
$2 \cdot 10^{-2}$ m²/s. The concentration distribution is shown in Figure 4 for the
shortest flow path. The elements representing the source and sink are surrounded by eight elements and four nodes each. The mean of the concentration
at these four nodes is plotted in Figures 6, 7, 8, and 9 as a function of
time, together with the breakthrough time for various dispersion coefficients.
Oscillation does not occur for values of D greater than 10^{-3} m²/s (Figs. 8
& 9). The influence of dispersion on the breakthrough time can be seen in
Figure 10. It influences the breakthrough time for $D > 10^{-3}$ m²/s. The relationship can be seen in Figure 11 for an injection rate of 6 L/min. Variation
of the potential field in the fracture shows that the breakthrough time is a
hyperbolic function of the injection rate (Fig. 12). The breakthrough curves
for $D = 0$ and 10^{-3} m²/s are shown in addition to the in-situ data in Figure 12. The agreement is good for flow rates > 6 L/min but the deviation is
large for low flow rates, which is to be expected when the Gitter-Peclet number (Eq. 8) is examined more closely. For an injection rate of 6 L/min and
$D = 0.001$ m²/s, the concentration distribution perpendicular to the direction
of flow as a function of time is shown in Figure 13; for an injection rate of
15 L/min and $D = 0.001$ m²/s, the concentration distribution as a function of
time in the direction of flow and perpendicular to it is shown in Figures 14
and 15. The concentration distribution is shown in Figure 16 for the same
conditions. The lines show a concentration of 10 % as a function of time. A
comparison of Figures 14 (15 L/min) and 16 (1 L/min) clearly shows the influence of the dispersion coefficient. The breakthrough time is halved by the
relatively large dispersion coefficient only due to advection (Fig. 12).

5.   CONCLUDING REMARKS

   The parameter study described above is a part of a research project to
be concluded by 1990. The objective is to determine the rate of propagation
of contaminated water in fractured rock. The validity of the numerical models
is to be established by in-situ tests. The understanding of the hydraulic behavior of fractured rock is to be expanded by extrapolation. The following
aspects are being emphasized:

      -- locating fracture systems,
      -- measurement of fractures,
      -- aperture width,
      -- filling material,
      -- extent of fracturing,
      -- extent to which fractures intersect,
      -- stresses and their orientation with respect to the discontinuities,

-- development of hydraulic test equipment,
-- development of numerical models for non-steady-state flow in unsaturated parts of the fluid, gas flow, permeability as a function of stress, and
-- parameter studies and comparisons with in-situ tests.

6. REFERENCES

[1] LIEDTKE, L. & PAHL, A.: "Transport of Dissolved Substances in Fissured Granite", 87 DOE/AECL Geostatistical Sensitivity and Uncertainty Methods for Ground-Water Flow and Radionuclide Transport Modelling, Sept. 15 - 17, 1987, San Francisco, California.

[2] LIEDTKE, L.: "Stofftransport im geklüfteten Granit", Festschrift aus Anlaß des 75. Geburtstags von Prof. Dr.-Ing. Erich Lackner, Eigenverlag, Hannover, 1988.

[3] GÄRTNER, S.: "Zur diskreten Approximation kontinuumsmechanischer Bilanzgleichungen", Institut für Strömungsmechanik und Elektronisches Rechnen im Bauwesen der Universität Hannover, Nr. 24/1987, Hannover, 1987.

[4] KINZELBACH, W.: "Numerische Methoden zur Modellierung des Transports von Schadstoffen im Grundwasser", R. Oldenburg Verlag, München, Wien, 1987.

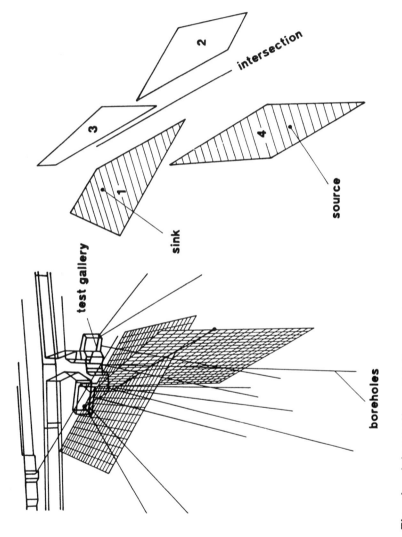

Fig. 1   Intersection of 2 fissure planes

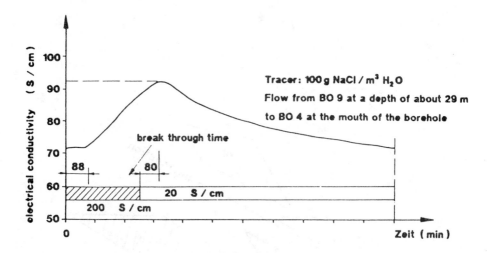

Tracer: 100 g NaCl / m³ H₂O

Flow from BO 9 at a depth of about 29 m
to BO 4 at the mouth of the borehole

break through time

88

80

200 S / cm

20 S / cm

Fig. 2 Electrical conductivity as a function of time [1]

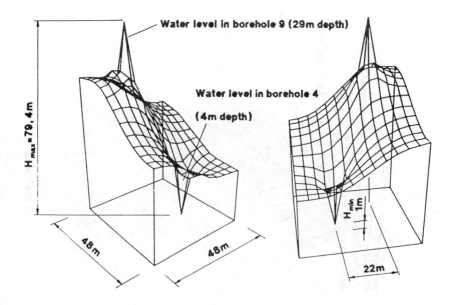

Water level in borehole 9 (29m depth)

Water level in borehole 4
(4m depth)

$H_{max} = 79,4 m$

H_{min} 1m

48m

48m

22m

Fig. 3 Potentiometric surface in a joint, joint width 2.2 mm [1]
$q = 15 l/min$, $k_k = 2 \cdot 10^{-3}$ m/s, $S = 10^{-8}$

- 110 -

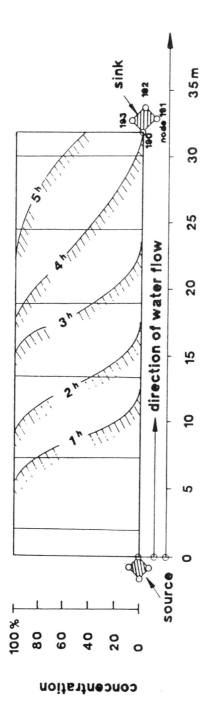

Fig. 4 Concentration distribution along the shortest distance of flow
$D = 2 \cdot 10^{-4} \, \text{m}^2/\text{s}$, fissure area $48 \times 48 \, \text{m}$, $k = 2.3 \cdot 10^{-2} \, \text{m/s}$, $S = 0$
width of fissure opening $0.002 \, \text{m}$, $Q = 6 \, \text{l/min}$, $\triangle t = 300 \, \text{s}$

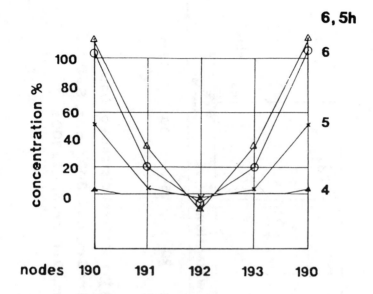

Fig. 5   Concentration distribution at the sink

$$q_{nodes} = 0.000025 \ m^3/s$$
$$\Sigma_{qi} = Q = 10^{-4} \ m^3/s \ \widehat{=} \ 6 \ l/min$$

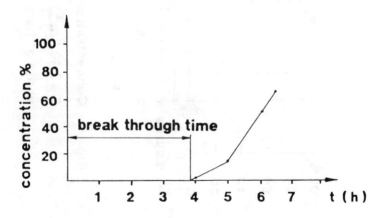

Fig. 6   Concentration at the sink

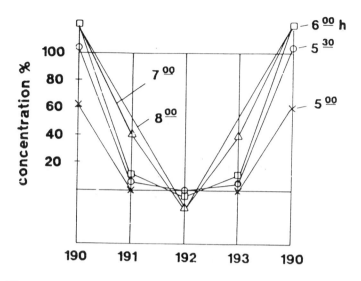

Fig. 7   Concentration at the sink

$$Q = 10^{-4} \, m^3 / s, \quad D = 2 \cdot 10^{-5} \quad m^2 / s$$

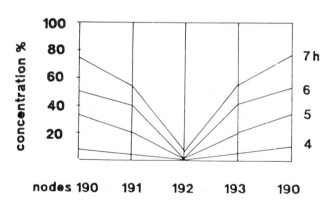

Fig. 8   Concentration at the sink

$$D = 10^{-3} \, m^2 / s$$

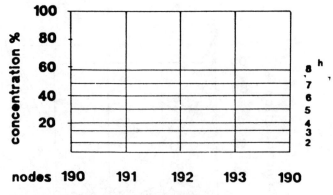

Fig. 9   Concentration at the sink

$$D = 2 \cdot 10^{-2} \, m^2 / s$$

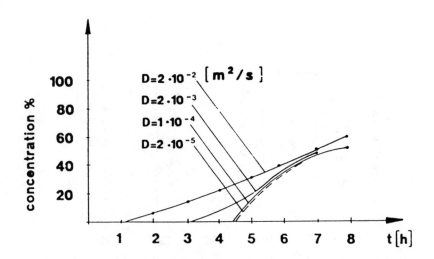

Fig. 10   Concentration distribution at the sink as a function of time

$$Q = 10^{-4} \, m^3 / s, \quad \triangle t = 300 \, s$$

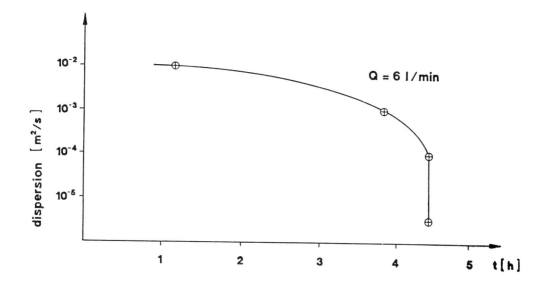

Fig. 11   Breakthrough time as a function of dispersion

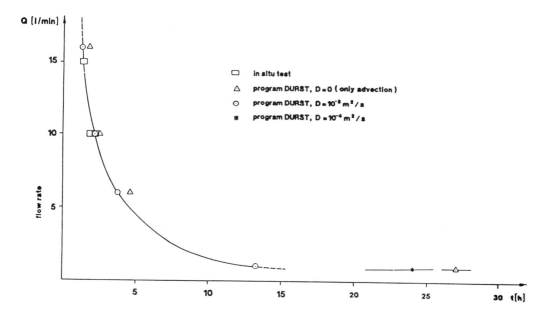

Fig. 12   Break through time as a function of flow rate

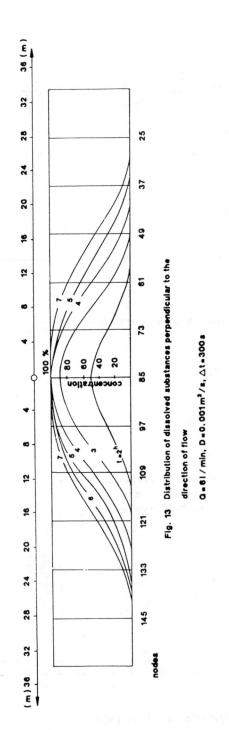

Fig. 13 Distribution of dissolved substances perpendicular to the
direction of flow

$Q = 6 \, l / min$, $D = 0,001 m^2/s$, $\Delta t = 300 s$

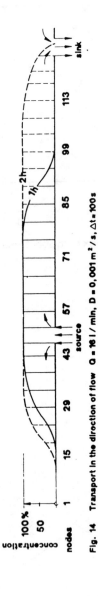

Fig. 14 Transport in the direction of flow $Q = 16 \, l / min$, $D = 0,001 m^2/s$, $\Delta t = 100 s$

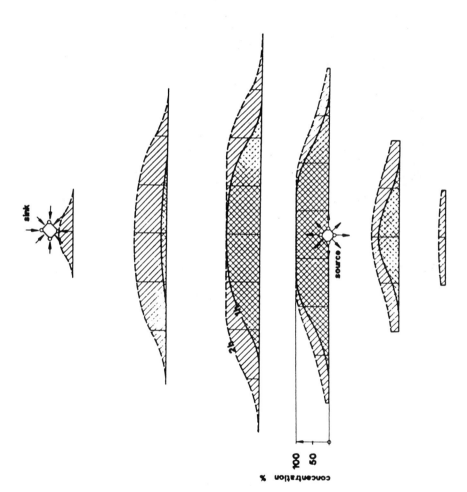

Fig. 15   Distribution of dissolved substances perpendicular to the direction of flow   Q = 18 l/min, D = 0.001 m²/s,   Δt = 100 s

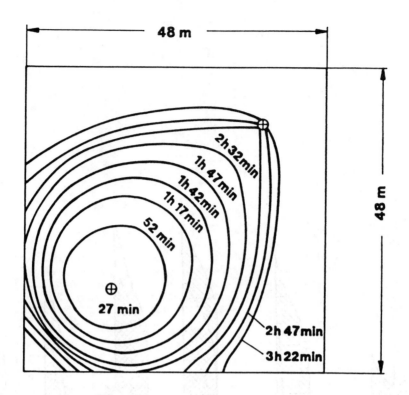

Fig. 16   Concentration distribution as a function of time

Concentration   C = 10 %

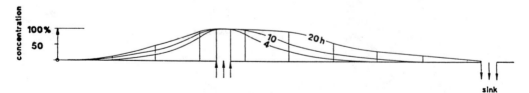

Fig. 17   Transport in the direction of flow   Q = 1l / min, D = 0.001m²/s, △t= 1200s

STRESS AND TEMPERATURE EFFECTS ON THE PERMEABILITY IN GRANITIC ROCK AROUND A GALLERY EXCAVATED BY CONTINUOUS MINING

J. Schneefuß, H. Kull and Th. Brasser
Gesellschaft für Strahlen- und Umweltforschung mbH München
Institut für Tieflagerung
D-3300 Braunschweig, Federal Republic of Germany

ABSTRACT

Several in situ tests are being performed at the Grimsel Test Site (CH) by GSF to investigate the hydraulic and mechanical reaction of the granite to excavation, heat input and ventilation. The ventilation test serves to investigate the hydrological conditions in the direct environment of a drift to determine the macropermeability of a fractured low permeable granitic rock. In the Heater Test the influence of heat input into the granite on temperature distribution, deformation and stress field are examined. Test results concerning stress and temperature effects on the permeability of the granitic rock will be demonstrated and interpreted.

EFFETS DES CONTRAINTES ET DE LA TEMPERATURE SUR LA PERMEABILITE D'UNE FORMATION GRANITIQUE ENCAISSANT D'UNE GALERIE CREUSEE FAISANT L'OBJECT DE TRAVAUX CONTINUS D'EXCAVATION

RESUME

Plusieurs expériences in situ sont exécutées dans le laboratoire souterrain de Grimsel en Suisse par la Gesellschaft für Strahlen-und Umweltforschung (GSF) pour étudier les effets sur les propriétés hydrauliques et mécaniques du granite les travaux d'excavation, de l'échauffement et de la ventilation. Les essais de ventilation ont pour but d'analyser les conditions hydrologiques dans le champ proche d'une galerie en vue de déterminer la macroperméabilité d'une formation granitique fissurée faiblement perméable. L'essai de chauffage vise à étudier l'influence de l'apport de chaleur dans le granit sur la distribution de la température, la déformation et le champ des contraintes. Les résultats des essais concernant les effets des contraintes et de la température sur la perméabilité de la roche granitique seront démontrés, puis interprétés.

1. Introduction

On the basis of a Swiss-German cooperation agreement which was signed in 1983 by the National Cooperative for the Storage of Radioactive Waste (NAGRA), the Research Centre for Environmental Sciences (GSF) and the Federal Institute for Geosciences and Natural Resources (BGR) the Federal Republic of Germany participates in the Grimsel Rock Laboratory Test Programs. Within the scope of development of geotechnical, geophysical and hydrogeologic investigation techniques as well as the characterization of fissured rock as a geological barrier for the safe disposal of radioactive waste GSF performs several in situ tests, namely

- Tiltmeter Measurements,
- Heater Test,
- Ventilation Test.

Today all these tests are running and supply test-specific results after extensive preparatory work during the years 1983 - 1985. Whereas the Swiss excavation test at Grimsel test site has been conceived especially for studying stress changes, rock deformations etc. due to the excavation of tunnels and caverns, the GSF tests have different test objectives, namely:

- the determination of periodic and non-periodic displacements of a rock mass resulting from e.g. tectonic movements or man-made events by measurements with tidal TILTMETERS,

- the determination of the thermomechanical behaviour of granite as well as the corresponding rock properties including the temperature dependent release of formation water by the HEATER TEST and

- the calculation of the hydraulic conductivity of low permeable fissured granitic rock with the help of a VENTILATION TEST.

However, within the scope of these tests data concerning stress and temperature effects on the permeability of granitic rock around a drift are also obtained.

The results allow statements on the permeability of the host rock in the vicinity of engineered tunnels and the near-field hydrology so that these may support further tests and facilitate the interpretation of excavation effects.

2. Ventilation effects

Water circulation in crystalline rock, such as in the Grimsel granodiorite, mainly occurs in communicating fissures. Within the region of tunnels excavated by mining the hydraulic conditions may possibly be disturbed in the area of excavation zone. In order to keep interference with the normal hydrogeological conditions as small as possible the ventilation drift was excavated using a full-thickness heading machine. The actual test area has a length of 74 m and a diameter of 3,5 m corresponding to a surface area of approx. 815 m^2 and to a cavity volume of about 715 m^3 respectively.

The objective of the Ventilation Test is to determine the hydraulic conductivity of the low permeable fissured crystalline rock mass [1, 2, 3]. The investigation procedure is similar to the in situ tests carried out previously at Stripa [4] and at Konrad [5].

Within this context the ventilation drift, being separated from the remaining tunnel system by means of mining cushions, is ventilated by an air-conditioning system. The intruding moisture is flushed out with the air stream, and the amount of condensed water is measured as a function of time in order to assess the water seepage into the ventilation drift.

In situ measurements of hydraulic pressure distribution in the direct surroundings of the ventilation drift are carried out in boreholes (with altogether 32 measurement sections) situated at a distance of 2,75 m and 5,25 m to the axis of the ventilation drift respectively (Fig. 1). By varying the ventilation temperature the influence on the liberated rock moisture and the prevailing hydraulic pressure distribution is determined.

2.1 Hydrogeological conditions

The ventilation drift is located in the massive, vertically structured Grimsel granodiorite. Stress areas are the more strongly textured, foliated areas as well as lamprophyre veins and mylonite bands. At the beginning of excavation the entire area was water saturated. Emanating rock water stems from dripping water from fissures and wet surfaces of confined areas. The main water bearing zone is a shear zone, designated as S_3, of approx. 70 cm thickness which forms a hydraulic system with the fissure system, named S_2, in the region of the ventilation drift. The observed hydraulic potential varies strongly and exhibits a heterogeneous distribution. The hydrostatic pressure of 4 MPa is not reached in the measurement areas close to the VE-drift. The rock temperature measurement initially was about 12 °C.

2.2 Ventilation tests

The first chamber of the subdivided ventilation drift was ventilated stepwise with temperatures of 13 °C, 17 °C, and 23 °C and was subsequently kept in an unventilated condition. The ventilation tests were terminated when a constant temperature and pressure in the measurement sections were observed. The amount of water seepage was also constant at this time. Apart from the change of temperature the relative humidity was varied several times. The ventilation tests were supplemented by pressure measurements in the boreholes.

2.3 Test results

The rates of water seepage from the first chamber were determined to be within 25 ml/min. over the entire test period. This finding was independent of the ventilation temperature and relative humidity (Fig. 2). However, a short-term increase of condensed water seepage was observed. Water inflow was limited to the shear zones S_2 and S_3. Analysis of the temperature distribution at the tunnel surface [6, 7] indicated channeling processes within the shear zones (Fig. 3). The contribution of the unfissured homogeneous matrix area in an unventilated condition is considered to be negli-

gible with respect to a quantitative mass balance. In the observation boreholes the measured potential varied between 0 - 1,2 MPa and so far no change
during the test period is indicated. In the shear zone the pressure values
increased shortly from 100 to 160 kPa and from 400 to 440 kPa, respectively.

The significant increase lasted for about 2 months. The different temperature steps in the drift caused an increase of the temperature level in
the rock. At a distance of one meter from tunnel face the temperature increased from initially 12 °C to 17 °C and up to 14 °C at 3,5 meters distance, respectively.

In the shear zone S_3 the influence of temperature changes was less
pronounced and differed by 1.5 °C. The increase of the rock temperature had
no measurable effect on the seepage level.

2.4 Determination of the permeability of the rock

The hydrogeological conditions in the test area may best be described by
a dual porosity and permeability system for which the Darcy Law can be applied. The permeability calculations carried out according to Thiem [8]
establish the stress zones S_2 and S_3 as areas with a hydraulic conductivity of K = 9 E - 10 m/s [90 µD]. Model calculations, carried out with a
simulation program, confirm the order of magnitude of the effective hydraulic conductivity. Also, an initial indication of the geometric dimensions of the joint network was obtained.

By contrast, the hydraulic conductivity of the matrix is much lower.
Permeability values of < 1 µD [K < 1E-11 m/s] were obtained in laboratory measurement on core samples. The pressure decrease measurements
(Fig. 4) in the corresponding measurement sections also rendered values
smaller than 10 µD [K < 1E-10 m/s].

3. Excavation effects and heat load

A further important influence on the host rock in the vicinity of the
drifts of a repository is the power input of canisters filled with heat producing waste into boreholes drilled perpendicular to the drifts. In borehole
emplacement the canisters are emplaced in areas which, according to their
distances to the drifts, are with and without excavation effects. The mechanical and hydraulical response of the host rock to this power input is investigated by means of Heater Tests [9, 10, 11].

3.1 Layout and instrumentation

The layout and instrumentation of the Heater Test at the Grimsel Test
Site is outlined in the vertical section in Fig. 5. Deformation of the rock
is measured in boreholes E, M and I with extensometers and inclinometers.
Rock stresses are investigated by hydraulic flat jacks in the boreholes "S".
Water inflow into the boreholes is registered in the boreholes "W". Holes
"B" are part of a seismic array to detect elastic waves caused by sudden re-
laxations of stresses in the rock. In all of these holes and also in those
named "T" the spread of temperature is measured.

The waste canisters are simulated by two electrical heaters in boreholes H1 and H2 at depths from - 12 m to - 18 m. Their power output can be controlled between 0 and 24 kW. Heater H1 is placed in an unfractured part of the "Zentrale Aaregranit" directly beside a strongly fractured fault zone, heater H2 directly in the centre of this zone. This arrangement represents the two marginals of the borehole emplacement with poor and highest faulting of the rock matrix. All reactions of the rock affected differently by excavation are within these limits.

3.2 Schedule of the test

After 1,5 years of investigating without heating ("Zero-State") the power input began with the operation of heater H1 in August 1986, followed by heater H2 in June 1987 (Fig. 6). According to the Swiss concept for final disposal the power output is controlled in such a way that the rock water remains below boiling point, i.e. 90 °C at the elevation of the test site. The power output required for heating with the necessary temperature is approximately 3 kW after one year (Fig. 7).

3.3 Temperature field

The spread of temperature after 1,5 years of heating is shown in Fig. 8. The maximum temperature at a distance of 0,5 m from the heater is less than 55 °C. More than 8 m away from the heater no significant rise in temperature could be registered.

3.4 Deformations

One result of the calculations carried out beforehand for the layout of the instrumentation was that the deformations in the rock would remain very small as a consequence of the prevention of expansion by the surrounding rock mass. This fact was verified by the measurements because the measured values were only slightly higher than the accuracy of the devices.

3.5 Rock stresses

Another decisive result of the precalculation was that the strong prevention of any expansion of the heated rock would produce rather high rock stresses which are to be detected by the stress monitors in the holes "S". The development of radial stresses at different distances from the heater is shown in Fig. 9. At a distance of 0,5 m they reach approx. 10 MPa. The triaxial compressive strength of the Grimsel granite has been found to be more than 250 MPa. The compressive strength of heated granite (100 °C) is approx. 90 %, that is 225 MPa, respectively.

The lithostatic pressure at a depth of perhaps 1000 m is approx. 30 MPa. The resulting stresses due to overburden and heating - but without tectonic influences - of approx. 40 MPa have a sufficiently safe distance from the strength of the granite like that at Grimsel even if the rock is affected by excavation effects.

3.6 Water conductivity

The water conductivity and its alteration due to artificial heat input has already been investigated by the Stripa heater experiments [12]. In these experiments the discharge increased at the start of heating, after that it decreased slowly until it almost stopped. Some time after shutdown of the heaters the discharge increased again. A similar reaction could not be registered in the Grimsel heater test (Fig. 10). A specific interrelation between power input and water inflow cannot be recognized. However, it must be considered that at Stripa, on the one hand, the power output was much higher and, on the other hand, the Stripa rock is mostly fractured and very wet while at Grimsel the rock is poorly fractured and rather dry. As neither in the solid granite nor in the faulted zone the water inflow was affected, it is obvious that the power input in parts of rock with excavation effects will be without influence on the permeability of the rock.

3.7 Pore water pressure

The pore water pressures at the bottom of the holes are plotted in Fig. 11. The rapid increase of temperature induces pore water pressures next to the heater of 1,0 MPa, at a distance of 1,0 m of 0,4 MPa and at a distance of 2,0 m of 0,05 MPa. The pressure also relaxes very quickly according to the permeability of the rock. A small increase of temperature as in S2 has only a negligible effect. Generally it can be said that the permeability of the granite only allows for the development of locally and temporarily limited pore water stresses. In case of zones affected by excavation the rise of pressure will be less than in the solid granite mentioned above.

4. Conclusions

During the Ventilation Test at Grimsel Test Site the test sections of VE-tunnel will be ventilated over long periods of time and with different air-humidity and temperature values, the last ranging from 13 °C to 23 °C.

The main results concerning the theme of this paper are the following

- Dominant water seepage into the tunnel is restrained to shear zones.

- The hydraulic pressure distribution around the gallery did not change significantly during the whole test period.

- The influence of temperature increase during ventilation became noticeable merely in the form of a slight pressure rise (0,04 -0,06 MPa) within the observation boreholes only in the the area of shear zones.

In the Heater Test at the Grimsel Test Site the rock has been heated for approximately 1.5 years with a temperature of 90 °C at the surface of the heaters and a power input ranging between 3,0 kW and 3,5 kW

each. Two heaters were placed in the solid granite as well as in a heavily disturbed zone. The essential results are:

- The deformations are very small. They are negligible with respect
  to the stability of a repository.

- The amount of rock stresses is about 10 MPa. Under the present
  conditions the total stresses are much smaller than the compressive strength of the rock.

- Under the present conditions there is no influence of heating on
  the water inflow into the dewatering holes.

- The rise of pore water pressure remains below 1 MPa.

The mechanical and hydraulical conditions range between the marginals (solid rock - disturbance) chosen for this test.

To sum up it may be confirmed that by means of the results obtained
from Ventilation Test and Heater Test significant stress and temperature
effects on the permeability in granitic rock around a gallery excavated
by continuous mining could not be determined.

References:

[1] Brewitz, W., Kull, H., and Tebbe, W.: "In situ experiments for the determination of macropermeability and RN-migration in faulted rock formations such as the oolithic iron ore in the Konrad Mine", Proc. Joint
Workshop Design and Instrumentation of in situ Experiments in Underground Laboratories for Radioactive Waste Disposal, Brussels, May 15 -
17, 1984, CEC/NEA (Ed.: B. Côme), Rotterdam (Balkema), 1984, 289 - 302

[2] Brewitz, W., Sachs, W., Schmidt, M. W., Schneefuß, J. and Tebbe, W.:
"Layout and Instrumentation of in situ experiments for determination of
hydraulic and mechanic rock mass properties of granite with and without
heat load in the underground laboratory Grimsel, Switzerland", Proc.
Joint Workshop Design and Instrumentation of in situ Experiments in Underground Laboratories for Radioactive Waste Disposal, Brussels,
May 15 - 17, 1984, CEC/NEA (Ed.: B. Côme), Rotterdam (Balkema), 1984,
289 - 302

[3] Brewitz, W. and Pahl, A.: "Participation of the Federal Republic of
Germany in the Grimsel Underground Rock Laboratory in Switzerland - Ob-
jectives and methods of in situ experiments in granite for radioactive
waste disposal", Proc. Int. Symp. on Siting, Design and Construction of
Underground Repositories for Radioactive Wastes in Hannover (FRG),
March 3 - 7, 1986, IAEA, Vienna, 1986, 477 - 490

[4] Wilson, C. R., Witherspoon, P. A., Long, J. C. S., Galbraith, R. M., DuBois, A. O. and McPherson, M. J.: "Large-scale Hydraulic Conductivity Measurements in Fractured Granite", Int. J. Rock Mech. Min. Sci & Geomech. Abstr., 20(6), (1983), 269 - 276

[5] Brasser, Th.: "Entwicklung hydrogeologischer Methoden zur Erkundung von Tiefengrundwasser in geringpermeablen Gesteinen", Brunnenbau, Bau von Wasserwerken, Rohrleitungsbau, 37(8), (1986), 283 - 286

[6] Brasser, Th. and Kull, H.: "Subsurface Application of an Infrared Thermometer to determine Zones of preferred Pathways", Int. Symp. Detection of Subsurface Flow-phenomena by self-potential, geoelectrical and thermometrical methods, Karlsruhe (Fed. Rep. of Germany), March 14 - 18, 1988

[7] Brasser, Th. and Kull, H.: "Untertägiger Einsatz eines Infrarot-Strahlungsthermometers zur Identifizierung und Abgrenzung wasserführender Kluftzonen", Felsbau (1988 - in print)

[8] Brasser, Th.: "Der Ventilationstest, ein untertägiger Versuch zur Ermittlung der Gebirgspermeabilität", Bergbau, 38(12), (1987), 523 - 526

[9] Carlsson, H.: "A Pilot Heater Test in the Stripa Granite", LBL-7086/SAC-06, (1978)

[10] Cook, N. G. W. and Hood, M.: "Full-Scale and Time-Scale Experiments at Stripa: Preliminary Results", LBL-7072/SAC 11, (1978)

[11] Schrauf, T., Pratt, H., Simonson, E., Hustrulid, W., Nelson, P., DuBois, A., Binnall, E. and Haught, R.: "Instrumentation Evaluation, Calibration and Installation for Heater Tests Simulating Nuclear Waste in Crystalline Rock, Sweden", LBL-8313/SAC-25, (1979)

[12] Nelson, P. H., Rachiele, R., Remer, J. S. and Carlsson, H. S.: "Water Inflow into Boreholes During the Stripa Experiments", LBL-12547/SAC-35, (1981)

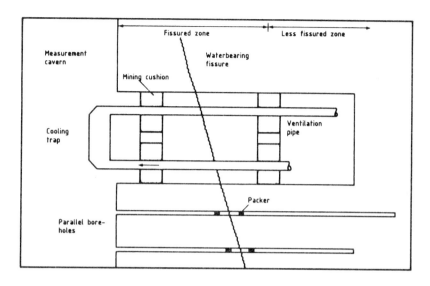

Fig. 1: Layout of the ventilation test in Grimsel Rock
Laboratory.

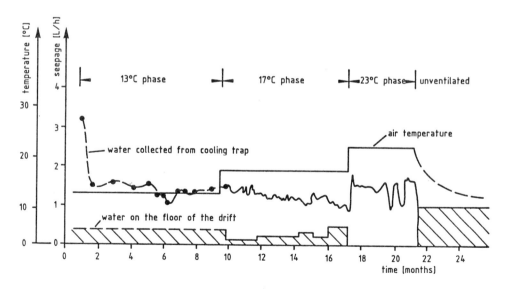

Fig. 2: Water seepage from the first chamber during ventilation.

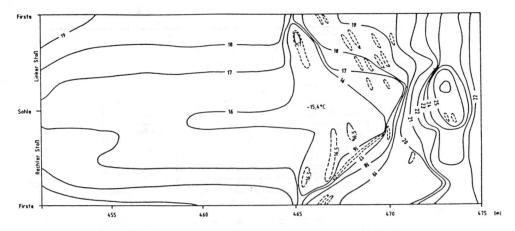

Fig. 3: Map of temperature isotherms in the first chamber at an
air temperature of 17 °C.

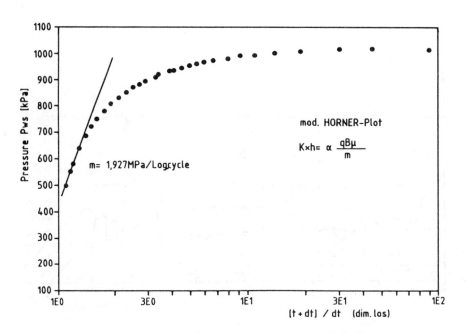

mod. HORNER-Plot

$$K \times h = \alpha \; \frac{qB\mu}{m}$$

m= 1,927MPa/Logcycle

Fig. 4: Example of the interpretation of a borehole pressure
test (HORNER plot).

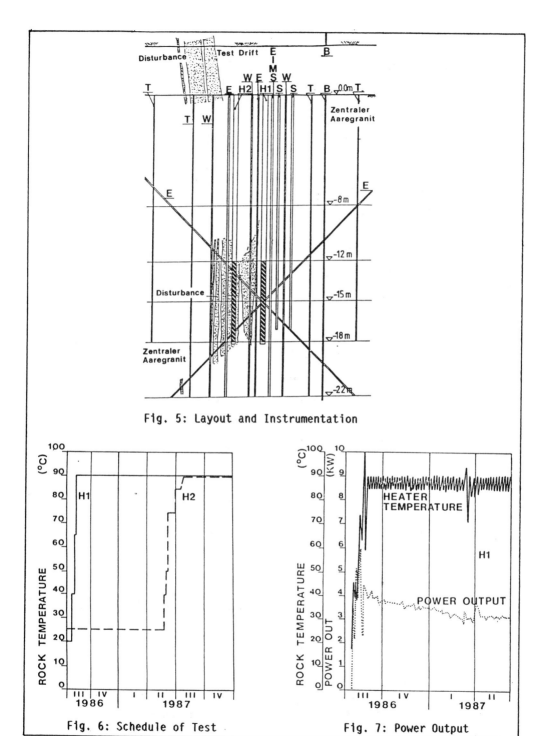

Fig. 5: Layout and Instrumentation

Fig. 6: Schedule of Test

Fig. 7: Power Output

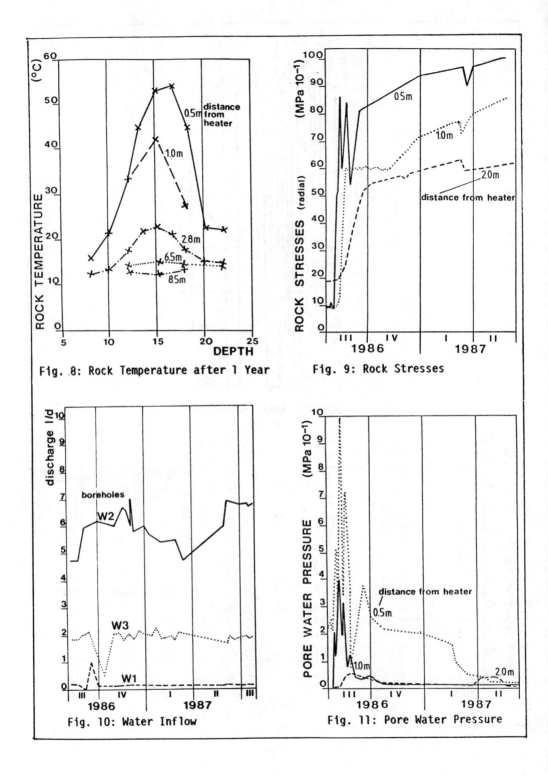

Fig. 8: Rock Temperature after 1 Year

Fig. 9: Rock Stresses

Fig. 10: Water Inflow

Fig. 11: Pore Water Pressure

EVALUATION OF THE SCALE AND CONSEQUENCES OF EXCAVATION EFFECTS ON THE GEOLOGICAL ENVIRONMENT

EVALUATIONS DE L'AMPLEUR ET DES CONSEQUENCES DES EFFETS DE TRAVAUX SUR LE MILIEU GEOLOGIQUE

C. DEVILLERS ; P. ESCALIER des ORRES
Commissariat à l'Energie Atomique
IPSN/DAS - B.P. 6
F92260 FONTENAY AUX ROSES - FRANCE

J.M. HOORELBEKE
Commissariat à l'Energie Atomique
IRDI/DERDCA/DIP - B.P. 171
F 30205 BAGNOLS SUR CEZE - FRANCE

Agence Nationale pour la Gestion des Déchets Radioactifs -
ANDRA - CEA
31-33, rue de la Fédération
F 75015 PARIS - FRANCE
(à partir du 1er Avril 1988)

A B S T R A C T

The Commissariat à l'Energie Atomique (French Atomic Energy
Commission) has conducted a bibliographic study of the
damages in the rock caused by the construction of a
repository, and several hydraulic simulations, to appreciate
the influence of these damages on the safety of the
repository.

These studies have led to the proposal of construction
techniques in accordance safety requirements and industrial
feasibility.

R E S U M E

Le Commissariat à l'Energie Atomique a effectué une synthèse
bibliographique des effets induits par la création du
stockage et à un ensemble de simulations hydrauliques, afin
d'apprécier leur incidence sur la sûreté du stockage. Ces
travaux aboutissent à la proposition de méthodes d'exécution
satisfaisant à la fois les exigences de sûreté et la
faisabilité industrielle.

1/ OBJECTIFS DES ETUDES

Le Commissariat à l'Energie Atomique a effectué une analyse générale des effets induits sur la barrière géologique par la construction d'un stockage, dans le cadre :

- d'une part du groupe de travail présidé par le Professeur GOGUEL chargé par le Gouvernement Français d'établir des critères techniques de sélection de sites de stockage [1] ;

- d'autre part, de travaux cofinancés par la Commission des Communautés Européennes [2 - 3].

Ces études comprennent une synthèse bibliographique des connaissances sur les dégradations dues à l'excavation [4], une approche des phénomènes thermomécaniques induits par les déchets exothermiques et un ensemble de modélisations des écoulements afin d'apprécier les conséquences des travaux sur la sûreté du stockage. Les études ont conduit à proposer des méthodes d'éxécution satisfaisant à la fois aux objectifs de sûreté et à la faisabilité industrielle du stockage. Bien qu'elles aient porté sur les quatre milieux géologiques envisagés en FRANCE (granite, schiste, sel, argile), la présentation rapide qui en est faite ici sera limitée au cas des sites cristallins.

2/ DESORDRES INDUITS PAR L'EXCAVATION

2.1) Méthodes de creusement

Pour le creusement de puits verticaux, ou inclinés, de galeries horizontales ou de très grandes cavités, on distingue essentiellement l'abattage traditionnel par tir à l'explosif et l'abattage mécanique basé sur le travail d'outils désagrégeant la roche. L'abattage traditionnel, de mise en oeuvre très simple, ne connaît aucune limite technologique vis à vis des ouvrages à réaliser. En revanche, l'énergie nécessaire à l'abattage d'une roche très dure comme le granite y limite les possibilités de l'abattage mécanique ; seules les machines à attaque globale, très puissantes et encombrantes, telles que les tunneliers, peuvent être utilisées, impliquant :

- des contraintes sur la géométrie des cavités : profil circulaire, de diamètre limité à moins de dix mètres pour les galeries, de quatre mètres environ pour les puits forés en montant et de deux mètres pour les puits forés en descendant ; rayons de courbure des galeries très grands ...

- des conditions économiques d'utilisation particulières, en particulier l'amortissement obligatoire des tunneliers sur plusieurs kilomètres de galeries à creuser sans discontinuer.

2.2) Effets de l'abattage à l'explosif

La rupture de la roche est due à la propagation d'une onde de choc, qui se traduit en chaque point par la succession de contraintes de traction et de compression. L'amortissement radial de l'onde de choc entraîne la distinction de plusieurs zones cylindriques autour de chaque charge : zone broyée, zone fissurée suivant plusieurs directions, zone fissurée radialement, zone non endommagée. L'extension de la zone endommagée augmente quand la charge d'explosif croît, quand la contrainte géostatique moyenne, donc la profondeur, diminuent et quand la résistance mécanique de la roche décroît. L'abattage en masse procède du tir de charges distribuées sur l'ensemble du front de taille.

Afin de limiter les dégradations induites dans le parement, plusieurs techniques ont été développées depuis trois décennies ; ces techniques sont plus coûteuses que l'abattage sans ménagement, car elles requièrent une longueur de trou foré supérieure :

- le prédécoupage consiste à réaliser préalablement à l'abattage une coupure matérielle entre le futur parement et la roche à enlever, par le tir simultané de faibles charges placées dans des trous très rapprochés, suivant le profil de la cavité ;

- le tir ménagé implique aussi la foration de trous rapprochés suivant le profil ; les charges faibles qui y sont placées, découplées de la roche, sont tirées toutes ensemble après la détonation des autres charges du front de taille ; ce tir périphérique détache de la roche saine les blocs fissurés par le tir principal.

L'extension des zones dégradées, observée par plusieurs auteurs, atteint généralement :

- en roche décomprimée de surface : quelques mètres,

- en souterrain, pour un tir non ménagé : un peu plus d'un mètre,

- en souterrain, pour un tir "adouci" : entre quelques décimètres et un mètre selon la qualité du tir.

De par son principe, le prédécoupage amène une meilleure régularité du profil du parement que le tir ménagé. En revanche, la difficulté de sa mise en oeuvre s'accroît avec la profondeur, la propagation de la fissure périmétrique étant entravée par la contrainte géostatique. De plus, le prédécoupage entraîne un ébranlement des roches plus important que le tir ménagé, d'où une densité de fractures induite dans la zone dégradée probablement supérieure ; ce point demande toutefois à être confirmé expérimentalement.

Si les méthodes d'abattage adoucies développées à ce jour préservent convenablement l'intégrité de la roche en voute et en piédroit d'une galerie, en revanche la protection de la sole n'a pratiquement jamais été recherchée dans les chantiers souterrains. Le découpage soigné de la sole à l'explosif entraîne des difficultés techniques particulières ; l'alternative d'un découpage par havage mécanisé, tel qu'il est souvent mis en oeuvre dans les roches tendres, paraît inadapté à la dureté du granite dans des conditions d'emploi industrielles.

2.3) Effets de l'abattage mécanique

L'abattage mécanique entraîne des désordres dans le parement rocheux beaucoup moins importants que l'explosif. Leur extension ne doit pas excéder quelques centimètres, la pénétration des outils de coupe étant de l'ordre de quelques millimètres. Par ailleurs, l'abattage mécanique se caractérise par un avancement régulier du front d'abattage, permettant aux mouvements de détente du massif de se dérouler progressivement.

2.4) Redistribution des contraintes - confortement

L'ouverture d'une cavité entraîne la redistribution des contraintes autour de la cavité. Cette redistribution s'accompagne nécessairement d'un accroissement du déviateur.

La prédiction de l'état de contraintes induit est impossible sans la mesure des contraintes géostatiques in situ, car les formations granitiques connues présentent une forte dispersion de la valeur des contraintes géostatiques horizontales.

Cependant, sauf cas de figure exceptionnel, la grande résistance mécanique du granite interdit toute rupture à une profondeur de l'ordre de cinq cent mètres. Le seul risque à prendre en compte est l'ouverture de discontinuités naturelles préexistantes ou induites par l'abattage, ainsi que la chute de blocs désolidarisés du massif.

Pour prévenir ces éventuelles chutes de bloc, un boulonnage de sécurité pourrait être mis en place. Le boulonnage amène un renforcement de la cohésion des roches, mais il implique la présence de trous forés radialement à partir de la cavité. La durée de vie du confortement est liée à la corrosion des tiges sous contrainte, à la fatigue de la roche au niveau de l'ancrage (pour un ancrage ponctuel) et au vieillissement du matériau de scellement (pour un ancrage réparti).

3/ EFFETS THERMOMECANIQUES

Parallèlement aux dégradations mécaniques liées à l'excavation, il apparaît indispensable de prendre en compte les éventuels remaniements d'origine thermomécanique affectant les mêmes volumes de roche du site. Autour des cavités de stockage de déchets de très haute activité, on distinguera les effets thermomécaniques intervenant à l'échelle de l'édifice polycristallin de la roche, et les effets macroscopiques. Les premiers sont liés à la dilatation thermique différentielle des grains de quartz, mica et feldspath, entraînant le développement de microfissures intergranulaires ; l'élévation de température du granite pourra être limitée à moins de 100°C, ce qui ne devra pas entraîner de dégradation significative. Les effets macroscopiques sont associés à l'élévation de la contrainte mécanique due à la dilatation confinée de la roche. Plusieurs calculs élasto-plastiques, fondés sur une élévation maximale de la température de l'ordre de 100°C, montrent qu'un granite sain se trouve encore très loin de l'état de rupture. La contrainte tangentielle la plus élevée est de l'ordre de 80 à 105 MPa [5]. Comme pour la redistribution de contrainte accompagnant l'ouverture des cavités, seules les discontinuités préexistantes pourront être remaniées par l'élévation de température ; dans la zone très proche des déchets, où subsiste un déviateur, on peut craindre le risque d'ouverture de certaines discontinuités ; en revanche, au delà de un à quelques mètres, le dégagement thermique entraîne un accroissement de la compression moyenne tendant à refermer les discontinuités.

Les phénomènes thermomécaniques apparaissant en champ lointain et, en particulier, à proximité du sol, sortent du cadre de cette présentation.

4/ INCIDENCES SUR LES ECOULEMENTS

4.1) Scénario d'évolution normale

Les définitions adoptées ici pour les scénarios sont celles retenues dans le cadre du programme de la CCE "PAGIS" (Performance Assessment of Geological Isolation System). On entend par scénario, soit un événement ou une séquence d'événements non engendrés par la présence des déchets, soit la prise en compte de défaillances des barrières naturelles ou ouvragées.

Le scénario dit "normal" est le plus probable : pour
une formation granitique, on considérera ainsi une
circulation d'eau dans un milieu microfissuré assimilé à un
milieu continu poreux équivalent obéissant à la loi de
Darcy, extrapolée à partir des tendances hydrogéologiques
actuelles du site et menant à un relâchement des
radionucléides.

Les scénarios altérés sont tous ceux qui diffèrent du
scénario normal : on pourra y trouver des événements déjà
présents dans le scénario normal, mais avec une amplitude ou
une date d'occurrence différentes ou des événements
nouveaux, de caractère aléatoire et de plus faible
probabilité.

Afin d'évaluer grossièrement l'incidence d'une zone
dégradée autour d'une cavité de stockage sur les écoulements
locaux, on peut utiliser un modèle analytique simple : la
cavité est représentée par un cylindre infini (rayon :
3,50 m), rempli d'une barrière ouvragée de perméabilité K1.
Une épaisseur de roche endommagée entoure la cavité de
perméabilité K2, supérieure à la perméabilité de la roche
saine supposée égale ici à $10^{-10}.m.J^{-1}$ (figure 1). La
vitesse de l'eau à une distance suffisante de la cavité est
supposée uniforme et inchangée par la construction du
stockage.

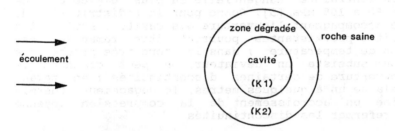

Figure 1 : représentation de la cavité

D'une manière générale, une augmentation de l'épaisseur
de la zone endommagée, ou de sa perméabilité, diminue les
débits entrant en cavité de stockage (figure 2), alors que
le débit total transitant dans la zone proche, assimilée à
un cylindre de rayon 10 mètres, ne croît que très légèrement
(figure 3).

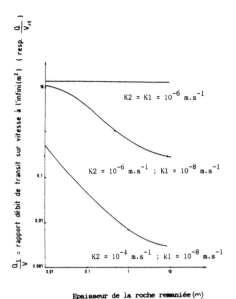

Figure 2

Variations du débit traversant
la cavité de stockage

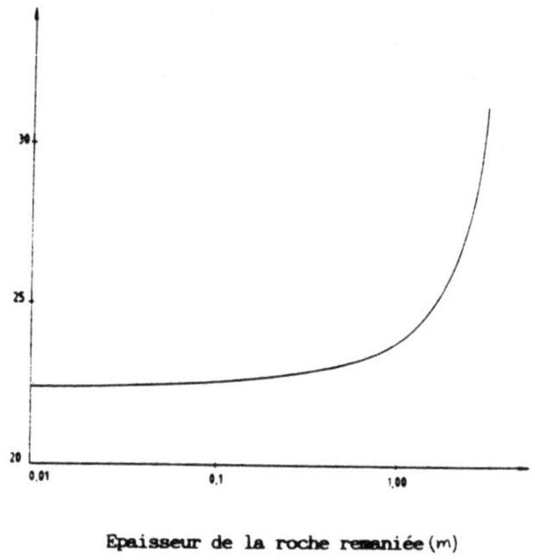

Figure 3

Variations du débit traversant
la zone proche

Les exemples de simulations numériques proposés dans la suite se rapportent au site de référence de l'option granite du programme PAGIS (Site dans le Massif Central en FRANCE). Rappelons qu'il s'agit d'un site expérimental pour la recherche sur lequel ont été réalisés deux forages profonds (1000 m et 500 m).

Une évaluation de l'incidence sur les écoulements de la perméabilisation du stockage a été effectuée, avec les hypothèses suivantes :

- représentation du stockage par un élément monodimensionnel d'emprise 30 m

- attribution d'une perméabilité de 3,5 10^{-9}m/s au stockage pour tenir compte des effets des travaux et d'un rebouchage imparfait des galeries (milieu ambiant, $K = 10^{-12}$ m/s).

Les calculs, effectués à l'aide du code METIS, ont
permis d'établir la carte des équipotentielles hydrauliques
avant creusement du dépôt (figure 4 - haut).

La figure 4 (en bas) montre l'incidence de la
perméabilisation du stockage sur les équipotentielles ; ces
dernières sont nettement modifiées au voisinage du dépôt.
Elles tendent à former une croix centrée sur la zone
perturbée, celle-ci évoluant vers une surface
équipotentielle. En prenant le cas limite d'une
perméabilisation infinie de la zone perturbée, on observe
une configuration identique.

La comparaison des champs de vitesses avant creusement
et en prenant en compte le dépôt perméabilisé fait apparaître un drain horizontal traversant ce dernier.

Les vitesses calculées sont toutefois, en raison du
faible gradient hydraulique, du même ordre de grandeur
autour du stockage et dans la zone perméabilisée.

En conclusion, on constate que les perturbations
entraînées par la perméabilisation du stockage sont d'extension limitée. Elles ne modifient pas sensiblement les
conditions des écoulements souterrains.

Une étude de sensibilité aux paramètres a été effectuée
en utilisant un modèle simplifié monodimensionnel : la coupe
bidimensionnelle utilisée précédemment est remplacée par un
tube de courant, tronçonné en trois zones de perméabilités
différentes. Les résultats obtenus mettent en évidence
l'influence prépondérante des paramètres liés à la
géosphère : la perméabilité, la dispersivité longitudinale
et le coefficient de retard du radioélément considéré. Le
tronçon le plus influant est celui représentant le trajet le
plus long, associé à la perméabilité la plus faible. Le
temps de transfert dans cette zone contribue le plus à la
durée totale du transport. En conclusion, ces calculs
tendent à montrer le rôle important du champ lointain
(géosphère); ils vont dans le même sens que ceux décrits
plus hauts. Il faut toutefois faire remarquer qu'ils ne
concernent que le scénario normal et un nombre limité de
paramètres ; en particulier, il n'a pas été différencié de
zone perturbée par les travaux ou un rebouchage imparfait
des galeries.

4.2) Scénarios altérés

Une étude d'un scénario d'intrusion humaine a été réalisée à partir de l'hypothèse de l'exploitation d'un gisement minier en forme d'amas dans le massif granitique. Cette exploitation se traduit par la création d'une cavité cubique de 100 m d'arête, supposée extérieure au dépôt, à 50 m de son extrémité Sud, au même niveau.

Les simulations montrent les modifications des équipotentielles provoquées par l'existence de la cavité d'une part dans les conditions d'écoulement naturelles, d'autre part en prenant en compte le stockage perméabilisé ($k = 3,5 \ 10^{-9}$ m/s) (figure 5). Il apparaît un drain horizontal, comme nous l'avions vu plus haut, qui entraîne des changements importants dans la répartition des débits d'exhaure selon les différentes parois de la cavité. Ainsi, la paroi la plus proche du dépôt représente alors 99% des débits entrant dans l'exploitation, contre 5% auparavant, 90% de l'exhaure provenant, en conditions naturelles, de la face au toit de la cavité. Le débit total, par contre, est peu modifié, 5% environ, par la prise en compte de la perméabilisation du stockage.

Les calculs des conséquences de l'intrusion ont été effectués en tenant compte de cette perméabilisation. Notons qu'il a été négligé le rôle de l'aérage qui pourrait conduire à une évaporation pour des débits d'exhaure faibles.

4.3) Phase d'exploitation

Outre les perturbations mécaniques infligées à la roche, la création des cavités entraîne des modifications du régime hydraulique. L'étude de P. GOBLET [2] a montré que le creusement des cavités du stockage ne devrait pas entraîner de dénoyage du massif, en raison de la très faible perméabilité du milieu-hôte au niveau du dépôt : ce dernier tend ainsi à être isolé des zones plus superficielles.

La détermination de la durée de la phase transitoire du régime hydraulique dépend de la valeur du coefficent d'emmagasinement spécifique. Ce paramètre est difficile à obtenir avec précision à partir des essais d'eau. La fourchette retenue dans cette étude porte sur quatre ordres de grandeur, ce qui entraîne une durée s'étalant entre 20 et 20 000 ans pour le transitoire dû au creusement des cavités de stockage. Une incertitude du même ordre de grandeur affecte la durée du retour à la normale après scellement du dépôt.

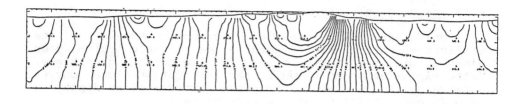

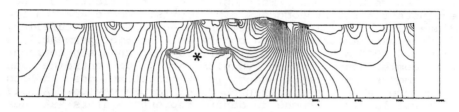

* Stockage perméabilisé

Figure 4 : Scénario normal :
équipotentielles initiales (en haut)
et après perméabilisation du stockage (en bas)

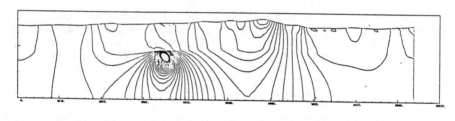

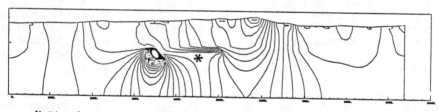

* Stockage perméabilisé

Figure 5 : Scénario altéré :
équipotentielles sans (en haut)
et avec (en bas) perméabilisation du stockage

5/ CONCLUSIONS

Les résultats obtenus par ces différents calculs tendent à montrer le rôle prépondérant du champ lointain de la barrière géologique, dans le cas d'un milieu très peu perméable vis à vis des éléments du champ proche, ce qui se traduit par une incidence modérée des dommages localisés au voisinage immédiat et à l'intérieur des cavités de stockage. Aussi le recours à des méthodes d'exécution des ouvrages minimisant à tout prix les désordres induits ne paraît pas s'imposer. De plus, on peut remarquer que les matériaux de remplissage des cavités et les colis de déchets eux-mêmes, présenteront probablement une porosité significative.

Toutefois, le rôle des éléments du champ proche devient plus important dans les scénarios accidentels, comme ceux, par exemple, d'intrusion humaine ou de défauts initiaux de certaines barrières de remplissage et de scellement ainsi que pendant les périodes pour lesquelles la barrière géologique n'est pas pleinement efficace, en particulier pendant la phase d'exploitation du stockage.

Ceci conduit à retenir des méthodes de creusement adoucies (tir ménagé) et de rebouchage soigné des cavités de stockage ; ces méthodes amèneront une limitation raisonnable de la porosité induite, dans des conditions techniques et économiques d'exploitation satisfaisantes. On visera en particulier l'indépendance mécanique et hydraulique des désordres associés à des cavités de stockage voisines de manière à éviter la création de cheminements préférentiels.

Par ailleurs, une exécution soignée limitant les hors profils favorisera la qualité du rebouchage.

Dans les milieux fluants, la porosité induite autour des ouvrages et dans le matériau de bouchage se cicatrise en tout ou partie à moyen terme.

REFERENCES BIBLIOGRAPHIQUES

[1]   Stockage des déchets radioactifs en formations géologiques - Rapport du Groupe de travail présidé par le Professeur GOGUEL ; Ministère de l'Industrie, des P. et T et du Tourisme - PARIS, France (1987)

[2]   Programme PAGIS - Option Granite
      Rapport final CCE - à paraître

[3]   GOBLET, P : "Simulation de la vie d'un site d'évacuation de déchets radioactifs à différents stades de son évolution" - Vol. 1
      Rapport CCE EUR 105 09 FR (1986)

[4]   HOORELBEKE, J.M. : "Incidences des travaux sur la distribution des perméabilités du site d'accueil"
      Rapport CEA R 53.99 - PARIS, France (1987)

[5]   HOORELBEKE, J.M. ; DOURTHE, M. : "Mechanical behaviour of host rock close to H.L.W. disposal cavities in a deep granitic formation",
      Proc. Scientific Basis for Nuclear Waste Management IX, MRS, STOCKHOLM, 1985, 791-798

ALTERATION OF THE CONDUCTIVITY OF ROCK
FROM EXCAVATIONS AS INDICATED BY THE
STRIPA PLUGGING AND BACKFILLING TESTS

R. Pusch* and A. Bergström**
*Clay Technology AB, and Lund University of
Technology and Natural Sciences, Lund, Sweden
**SKB, Stockholm, Sweden

ABSTRACT

Three research activities, forming part of the Stripa Project, showed disturbance on excavation of rock affecting its hydraulic conductivity. Ways of compensating for it by sealing were identified.

The Buffer Mass Test area had previously been used for a "macropermeability experiment", which had given a gross permeability value of 10^{-10} m/s of the rock. It was concluded from the Buffer Mass Test, that the axial conductivity of the disturbed zone was orders of magnitude higher than that. The Shaft plugging test demonstrated that a few major fractures, widened by blasting or stress relief, formed a disturbed zone with a significant axial hydraulic conductivity. Plugging with highly compacted Na bentonite blocks effectively blocked the flow. The Tunnel plugging test demonstrated that even careful blasting produces fissuring to a few decimeters depth from the free surface. The evaluated enhanced axial conductivity was concluded to result from an improved interconnectivity and widening of a few groutable fractures. Proper engineering design and strategic location of plugs and sealing through grouting can create stagnant water regimes in repositories.

MODIFICATION DE LA CONDUCTIVITE D'UNE FORMATION ROCHEUSE
SOUS L'EFFET DE TRAVAUX D'EXCAVATION : RESULTATS OBTENUS DANS LE CADRE
DES ESSAIS D'OBTURATION ET DE REMBLAYAGE MENES A STRIPA

RESUME

Il ressort de trois activités de recherche menées dans le cadre du projet de STRIPA que des travaux d'excavation menés dans une formation rocheuse entraînent une perturbation de la conductivité hydraulique de cette roche. On a déterminé des moyens de corriger cet inconvénient par des opérations de scellement.

La zone de l'essai de masse tampon avait déjà été utilisée pour mener une expérience de "macroperméabilité" qui a permis d'attribuer une perméabilité de 10^{-10} m/s à l'ensemble de la formation. Les auteurs ont conclu de l'essai de masse tampon que la conductivité axiale de la zone perturbée dépassait cette valuer de plusieurs ordres de grandeur. L'essai d'obturation du puits a démontré que quelques fissures importantes, élargies par le travail aux explosifs ou le relâchement des contraintes, formaient une zone perturbé ayant une conductivité hydraulique axiale importante. Une obturation au moyen de blocs de bentonite compactée au sodium arrête effectivement l'écoulement. L'essai d'obturation de la galerie a démontré que même si on prenait des précautions le travail aux explosifs entraînait la formation de fissures sur une profondeur de quelques décimètres sous la surface. Les auteurs ont conclu que l'augmentation évaluée de la conductivité axiale découlait d'une meilleure interconnexion et d'un élargissement de quelques fissures susceptibles d'être cimentées par injection. Une conception technique adaptée et une localisation judicieuse des dispositifs d'obturation et de scellement mis en place par injection peuvent entraîner la formation de régimes hydrauliques stagnants dans les dépôts.

1. INTRODUCTION

Experiments with backfilling and plugging of excavations in the Stripa mine have given evidence not only of how low-permeable clay-based sealings can be made, but also of how water flows in the rock close to such sealings. In turn, this has shown that tunnel and shaft excavation generates a very substantial increase in hydraulic conductivity of the rock in the axial direction of such excavations. This strongly supports the idea of a "disturbed" zone around them as outlined in this paper. Examples are given of how it can be effectively blocked by applying a suitable engineering design by selecting suitable sites for plugs, and by applying rock sealing.

2. THE STRIPA TESTS

2.1 General

Three Stripa experiments based on the use of Na bentonite clay have a bearing on the alteration of the hydraulic conductivity that results from the excavation of tunnels and shafts in granite. A major one is the so-called "Buffer Mass Test", (BMT) which comprised backfilling of a section of a drift, equipped with simulated canister deposition holes, while the other two experiments had the form of inserting clay plugs in a shaft and drift, respectively. The test sites, which are all located 350-380 m below the ground surface, are shown in Fig 1.

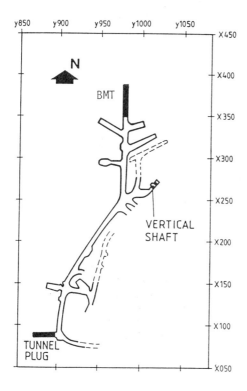

Fig 1.   Location of the tests

2.2 The Buffer Mass Test

2.2.1 *Macropermeability study*

The Buffer Mass Test, which was conducted in the Stripa mine in 1981-1985, gave a good view of the altered hydraulic conductivity that results from the excavation of a drift in granite. The conductivity of the virgin rock had been investigated with respect to the water inflow in the so-called Macropermeability Test, which was conducted by the Lawrence Berkeley Laboratory (LBL) and which preceded the Buffer Mass Test [1]. This experiment indicated that the gross hydraulic conductivity of the rock mass is about 10^{-10} m/s and that the piezometric head is 0.4-0.9 MPa at about 2 m distance from the periphery of the drift and 0.8-1.3 MPa at 10 m distance. It was concluded from the radial pressure gradient that a "skin" of lower radial permeability surrounds the drift and that this skin has a hydraulic conductivity of about $4 \cdot 10^{-11}$ m/s if it extends from the drift wall and 2.5 m outwards, or lower than that if the extension is smaller. The water inflow was measured as the net moisture pickup of the ventilation system inside a sealed portion of the drift, which had a temperature that was 20°C-45°C in various test periods while the ambient rock temperature is only 11-13°C. The heating may have tended to reduce the aperture of all fractures that extend to the periphery of the drift and may thus partially have been responsible for the skin effect.

2.2.2 *BMT*

The purpose of the Buffer Mass Test was to investigate the rate of water uptake and development of swelling pressures in 6 large holes with highly compacted bentonite embedding electrical heaters, and in the 12 m long inner part of the 35 m long drift that was backfilled with bentonite/sand mixtures [2], cf. Fig 2. This back-

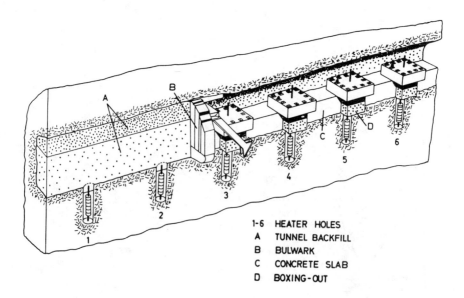

1-6	HEATER HOLES
A	TUNNEL BACKFILL
B	BULWARK
C	CONCRETE SLAB
D	BOXING-OUT

Fig 2.   Main features of the Buffer Mass Test. Figures refer to the heater holes

filling absorbed water that flowed in from the surrounding rock and the prediction was that the pore pressures, as measured by piezometers at the rock/backfill interface and in the shallow rock, would rise significantly in the course of the successive saturation of the backfill. The reason for this was simply that water would tend to flow more rapidly to the backfill than it could adsorb in the later part of the experiment.

Since the piezometric heads were known to be at least 0.4 MPa a couple of meters from the tunnel walls, pressures on the same order of magnitude were thus expected at the rock/backfill interface 2-3 years after the application of the backfill. It turned out, however, that the pore water pressure never rose to more than 0.03-0.05 MPa, which was currently explained by a severe overestimation of the rate of water uptake by the backfill. When the mass was finally excavated some four years after the application it turned out to be largely saturated as had been originally predicted, which implies that water flowing towards the drift must have been effectively discharged through the rock in the axial direction of the drift in the later part of the experiment. This would be in agreement with the observation that there was a slight pressure gradient from the inner to the outer end of the backfilled part, an approximate average figure being 0.2. The walls of the outer part of the drift constantly gave a dry impression and the increased water outflow from the backfilled inner part must therefore mainly have taken place below the tunnel floor. This was also evidenced by measuring the inflow in the emptied heater hole no 3 immediately outside the bulkhead. The heater experiment in this hole ended half a year before the bulkhead was opened and the backfill removed, and we see from Fig 3, that the inflow actually

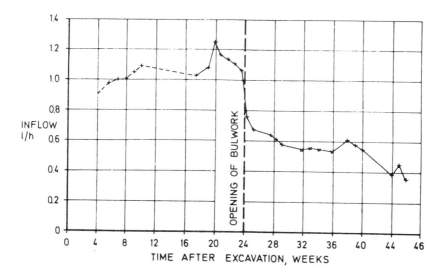

Fig 3.   Measured discharge from the inner part of the BMT drift into hole no 3 before and after emptying the drift

increased prior to this latter event. After removing the backfill, the water could again be discharged from the rock surface in the inner part of the drift, by which the inflow in the heater hole logically dropped to a small value. Because the hole had served as a major sink since it had been emptied, a large part of the water that was discharged under the low gradient operating in the axial direction of the drift accumulated in the hole. It was concluded from the measurements that this discharge was at least 0.8 l/h and this offers a possibility of estimating the hydraulic conductivity in the axial direction of the drift. Thus, making the conservative assumption that the axial discharge took place through a 0.5-1 m wide peripheral zone, we find that its average hydraulic conductivity is in the range of 10^{-7} to 10^{-6} m/s.

The rather odd rock stress situation, characterized by a major principal stress of 20 MPa perpendicular to the drift and a minimum principal stress of 4 MPa oriented vertically, is assumed to keep subhorizontal fractures in the central part of the walls open and the steep ones in the roof and floor closed. Steep fractures close to the walls are also expected to widen and propagate so that a disturbed zone with largely increased hydraulic conductivity in the axial direction is formed by stress relief. The distribution of water in the backfill indicated that it had been most effectively fed with water from the floor. This suggests that while stress relief may have had an effect on fracture apertures and rock conductivity in the walls, blasting had a dominant effect on the hydraulic conductivity of the floor.

2.3   The Shaft Plugging Test

2.3.1 *General*

A 14 m deep shaft with a diameter of 1-1.3 m and both ends open had been excavated by use of careful blasting and slot drilling as part of a preceding LBL test. The main geometrical features, which are shown in Fig 4, suggested that plugging of the shaft by expansive, highly compacted bentonite could have different effects on the flow of water in the adjacent rock. Thus, flow along the plug on the blasted side was assumed to exceed that on the slot-drilled side when applying water overpressure in the sand-filled chamber in the center of the plug as shown in Fig 5 [3]. A reference test with concrete plugs (expansive cement) preceded the main test in which the plugs consisted of well fitting blocks of highly compacted Na bentonite.

2.3.2 *Hydraulic characterization of the rock*

Careful fracture mapping and "crosshole" flow tests using 16 ϕ56 mm observation holes drilled parallel to and around the shaft gave the pattern of major hydraulically active features that is shown in Fig 6, and specified in Table I. The fracture-rich zone C is assumed to have resulted from the blasting, while the other ones are natural features. The natural piezometric head at the plug level was practically zero.

The rock stress situation in the slab is concluded to be characterized by a maximum horizontal stress of about 20 MPa oriented W/E to NW/SE, while the vertical stress is practically zero. The intermediate principal stress is estimated at about 10 MPa. Since all the structures are steeply oriented and have a long extension also small changes in aperture induced by the changes in stress would have a strong effect on their conductivity. The mapping showed that several fractures on the blasted side had apertures that were locally up to 2 mm.

2.3.3 *Test results*

The outflow of injected water from the saturated injection chamber at a pressure 0.1 MPa was expected to take place through the five discrete structures A-E at a maximum predicted rate of about 10 l/h, which would correspond roughly to a 20 cm wide zone with an equivalent hydraulic conductivity of about $5 \cdot 10^{-8}$ m/s

Table I. Major water-bearing features in the shaft

Code	Strike	Dip, °	Character
A	WNW/ESE	60	Fracture-rich zone
B	W/E	90	Large open joint
C	NW/SE	70-90	Fracture-rich zone
D	NW/SE	65	Plane fracture
E	WNW/ESE	60	Plane fracture

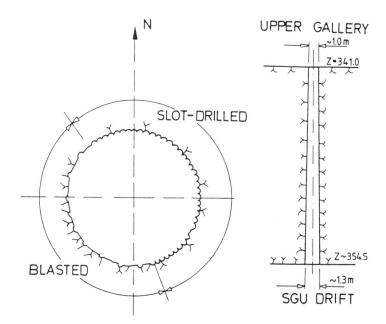

Fig 4.   Geometrical features of the shaft

applying the concept of flow through porous media. The actually observed steady state outflow in the reference test was found to be 8-9 l/h, the major flow paths being the B, A and E structures as concluded from tracer tests.

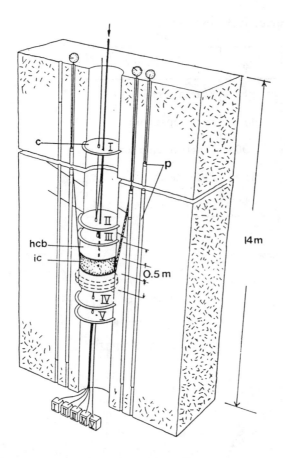

Fig 5.   Test arrangement in the shaft. (ic) is sand-filled chamber in the center of
          the plug consisting of highly compacted bentonite (hcb) in the main test
          and of concrete in a preceding reference test. (c) are collectors, numbered
          I-V for collecting water from the rock. (p) represents packers in surround-
          ing observation holes, arranged so that water could be sampled in isolated
          sections

          In the main test the concrete plugs had been replaced by sets of blocks of
highly compacted bentonite, and a 20 cm deep and high slot had been sawed around
the lower plug and filled with bentonite blocks. By this, the B-structure was expected
to have been cut off effectively. The initially recorded outflow was practically identical
with that of the reference test, indicating that substantial leakage initially took place
along the interface between the rock and the plugs. Already after about 3 weeks the
flow had become significantly lower than in the reference test, which is ascribed to the
cutoff of the B-structure. In the course of the test the flow continued to drop and
approached about 0.1 l/h after a couple of months. The injection pressure was then
increased to 0.2 MPa, by which the outflow first increased to about 1 l/h and then
dropped to less than 0.1 l/h after slightly less than two months.

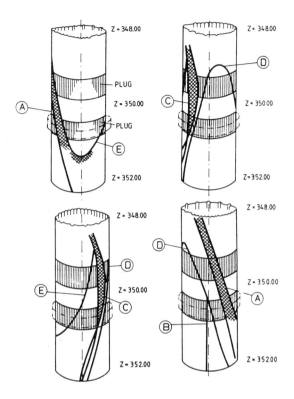

Fig 6.   The major water-bearing structures in the shaft

Measurement of the swelling pressure by use of Gloetzl cells in the plugs indicated that the expansion became substantial after about 1 month, at which stage the average recorded pressure was 0.3 MPa. This coincides in principle with the time at which the outflow from the injection chamber began to drop below about 1 l/h. After 3 months, when the 0.2 MPa injection pressure had been applied for 1 month, the average swelling pressure had increased to about 0.6 MPa, which may have had an enhanced effect on the outflow. One reason for the time-dependent sealing effect would thus logically be the tendency of closure of the subparallel fractures D and E. However, using common relationships between fracture aperture and normal pressure, the hydraulic conductivity of these fractures would not be decreased by more than 10-50 %, which indicates that other time-dependent sealing effects dominated. The major ones turned out to be 1) that the contact between the clay plug and the rock became perfectly tight by the swelling pressure which pressed the clay against the rock, and 2) that clay was released into the fractures. Both effects were clearly identified at the excavation of the plugs, the penetration amounting at several centimeters in fractures with an aperture of 1-2 mm.

It is concluded that the initially observed reduction of flow was caused by replacing high-permeability rock close to the walls by clay, and by intercepting continuous longitudinal fractures. The major part of the time-dependent flow reduction after the initial phase is concluded to have been caused by the development of a perfectly tight plug/rock contact and by clay penetration into fractures.

2.4   The Tunnel Plugging Test

2.4.1 *General*

The Tunnel Plugging Test [4] can be considered as a magnified version of the shaft sealing test with the aim of investigating how a strongly water-bearing zone can be isolated from a repository tunnel. The principle of testing was similar, i.e. water was injected into a sand-filled chamber separating two plugs, the main difference being that the confining rock could resist much higher injection pressures, up to 3 MPa, without risk of hydraulic fracturing. The plugs, which consisted of concrete bulkheads equipped with "O-ring" type sealings of highly compacted bentonite, were connected by a 1.5 m diameter steel tube which simulated a casing that would allow for through-transport during excavation and waste application in an actual repository, cf. Fig 7.

The major aim of the test was to find out how, and to what rate and extent that bentonite can produce a sealing effect by interacting with the rock.

2.4.2 *Hydraulic characterization*

Careful fracture mapping showed that no significantly water-bearing structural features existed at the inner plug but that strong outflow from the injection chamber was expected at the outer plug through a 1 m wide pegmatite zone at the crown and through the set of major fractures in the floor that is shown in Fig 8.

The piezometric head at the test site was estimated at 1-1.5 MPa before the rock excavation started and these initial pressure conditions were assumed to be preserved at more than about 3-5 m distance from the rock surface throughout the test. Within about 1 m distance the pressure had probably dropped to less than 0.5 MPa before the test started.

The local rock stress situation is not known but using i.a. the values for the BMT area one would assume a relatively isotropic stress condition and a maximum principal stress of about 10 MPa. This is expected to yield a smaller increase in axial conductivity caused by stress relief than in the BMT, and therefore a lower net hydraulic conductivity of the disturbed zone.

Assuming that the disturbed zone extends 1 m into the rock from the approximately 3 m wide and high drift, and that it is characterized by an average hydraulic conductivity of 10^{-7} to 10^{-8}, applying the concept of flow through porous media, it was expected that there would be practically no outflow from the injection chamber at pressures up to 1 MPa provided that no leakage would take place along the rock/plug interface. At 2 MPa pressure the outflow would be 50-500 l/h, while 3 MPa pressure would theoretically yield an outflow of 100-1000 l/h. Leakage through the bentonite was not expected once this material had become saturated to within a few centimeters from the rock surface. Based on experience from laboratory tests it was assumed that such a matured contact would be fully developed in a week after the onset of the test. Considerable leakage at the plug/rock interface was predicted for the first week after stepping up the pressure to 0.1 MPa.

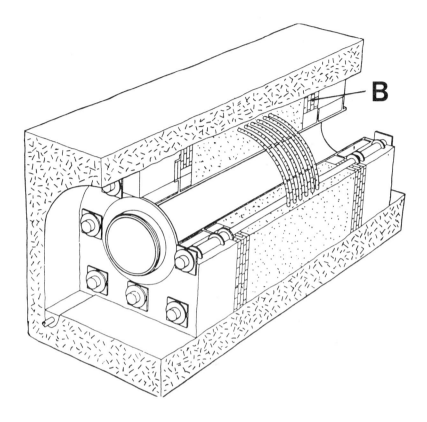

Fig 7.   General view of the test arrangement with the central steel casing and
          tie-rods passing through the concrete plugs with bentonite sealings (B) at
          the ends, and the sand-filled chamber, which simulates a richly water-
          bearing rock zone

Successive saturation of a larger part of the bentonite was expected to yield
a swelling pressure that would compress fractures that are oriented more or less
parallel to the plug axis. The major, potentially water-bearing steep fractures in the
tunnel floor and the pegmatite zone would not be affected by this while a number of
oblique fractures in the walls at the outer plug end would get their apertures reduced.
To that comes the effect of bentonite penetrating into fractures that transport water
along the bentonite/ rock interface, and these two sealing effects were estimated to
bring down the leakage through the rock significantly as in the shaft sealing test.
Thus, at 3 MPa injection pressure the assumed initial 100-1000 l/h flow was predicted
to be reduced by about 20-40 % after 1 year, the rate of reduction of the inflow being
very much dependent on the bentonite saturation rate. The minimum estimated leakage at the end of the 3 MPa pressure period was 60 l/h, while the maximum was about
800 l/h. The lower figure corresponds to a perfectly tight plug construction and a
conductivity of 10^{-8} m/s of the assumed, pressurized disturbed zone.

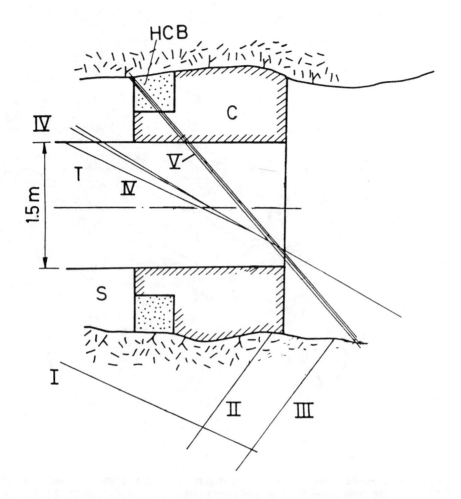

Fig 8.   Dominant water-bearing fractures at the outer end of the tunnel plug, all being steeply oriented. C represents concrete, HCB highly compacted bentonite and T the steel casing. S is the sand-filled injection chamber

2.4.3 *Test results*

The leakage turned out to be about 200 l/h at the application of 0.1 MPa water pressure early in the test but it dropped considerably in the course of the 20 months long test. During the final 3 MPa pressure period, which lasted for about 10 months, the leakage dropped from about 200 l/h to 75 l/h and this very significant reduction was found to be caused by three effects. The major one was the establishment of a very "intimate" contact between the rock surface and the bentonite, while the flow-reducing influence of the swelling pressure on certain rock fractures and the penetration of bentonite into fractures were less important but still of some significance.

It appears that most of the visible flow from the rock initially came from the pegmatite zone just above the outer plug and from a few steeply oriented fractures in the walls adjacent to this plug. These outflows decreased very much in the course of the test. Measurements showed that the outflow from the inner plug amounted to 10 % or less of the total leakage.

At the end of the test the fractures termed I-V in Fig 8 where sealed by grouting, using "dynamic injection technique" and Li-bentonite ($w = w_L = 500$ %), and the outflow then became reduced by about 50-75 % at pressure heads of 0.25-1 MPa while piping was initiated at 2 MPa pressure head and resulted in a reduction by only 20 % [5]. The fact that grouting gave such an obvious sealing effect demonstrates that the major water-bearing features had been identified and that they were groutable.

A very important observation concerning the influence of blasting was made at the core drilling for grouting. Thus, it became very obvious that the rock was richly fissured by the blasting to about 0.3-0.5 m depth not only in the floor, but also in the walls, and this suggests that there is a shallow disturbed zone with an enhanced isotropic hydraulic conductivity surrounding all blasted excavations.

3.   CONCLUSIONS

3.1   General

Although there is no safe quantitative figures for the hydraulic conductivity of the disturbed zone at the described test sites, it is concluded that it is highest at the BMT drift and somewhat lower at the Tunnel Plug drift, and even a bit lower at the Shaft Plug. An explanation for this may be the different rock stress conditions, the strongly anisotropic stress field with a high horizontal principal stress across the drift being a probable cause of the particularly high hydraulic conductivity at the BMT drift.

Blasting is concluded to yield a shallow pervious zone around the periphery of all blasted excavations. This is manifested by the fact that the backfill in the BMT was very uniformly wetted, and by the observation that the aperture of a fairly large number of exposed fractures is in the range of 0.1-2 mm.

What we learned from the Stripa tests is, thus, that since certain fractures that are subparallel to long-extending excavations will widen and propagate by stress relief at any initial rock stress situation, and since they will be connected to the shallow pervious zone by fractures that intersect the excavations, a continuous disturbed zone with a significantly enhanced hydraulic conductivity is unevitable when excavation is made by blasting.

3.2   Engineering design principles

A major outcome of the Stripa tests is that there are effective means of cutting off the continuous "superconductors" that surround shafts and tunnels in repositories, by which the beneficial state of stagnant water in the vicinity of waste packages can be obtained. Strategically located plugs, or other rock seals such as groutings, can also be used to separate different groundwater regimes, such as saline and fresh groundwaters, or surface water with low pH and deep groundwater with normal pH. Fig 9 illustrates such applications.

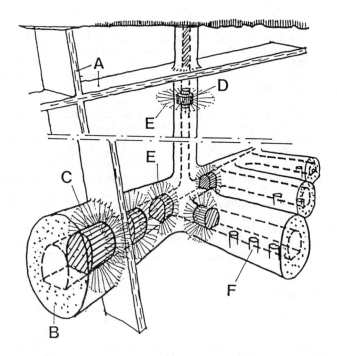

Fig 9.   Artist's view of possible means of cutting off superconductors in a
          repository
          A) Natural hydraulically active zones, B) Disturbed zone ("superconduc-
          tor") C) Clay plug, D) Clay plug with slot, E) Grouting, F) Deposition
          holes

4.   REFERENCES

1   Gale, J.E., Witherspoon, P.A., Wilson, C.R. and Rouleau, A.:
        "Hydrogeological Characterization of the Stripa Site. Proc. NEA Workshop
        on Geological Disposal of Radioactive Waste. In Situ Experiments in
        Granite, OECD, Paris, 1983

2   Pusch, R., Börgesson, L., and Ramqvist, G.: "Final Report of the Buffer
        Mass Test - Volume II: Test results". Stripa Project Technical Report 85-12,
        1985

3   Pusch, R., Börgesson, L., and Ramqvist, G.: "Final Report of the Borehole,
        Shaft, and Tunnel Sealing Test - Volume II: Shaft Plugging". Stripa Project
        Technical Report 87-02, 1987

4   Pusch, R., Börgesson, L., and Ramqvist, G.: "Final Report of the Borehole,
        Shaft, and Tunnel Sealing Test - Volume III: Tunnel plugging": Stripa
        Project Technical Report 87-03, 1987

5   Pusch, R., Erlström, M., and Börgesson, L.: "Sealing of Rock Fractures. A
        survey of Potentially Useful Methods and Substances. SKB Technical
        Report 85-17, 1985

CHARACTERISTICS OF THE EXCAVATION RESPONSE ZONE
AS APPLIED TO SHAFT SEALING

A.T. Jakubick*; R. Klein*; M.N. Gray**; L.D. Keil***
*: Ontario Hydro, Civil Research Department
**: Atomic Energy of Canada Limited, WNRE
***: Keil & Associates Ltd.

ABSTRACT

Effective sealing of an underground repository requires high quality
seals. Clay-based backfills in combination with cement plugs are the possible
candidates. An important component of the shaft sealing program is the
development of techniques to predict and monitor the flow in the excavation
response zone (ERZ) and at the seal-rock interface. The mechanical and
hydraulic performance of the seal and backfill and the flow characteristics of
the ERZ under in situ conditions will be tested in three experimental shafts
to be excavated in the URL. The use of the vacuum permeability technique for
characterisation of the hydraulic properties and of the permeability
distribution in the ERZ under various geomechanical conditions is discussed.

CARACTERISTIQUES DE LA ZONE AFFECTEE PAR L'EXCAVATION :
CAS DE SCELLEMENT D'UN PUITS

RESUME

Le scellement efficace d'un dépôt souterrain nécessite du matériel de
scellement de première qualité. Le remblayage à base d'argile combiné à des
bouchons en ciment est une option possible. La mise au point de techniques
pour prévoir et surveiller le débit dans la zone affectée par l'excavation
(ERZ) et dans l'interface roche matériel de scellement représente un élément
important du programme de scellement des puits. La tenue mécanique et
hydraulique du matériau de scellement et de remblayage et les caractéristiques
du débit d l'ERZ dans des conditions in situ seront vérifiés dans trois puits
expérimentaux qui seront creusés au laboratoire souterrain de recherche (URL).
L'exposé traite également de l'utilisation de la technique de perméabilité
sous vide pour la caractérisation des propriétés hydrauliques et de la
distribution de la perméabilité dans l'ERZ dans diverses conditions
géomécaniques.

1.0   INTRODUCTION

Research and development of sealing materials and approaches for a spent fuel waste repository is an ongoing activity at the Whiteshell Nuclear Research Establishment of the Atomic Energy of Canada. It involves both theoretical studies, laboratory investigations as well as large scale experiments at the Underground Research Laboratory at Lac du Bonnet, Manitoba. The URL provides the opportunity to conduct studies in a realistic geological environment.

During development of the URL, considerable effort was expended to document the engineering and geological properties of the site and their response to the excavation of the URL. This work is ongoing during the current period of deepening the shaft from 240 m to 440 m. Recently, a major grouting experiment, jointly funded by AECL and the United States Department of Energy, was initiated to investigate the use of cement based grouts for sealing. Some grouting was performed in May, 1987 around the main shaft to gain experience with the various components. The main grouting experiment will be undertaken in the fall of 1988 around the ventilation raise.

Research into vault sealing is proceeding on a multi-faceted program of materials research and engineering design [1]. Both are closely related to the geological environment into which the seals will be emplaced. Materials under consideration are both clay based and cement based. The geological environment issues pertinent to sealing include the properties (both physical and chemical) of the intact rock as well as any defects which occur either naturally (such as fracture zones) or may be induced due to repository construction. The latter category focuses on the zone of rock adjacent to the excavations which could undergo stress relaxation or experience deformations as a result of the excavation. Generally, this zone has been termed the "Excavation Response Zone" (ERZ).

2.0   SHAFT SEALING EXPERIMENTS

Shaft sealing experiments are proposed to be undertaken in the URL after 1990 during the operating phase of the facility. The three year planning and preliminary design activity for these experiments is jointly funded by AECL and the US/DOE as part of their joint Subsidiary Agreement No. 1 [2]. The planning focuses on three principal activities, as follows:

- materials development: both clay based backfills and cement based concrete plugs are proposed for the experiment. Laboratory studies are in progress to examine their engineering properties and behaviour, as well as their longevity to leaching or alteration in the geological environment.

- instrumentation development: extensive instrumentation will be required for the experiment and thorough checking of their performance, accuracy and durability is essential. In addition, some new instrumentation concepts are being developed to measure the moisture content of clay based materials in situ, remotely using resonant frequency generators, and to detect tracers using fiber optics.

-    experiment design: the design of the experiment involves a
careful review of the priorities and issues facing shaft sealing,
numerical modeling for parametric studies and performance
predictions, in full perspective of previous work by others and of
established concerns.

The shaft sealing experiment currently is planned to comprise three
shafts, each approximately 3 m diameter and 50 m long (Figure 1). The
Canadian concept envisages that all shafts will be either raise bored or
reamed prior to sealing. Thus two of the experiment shafts will be raise
bored. The third shaft is proposed to be excavated by drill and blast methods
both as a mechanism to exaggerate any damage to the rock by the excavation,
and in reflection that some countries are proposing blasted shafts. The
hydraulic and mechanical performance of shaft backfills and seals will be
investigated in these three shafts. All shafts will be located in a zone of
massive granite with a minimum of jointing, and where high in situ rock
stresses prevail. The rationale for this site selection was that in any
repository the shaft seals would be located in the more massive sections of
high in situ stresses for improved hydraulic performance. Also, a joint free
environment will provide better experimental control of seepage flows.

One shaft is proposed for a direct comparison of two alternative
concrete plugs. These could involve alternative concrete designs or, as
shown, alternative geometric designs. The test concept is to establish the
hydraulic aspects of the adjacent rock, particularly the zone immediately
around the shaft opening where excavation induced damage could have occurred.
In this regard, detailed definition of the location and extent of all jointing
is important, along with the measurement of the permeability of these joints.
While the permeability of individual joints is of interest, the connected
permeability of the jointed network over the plug length is paramount.

After characterization of the ERZ, the concrete plugs will be placed
together with the instrumentation to monitor their performance (Figure 2 and
3). The recharge zone will be pressurized and flow monitored until steady
state conditions are achieved. Careful attention will be paid to the design
of the instrumentation to isolate, as best as possible, flows through the seal
itself, flows along the seal/rock interface, and flows through the ERZ. On
completion of this first hydraulic test, grouting of the seal/rock interface
and the ERZ will be performed, and the hydraulic flow tests repeated to
demonstrate the improved performance. On completion of the second flow test,
the experiment will be carefully dismantled to observe the exact conditions
tested and to identify any defects which may have inadvertently been
incorporated into the experiment. Particular attention will be paid to the
condition of the seal/rock interface for evidence of mechanical bonds and
chemical interaction.

The third shaft is proposed for large scale examination of the
deformation performance of the shaft backfill material, as well as the
potential for hydraulic separation at the backfill/rock interface.
Instrumentation proposed (Figure 4) will focus on the stress and deformation
of the column of backfill for comparison with the predicted performance from
silo theory. Other instrumentation will monitor the saturation process and
hydrofacing tendency after pressurization of the recharge band.

Numerical modeling of the sealing systems describing the above concepts clearly shows that seepage through the intact seal materials will not be a concern. Flow at the seal/rock interface could be a concern depending on the concrete mix design and the curing conditions. Instead, the majority of seepage flow is expected to occur through the ungrouted disturbed zone of rock adjacent to the excavation opening. Grouting is expected to moderate these flows substantially - and indeed, the purpose of the experiment is both to quantify flows through the untreated disturbed zone, and to demonstrate the improvements possible by treatment such as grouting.

With the majority of the seepage flow expected to be through the excavation response zone, its detailed characterization will be a high priority item of the overall experiment. A number of researchers are investigating alternative approaches to identifying the presence of joints behind the excavation surface. Radar is showing promise, as are various geophysical tomographic approaches. Deformation measurements using extensometers can identify zones of relaxation and thereby indicate the presence of discontinuities. The method, however, which is capable of measuring the hydraulic properties, and particularly the existence of connected permeability along larger lengths paralleling the opening, will be the most valuable one in regards to the sealing of shafts or tunnels in a repository.

3.0 VACUUM PERMEABILITY TECHNIQUE

In Section 2.0, we established that the excavation response zone (ERZ) may serve as a pathway of preferential leakage around shaft seals. The available data are incomplete, but they show that the extent of the ERZ is in the range of .10 to 10 m. It follows that the permeability of this zone has to be measured on a comparable scale. Although large scale permeability tests are statistically more suitable in jointed rock, they fail to provide the details of permeability distribution in the ERZ. An additional characteristic of the ERZ is that, except for places of active water inflow, its joints tend to be unsaturated or partially saturated. Consequently, for characterization of leakage through the ERZ, small scale air permeability tests may be suitable. The vacuum permeability technique offers the special advantages of not pressurizing the ERZ and of having a drying effect in partially saturated fractures due to the lowering of the atmospheric pressure.

Similar to air injection testing of rock fractures, such as used by Bordman [3], Miller [4] and Montazar and Hustrulid [5], the vacuum testing is carried out by sealing a particular section of a drill hole and performing a small scale, high precision transient vacuum test, ie, a transient pressure test below the atmospheric pressure level. For our testing, we use an adaptation of Figg's vacuum permeability technique [6]. The technique was tested in the laboratory on large-size concrete cylinders with and without induced fractures [7] and in the ERZ of underground excavations in various types of rock (quartzite, limestone, gneiss, and granite) [8,9]. The approach is directed at testing unsaturated or partially saturated media. In excavations below the water table it is only suitable in the ERZ and not in the saturated rock mass.

3.1   Testing Procedure

A radial negative pressure gradient toward the tested section is created by means of a vacuum pump (Figure 5). The vacuum pump used has a free air displacement of 25 litres per minute with a linear performance in the range of 0.01 to 760 torr. The testing is done in EX (1.5") and NX (3") size holes at 25 cm intervals but it is easily adjustable to other sizes and test intervals. Both diamond drilled and percussion holes can be used. The information obtained during testing are the two ratios p/t and p/Q, where p is the pressure in the tested section, t is time and Q is the flow due to rock permeability at the specific pressure.

Accordingly, there are two approaches to testing, each convenient in different permeability ranges:

1)   at low permeabilities, monitoring the pressure recovery after evacuation

$$\bar{V}/(t_2 - t_1) = K_i / \ln (1 - \frac{p_2}{p_1}),$$

2)   at high permeabilities, pressure/flow equilibration

$$Q = Q_p (p_a - p_p),$$

where $\bar{V}$ is the volume of the test section, K_i is the air conductivity of the walls of the test section, p_1 and p_2 are the pressures at times t_1 and t_2, Q_p is the effective pumping rate, p_a is the achievable vacuum, and p_p, the end pressure of the pump. Testing in dense matrices or tight fractures produces as final output only the values p_a and p/t, and no measurable flow.

For a Darcian evaluation of tests based on pressure/flow equilibration, the validity of the linear flow assumption can be limiting. A considerable turbulent flow component was observed in fractures of the ERZ at flow rates above approximately 700 cm^3/min [8].

The limits of the lowest detectable permeability, when monitoring pressure recovery, are set by the leakage rate of the testing system. This leakage for the NX size packers under field conditions is:

Range [torr]	Leakage rate [torr/min]
1 to 20	0.2
20 to 100	0.05
100 to 150	0.045
150 to 200	0.04

Practically we attribute any leakage which results in a vacuum decay with a characteristic time of $\tau \geq (1.5 - 2) \times 10^4$ minutes to system leakage.

To fully define the test, in addition to pressure and flow, the temperature and air humidity (if required) in the tested section and outside the borehole are recorded. The measuring ranges and accuracy of the monitoring instruments are:

	Range	Accuracy
Pressure	0 to 1000 torr	0.05%
flow	1 to 50 cm^3/min	1%
	1 to 500 cm^3/min	1%
	1 to 10^4 cm^3/min	1%
temperature	0 to 100°C	0.1%
humidity	0 to 100% RH	4%

The air pumping is continued as long as is necessary under the specific conditions dictated by the permeability of the tested medium. In general, pumping is terminated after steady-state conditions are achieved. At this point, the pressure build-up is initiated. The duration of this build-up consistently depends on the air permeability of the tested medium.

The evaluation of the pressure recovery is conceptually based on utilization of the same variables and parameters as in transient pressure testing (drill stem testing) of gas wells. Using Darcy's law with the equation of state and rock compressibility relationship in the mass conservation equation, the flow equation for gases is derived. Numerical solutions can then be applied to solve the equation and to obtain the permeability value. However, for many tests an analytical solution is possible or acceptable as an approximation. In this case, the gas flow equation is partly linearized in respect to viscosity and compressibility of air, for instance, by using the pressure squared solution technique and the permeability derived [10].

Single continuous joints are approximated as two dimensional confined conductors. Provided the joint, not the area of influence (drainage area), intersects the excavation wall, analytical solutions of the radial flow equation provide reasonable results. A connection to the atmosphere is also assumed when a system of joints is intersected by the test hole. Accordingly, in both situations a constant pressure outer boundary condition applies. Details of the testing and evaluation procedure are discussed in [7].

As pointed out in the Introduction, it is important that the scale of the test be comparable to the scale of the ERZ.

In case of a single fracture, we can use the total volume of extracted air to calculate the area of influence of the vacuum permeability test. The storage volume (volume of the tested section plus dead volume of the measuring system under field conditions) is constant within ± 30 cm^3 standard deviation. Thus, any volume of air extracted during the test in excess of the storage volume can be assumed to originate from the tested formation. Because the maximum available pressure head is constant within limits of atmospheric pressure fluctuation, the drained fracture volume V_f for low permeability fractures (achievable vacuum $p_a \leq 40$ torr) is given by the difference of the integral of the flow, Q (t), recorded between the start of pumping at t_0 and the end of the flow period at t_1, and the storage volume Vs:

$$V_f = \int_{t_0}^{t_1} Q(t)\, dt - V_s. \tag{1}$$

For example, the average storage volume in the NX borehole URL-HG-El was Vs = 1159 cm^3 (Vs of the EX boreholes in the URL shaft collar was 515 cm^3). For the fracture identified at depth 1415 to 1440 cm in the HG-El borehole (extracted volume 1448 cm^3) follows:

$$V_f = (289 \pm 30)\ \text{cm}^3$$

This value is typical for fractures encountered in the URL shaft collar.

Using the "cubic law" derived from the ideal plane model and the estimated permeability of k = 1.7 x 10^{-11} cm^2 (17 mD), the hydraulic aperture, e, of the fracture is:

$$\begin{aligned} e^3 &= 12\ k/N \\ e &= 17\ \mu m \end{aligned} \tag{2}$$

where N is the joint frequency in the tested section. From (1) and (2), the radius of influence, r, follows:

$$r = \sqrt{V_f/\pi e}\ = 231\ \text{cm} \tag{3}$$

The corresponding area of influence is approximately 17 m^2. This means that the extent of the area drained during the test is comparable with the area of individual fractures in the near-field of the URL and the vacuum permeability test is appropriate for the ERZ range of scale.

3.2 Permeability Distributions

Darlington GS Water Intake Tunnel. As mentioned in Section 3, the vacuum permeability technique has been employed to determine the magnitude and extent of permeability disturbance under various geomechanical conditions. The measurements in the Darlington GS water intake tunnel, near Toronto, illustrate the expected permeability distribution quite well (Figure 6). The tunnel is oriented north-south approximately 60 m below the surface in horizontally bedded dense limestones with shaley interbeds. The in situ stress situation and the fracturing of the rock is given in Lukajic [11] and McKay [12].

As Figure 6 shows, the permeability distribution measured in three NX boreholes decreases with distance from the excavation wall. Because of the wide range of permeabilities, ka, the measurements are presented as decadic logarithms of 10^6 x ka. The zone of increased permeability around the excavation extends over 0.9 m deep in the sidewall and 0.4 m deep in the crown.   The shallow depth of the ERZ in the crown is attributed to higher stresses.   The increased permeability within the ERZ is probably due to lowering of radial stresses around the tunnel which resulted in expansion of the orthogonal sets of vertical and bedding joints.   Blast damage very likely contributed to the perimetral fracturation.   The joints beyond the ERZ remained tight due to vertical and horizontal stresses acting in the intact rock mass.   The sensitivity of permeability to stress is obvious from the comparison of measurements carried out before and after construction of the liner.   Due to stress redistribution after construction of the concrete liner, the permeability distribution became narrower and the magnitude of the permeability values decreased.

Colorado School of Mines.   The pattern of permeability distribution in the Colorado School of Mines' experimental mine in Idaho Springs is more complex than that in Darlington.   The rock around the ONWI test room is a precambrian medium to coarse grained granitic gneiss interspersed with lenses of pegmatite, biotite schist and quartz.   The Swedish cautious blasting technique was used to excavate the room which is 5 m wide and 3 m high and approximately 100 m below the surface.   The geology, fracturing and in situ stresses around the excavation are described by Montazar and Hustrulid [5].

The vacuum testing was done in NX boreholes drilled radial to the excavated room.   Our measurements were carried out mainly for purposes of calibration of the testing method [8] and the accuracy of the instruments used was slightly below that stated in Section 3.1.   There is no monotonically decreasing distribution of permeabilities away from the excavation wall.   The zone of increased permeability near the excavation perimeter extends to a depth of 0.5 to 1.15 m followed by vacuum tight zones.   However, considerable permeabilities exist beyond the tight zones in deeper parts of the rock mass (Figure 7).

While the radial permeability distribution alternated between permeable and nonpermeable zones, the longitudinal distribution showed discrete, permeable fractures rather than permeable zones.

Underground Research Laboratory.   During the tests in the pink gneissic granite of the shaft collar at AECL's Underground Research Laboratory (URL) in Lac du Bonnet, Manitoba, we did not find a continuous permeable zone around the excavation, only sparsely spaced discrete permeable joints.   The testing was done in six horizontal EX boreholes, drilled radial from the shaft collar at the 15 m level.   The depth of each hole was 15 m which well covered the excavation response zone.   The full length of the holes was tested with 25 cm intervals but only 15 single zones of permeability were encountered (Figure 8).   It is likely that these horizontal boreholes preferentially sampled vertical/subvertical joints.   In borehole 4, drilled at the east end of the north wall, joints with active water flow were detected.   In five of the six radial boreholes (Nos. 1,2,3,4 and 6) no discernible permeability was observed in the test sections close to the excavation wall (depth 8 to 33 cm).   The only exception was the high rock wall permeability of borehole No. 5 where a constant flow of 5150 cm^3/min was measured at the 712 torr level (5 torr below the atmospheric pressure during the test).

Evidently the main factor controlling the near field permeability at the URL is the wide spacing of pre-existing joints.   The excavation response at this shallow depth seems to have remained limited to stress perturbation and perhaps to displacement along the subhorizontal joints observed in the upper part of the shaft collar. A more pronounced excavation response, however, can be expected under higher lithostatic pressure conditions.

3.3   Discussion

Our investigations (Section 3.2) show that the permeability of the excavation response zone cannot be generally described in terms of a homogeneous anisotropic porous medium of defined thickness with a monotonically decreasing permeability distribution. Rather, the permeability distribution is a product of the superimposed effects of the fracture system geometry and of the locally dominant fracture deformation component leading to either shear-induced dilation or stress compression of fractures.

In the Darlington Tunnel, due to the high density of fractures, the evacuation during testing did not remain confined to a single fracture. Very likely the boundary conditions were violated as well. In some of the tests, the drainage area intersected the excavation wall. For these reasons the permeabilities measured at Darlington must be considered index values only.

In the case of the Colorado School of Mines, a stress-joint interaction effect is clearly superimposed on the permeability of the excavation response zone. As a result, alternating zones of permeable and nonpermeable fractures developed. As the permeability measurements cannot be clearly attributed to either individual fractures or to statistically representative parts of the fracture system, they can only be interpreted as apparent air permeabilities.

Due to the sparsely spaced fractures in the URL, it was possible to isolate and test individual fractures. As the hydraulic behaviour of the fractures is similar to that of a homogeneous anisotropic medium, the permeabilities of single fractures can be calculated with some confidence. Although they do not represent the tensor permeability of the rock mass, they do predict the actual hydraulic behaviour of individual fractures.

In order to obtain meaningful data on flow through ERZ, the proper location of the test boreholes has to be ascertained; this requires first establishing the three-dimensional geometry of the fractures through a statistically relevant number of small scale tests.

4.0   IMPLICATIONS

The current shaft sealing experimental concept (Figures 2 and 3) assumes a series of shallow holes will be diamond drilled radially out from the shaft. One ring of holes is located opposite the sand recharge zone. A separate ring of holes is located towards the downstream limit of the plug. These holes will be used for the following purposes:

- they are installed to act as recharge holes into the ERZ to initiate flow longitudinally in the rock across the shaft plug and as collection holes at the downstream limit of the plug. Tracer detection equipment will be installed in the latter.

- diamond drill core obtained from the drilling will document the rock type and condition and will identify any fissures or fractures. A borehole camera will be used to confirm the presence of fissures.

- these holes will also be used for a complete suite of in-situ testing including cross hole geophysics, pump in hydrogeologic testing, as well as the vacuum permeability testing.

Geophysical testing, although valuable for monitoring changes in mass properties within the ERZ, may be limited in ability to assign meaningful hydraulic properties to the disturbed zone. The vacuum permeability technique, despite the acknowledged difficulties in interpretation, is expected to achieve the following:

- document "vacuum tight" lengths of the borehole. Such zones are of no interest to seepage flows within the ERZ.

- document zones which have a measurable air permeability. Analysis of the test data would provide in the worst case a relative index value which would help focus the experiment considerations. Under favourable conditions, the test data will provide the aperture and extent of fissures. The test data would, of course, be correlated with the diamond coring, borehole photography, and geophysical information.

- document zones where cross hole responses were detected. Such interconnection would be subjected to extensive testing using all available techniques so as to maximize the information available prior to placement of the seal, and the performance of the classical hydrogeological testing.

Unlike the plug material, the conditions in the ERZ cannot be easily controlled by the engineering design. Therefore the careful documentation of the characteristics of this zone, and its hydraulic significance is essential to both predictive estimates and to the ability to affect remedial measures (should such be necessary). The Vacuum Permeability Technique promises to be a valuable tool for this task.

REFERENCES

[1]   Seymour, P.H. and Gray, M.N.: "Borehole and Shaft Seals - Design Considerations", Proceedings of the Nineteenth Information Meeting of the Nuclear Waste Management Program, Atomic Energy of Canada, 1985, TR-350.

[2]   Keil, L.D.: "Borehole and Shaft Sealing Experiment Planning Committee Meetings, Minutes", AECL-US/DOE Cooperative Program, Atomic Energy of Canada internal document, 1987-1988.

[3]   Bordman, C.R. and Skrove, J.W.: "Distribution in Fracture Permeability of a Granitic Rock Mass Following a Contained Nuclear Explosion", Journal of Petroleum Technology, 18, 5, 1966, 619-623.

[4] Miller, C.H., Cunningham, D.R. and Cunningham, M.J.: "An Air Injection Technique to Study Intensity of Fracturing Around a Tunnel in Volcanic Rocks", Bulletin of the Association of Engineering Geologist, XI, 3 1974, 203-217.

[5] Montazar, P.M., and Hustrulid, W.A., "An Investigation of Fracture Permeability in Metamorphic Rock Around an Underground Opening", ONWI Topical Report No. 5, 1981.

[6] Figg, J.W.: "Method of Measuring the Air and Water Permeability of Concrete", Magazine of Concrete Research, 25, 85, 1973, 213-219.

[7] Jakubick, A.T. and Klein, R.: "Multiparameter Testing of Permeability by Transient Vacuum Technique", in Coupled Processes Symposium Proceedings, Chiu-Fu Tsang, ed., Academic Press, New York, 1987.

[8] Jakubick, A.T.: "Vacuum Logging of the Integrity of the Near-Excavation Zone", Field Measurements in Geomechanics, Proc. of the International Symposium, Zurich, 1984, 163-175.

[9] Jakubick, A.T.: "Permeability Monitoring of the Near-Field of Underground Openings", Ontario Hydro Research Report No. 83-520-K, 1983.

[10] Dake, L.P.: "Fundamentals of Reservoir Engineering. Elsevier, Amsterdam, 1987.

[11] Lukajic, B.J.: "Geotechnical Experience with Tunnel Portal Construction", 14th Canadian Rock Mechanics Symposium, Vancouver, British Columbia, 1982.

[12] McKay, D.A.: "In Situ Stress Measurements at Darlington GS", Ontario Hydro Research Report No. 79-47-K, 1979.

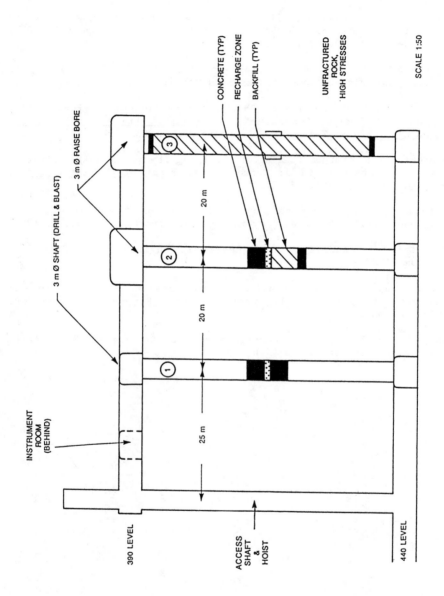

FIGURE 1

SHAFT SEALING EXPERIMENT

GENERAL ARRANGEMENT

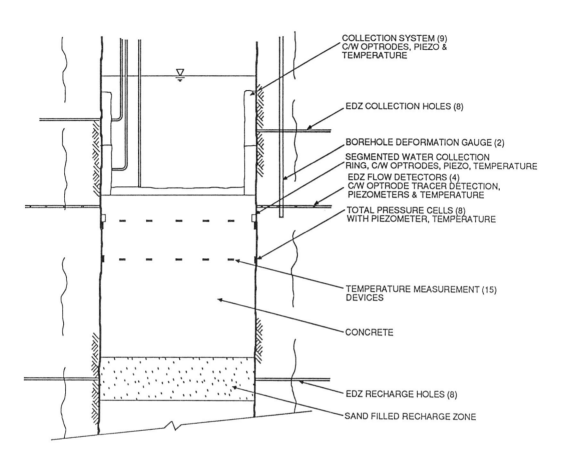

COLLECTION SYSTEM (9)
C/W OPTRODES, PIEZO &
TEMPERATURE

EDZ COLLECTION HOLES (8)

BOREHOLE DEFORMATION GAUGE (2)

SEGMENTED WATER COLLECTION
RING, C/W OPTRODES, PIEZO, TEMPERATURE

EDZ FLOW DETECTORS (4)
C/W OPTRODE TRACER DETECTION,
PIEZOMETERS & TEMPERATURE

TOTAL PRESSURE CELLS (8)
WITH PIEZOMETER, TEMPERATURE

TEMPERATURE MEASUREMENT (15)
DEVICES

CONCRETE

EDZ RECHARGE HOLES (8)

SAND FILLED RECHARGE ZONE

FIGURE 2
TYPICAL CONCRETE SEAL INSTRUMENTATION CONCEPT
1:5

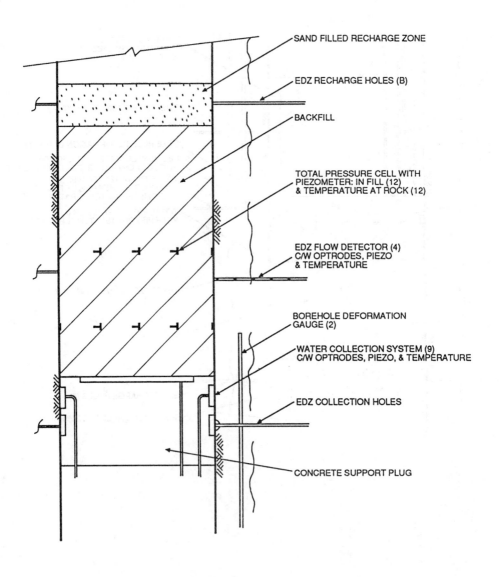

SAND FILLED RECHARGE ZONE

EDZ RECHARGE HOLES (B)

BACKFILL

TOTAL PRESSURE CELL WITH
PIEZOMETER: IN FILL (12)
& TEMPERATURE AT ROCK (12)

EDZ FLOW DETECTOR (4)
C/W OPTRODES, PIEZO
& TEMPERATURE

BOREHOLE DEFORMATION
GAUGE (2)

WATER COLLECTION SYSTEM (9)
C/W OPTRODES, PIEZO, & TEMPERATURE

EDZ COLLECTION HOLES

CONCRETE SUPPORT PLUG

FIGURE 3
BACKFILL SEAL INSTRUMENTATION CONCEPT
1:5

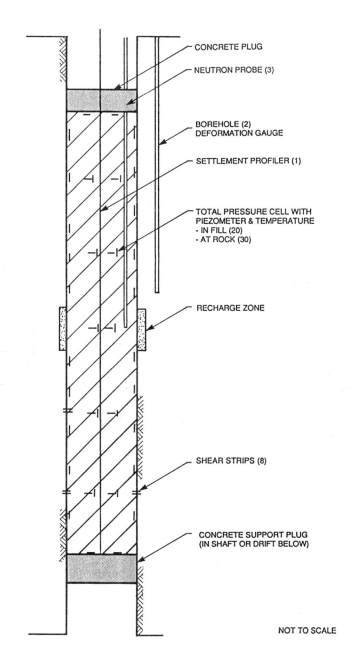

CONCRETE PLUG

NEUTRON PROBE (3)

BOREHOLE (2)
DEFORMATION GAUGE

SETTLEMENT PROFILER (1)

TOTAL PRESSURE CELL WITH
PIEZOMETER & TEMPERATURE
- IN FILL (20)
- AT ROCK (30)

RECHARGE ZONE

SHEAR STRIPS (8)

CONCRETE SUPPORT PLUG
(IN SHAFT OR DRIFT BELOW)

NOT TO SCALE

FIGURE 4
SHAFT BACKFILL INSTRUMENTATION CONCEPTS

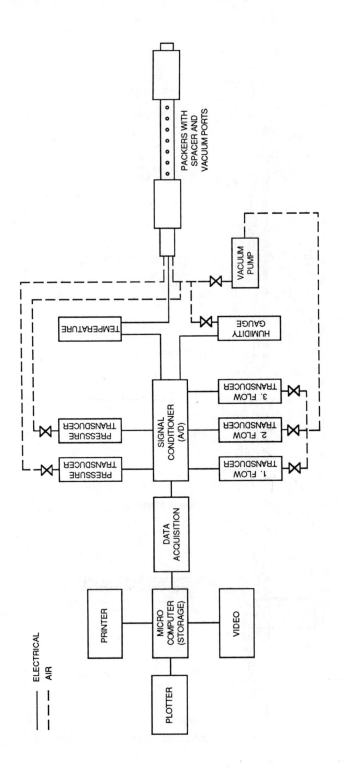

FIGURE 5

SCHEMATIC OF VACUUM PERMEABILITY TESTING SYSTEM

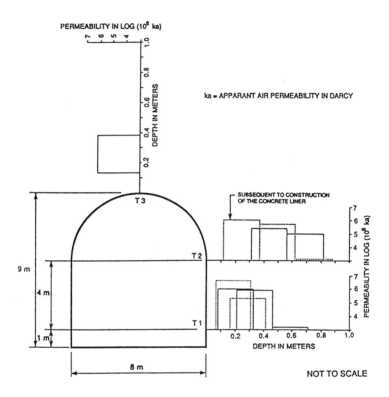

FIGURE 6
DARLINGTON GS WATER INTAKE TUNNEL

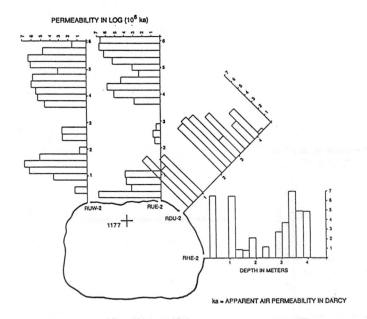

PERMEABILITY IN LOG (10⁶ ka)

ka = APPARENT AIR PERMEABILITY IN DARCY

FIGURE 7

COLORADO SCHOOL OF MINES' EXPERIMENTAL MINE
RADIAL BOREHOLE VACUUM PERMEABILITY
SECTION OF THE ONWI ROOM IN THE AREA OF THE
SECOND RING

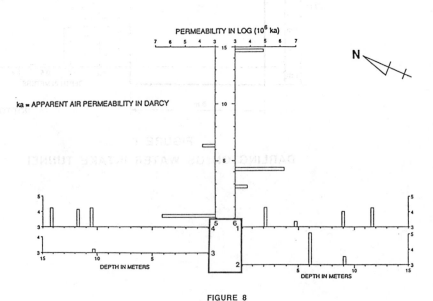

PERMEABILITY IN LOG (10⁶ ka)

ka = APPARENT AIR PERMEABILITY IN DARCY

FIGURE 8

UNDERGROUND RESEARCH LABORATORY, MANITOBA
SHAFT COLLAR PLAN VIEW
NEAR FIELD PERMEABILITIES AT THE 15 METER LEVEL

EXCAVATION EFFECTS ON TUFF - RECENT FINDINGS AND
PLANS FOR INVESTIGATIONS AT YUCCA MOUNTAIN*

T.E. Blejwas, R.M. Zimmerman, and L.E. Shephard
Sandia National Laboratories
Albuquerque, New Mexico, United States

ABSTRACT

Plans for site-characterisation testing and constructing an exploratory
shaft facility (ESF) at Yucca Mountain, Nevada, have been influenced by the
construction and monitoring of stable openings in G-Tunnel on the Nevada Test
Site. G-Tunnel provides access for testing in a thin bed of unsaturated
welded tuff that is similar to that at Yucca Mountain. The data from the
experiments in the ESF will be used to validate analytical methods for
predicting the response of underground openings to the excavation process and
to the heat generated by the waste.

EFFETS DES TRAVAUX D'EXCAVATION SUR LES TUFS :
RESULTATS RECENTS ET PLANS CONCERNANT
LES RECHERCHES ENTREPRISES A YUCCA MOUNTAIN

RESUME

On a mis à profit les renseignements obtenus lors de la construction
d'ouvertures stables dans le tunnel G du site d'essai au Nevada et de leur
surveillance dans le cadre de l'élaboration des projets d'essais de
caractérisation du site et de construction d'un puits d'exploration à Yucca
Mountain (Nevada). Le tunnel G permet de réaliser des essais dans un mince lit
du tuf soudé et non saturé, semblable à celui que l'on retrouve à Yucca
Mountain. Les données obtenus à la suite des expériences réalisées dans le
puits d'exploration serviront à valider les méthodes d'analyse. Celles-ci
permettront de prévoir la réaction des ouvertures souterraines au dégagement
de chaleur des déchets lors du processus d'excavation.

---

* This work was supported by the U.S. Department of Energy (US DOE) under
contract DE-AC04-76DP00789.

Consideration of Yucca Mountain as a repository for high-level nuclear
waste has been ongoing for several years by the Nevada Nuclear Waste Storage
Investigations (NNWSI) Project. In December, the Nuclear Waste Policy
Amendments Act of 1987 [1] was enacted and Yucca Mountain was designated as
the site to be characterized. Under the provisions of the Act, work ceased
at other sites at which characterization was planned. The current
understanding of the nature of Yucca Mountain and the U.S. Department of
Energy's (DOE) plans to characterize the site were released for comment in
January 1988 [2].

The stratigraphy of Yucca Mountain consists primarily of units of welded
and nonwelded tuff (Figure 1). The target thermal-mechanical unit into which
the waste would be emplaced, TSw2, is moderately to densely welded,
fractured, and on the order of 200 m thick. The porosity of the welded
matrix material is variable with some locations having spherical or
compressed lithophysal (gas-formed) cavities. Vitric and zeolitized
nonwelded tuff units underlie the waste-emplacement area at most locations.
The planned repository excavations include drifts and waste-emplacement
boreholes, which would be located in the welded units, and shafts and ramps,
which would pass through both welded and nonwelded units.

Yucca Mountain is located in an arid area that receives an average
precipitation of about 150 mm a year [3], of which only a small percentage
actually infiltrates the surface. At the planned horizon for the repository,
which is about 250 m above the water table and about 300 m below the surface
of the mountain, water is believed to be flowing primarily downward through
the matrix material with a flux of less than 1 mm per year [4]. At the in
situ partial saturation levels, which are estimated to be between 0.4 and
0.65 at the repository horizon [2], the fractures are not expected to
influence significantly the movement of liquid moisture; i.e., continuous
fracture flow is not expected in an unsaturated environment [5].

G-Tunnel is located in Rainier Mesa about 40 km from Yucca Mountain.
Although, like Yucca Mountain, Rainier Mesa is composed primarily of volcanic
tuffs, only a 12 m thick unit of welded tuff (including a rubble zone as
shown in Figure 2) is accessible from G-Tunnel. Lithophysae have not been
identified in G-Tunnel. The joint system in the available welded-tuff unit
at Rainier Mesa has predominately vertical joints with a frequency of about 3
to 5 per meter [6]. The joint system in the TSw2 unit at Yucca Mountain may
be similar but, because only data from vertical coreholes and surface-outcrop
mappings are available, the precise nature is uncertain.

The Site Characterization Plan for Yucca Mountain is scheduled for
release late this year following comment on the present draft version [2].
With this release date, construction of the Exploratory Shaft Facility (ESF)
would begin about the middle of 1989. This large underground facility, which
will be located in the northeast part of the repository, will include two
vertical shafts and about three kilometers of drifts at the repository
horizon. About one-half of the drift length will be along the route of
future drifts for the repository. On the basis of existing corehole and
surface data, the ESF site is considered to be representative of the entire

repository block [7]. Present plans also include a number of new explorations from and on the surface, including coreholes, drillholes, and trenches.

IMPORTANCE OF EXCAVATION EFFECTS

Understanding and predicting the behavior of underground openings during the operational (preclosure) period of a repository are important components of a possible license application. Excavations alter the in situ stress regime and modify the physical characteristics of the geologic medium into which waste will be emplaced. The heat generated by the waste both induces additional loads and further modifies the rock characteristics. Therefore, understanding excavation effects is an important component to understanding and predicting the total response of underground openings. However, the addition of heat is the single largest factor that differentiates the behavior of openings in a repository at Yucca Mountain from the behavior of typical mined openings.

Although the possible formation of new fractures and the activation of existing fractures caused by mining may modify the response of openings at Yucca Mountain, the fractures are not expected to change the local hydrologic and transport characteristics of the already fractured unsaturated medium. Possible exceptions to this include the impacts of bounding scenarios that would raise the local saturation around shafts to fully saturated. In conclusion, excavation effects at Yucca Mountain are important to preclosure performance but may be of limited importance to waste isolation following closure of the repository.

MINE-BY EXPERIMENT IN G-TUNNEL

Experiments for evaluating possible waste disposal in tuffs have been conducted in G-Tunnel since 1980 (summarized by Zimmerman and Finley [6]). The excavation activities were recently culminated with a mine-by experiment that included the monitoring of the mining of a drift with dimensions similar to large drifts planned for Yucca Mountain [8,9,10]. A cross-sectional view and plan view of the experiment are shown in Figures 2 and 3. The experiment began with the mining of an observation drift in nonwelded tuff. Multiple-point borehole extensometers (MPBXs) were installed in holes drilled from the observation drift toward the rock mass that would be excavated as part of the demonstration drift. Additionally, pre- and postmining borehole injection measurements using pressurized water between packers were made from other holes in the observation drift.

The demonstration drift was mined parallel to the observation drift and above and to the side of that drift (Figure 2). The locations for the two drifts were dictated in part by existing drifts and the stratigraphy. Although most of the demonstration drift is located in welded tuff, much of the floor is in a transition zone that has varying degrees of welding. Controlled blasting (smooth blasting), which is the technique planned for most of the repository drifts at Yucca Mountain, was used for mining the drift. Ground support consists of rockbolts and wire mesh. As the demonstration drift was mined, MPBXs and pins for cross-drift convergence measurements were installed close to the face of an advance. Once instruments were installed, they were used to monitor each successive round of blasting. After the mining was completed, the borehole injection measurements from the observation drift were repeated. These measurements could be made no closer than about one meter from the surface of the demonstration drift because of the limitations of the packer system.

During the mining of the demonstration drift, a dip-slip fault was encountered. Drilling and mining through the fault zone was more difficult, and a shear zone around the fault is evident in postmining examinations of the demonstration drift. The fault offset is on the order of 2.5 m, but the fault had not been noticed in the observation drift before its discovery in the demonstration drift.

A sample of the data collected during the mine-by experiment is presented to provide insight into the excavation effects of the G-Tunnel mining. A complete set of data collected during the mine-by experiment is presented in Reference 10. (Examples of some measurements are also presented in Reference 8). In Figure 4, the horizontal and vertical convergence measurements at the various stations in the demonstration drift are presented. Although the fault that intersects the drift influenced the magnitudes and rates of convergence in a manner that is not fully understood (e.g., the convergences at Stations B, C, D, and E), none of the observed convergences or rates suggest serious stability problems [8]. The MPBX measurements show similar trends and lead to the same conclusion.

Borehole injection measurements from before and after the mining of the demonstration drift are shown in Figure 5. Data are presented in terms of a hydraulic quotient (HQ), which is the ratio of the flow rate to the injection pressure. Near the fault at the stations D2 and D3, the HQ increased significantly after the mining. However, at other locations the differences are small, with many locations showing a decreased HQ following the mining. If new fractures near the measurement locations were induced by the mining, their presence is not evident from these measurements. The increased HQ near the fault is very likely a result of movement within the disturbed zone around the fault.

Because the G-Tunnel mining was conducted in a complex stratigraphy with some important differences from the expected geologic environment at Yucca Mountain, conclusions about excavation effects at Yucca Mountain must be qualified. Still, G-Tunnel provides the best available information on mining in welded tuff, and the influence of the G-Tunnel experience is evident in the plans for the ESF that are described below. In addition, some general conclusions on the approach to conducting experiments in the ESF are appropriate: 1) Geologic characterization of test locations before and during instrumentation are necessary to ensure that all of the appropriate data are obtained. 2) A high degree of instrument redundancy is required because no instrumentation set can be assured of operating without malfunctions. 3) Finally, to facilitate analytical modeling, the boundary conditions for any validation experiments must be carefully considered in the design of the experiments.

OTHER G-TUNNEL EXPERIMENTS

Other experiments conducted in G-Tunnel include a heated-block experiment, several borehole heater experiments, over-core stress

measurements, and pressurized slot tests. Results from these experiments
have been summarized by Zimmerman and Finley [6] and will not be described in
detail here. Some comparisons with analytical results are favorable [11],
suggesting that similar experiments at Yucca Mountain may provide suitable
data for validating the analytical models.

PLANS FOR CONSTRUCTION OF THE EXPLORATORY SHAFT FACILITY

Based in part on the experience with mining in G-Tunnel, the shafts and
drifts in the ESF will be mined using smooth blasting. Although differences
in stratigraphy make direct comparisons between G-Tunnel and the ESF
uncertain, present plans for the ground support in the ESF include rockbolts
and wire mesh that are similar to those used in the demonstration drift in G-
Tunnel. However, if the conditions are found to require more elaborate
ground-support systems (e.g., steel sets), back-up plans and procedures for
installation of such systems will be available. Also, analyses are presently
underway to determine whether the rockbolt system is suitable for the ESF
drifts that will become parts of the repository.

EXCAVATION-EFFECTS EXPERIMENTS IN THE EXPLORATORY SHAFT FACILITY

Several experiments that investigate excavation effects will be conducted
during and after the construction of the ESF. One is aimed at the
performance of vertical shafts, while the others will monitor the behavior of
drifts. Although not specifically excavation-effects experiments, several
other ESF experiments that include heaters to simulate the effects of
emplaced waste will also provide insight into the total response around
openings, which includes the excavation effects. In general, the excavation
investigations are designed to provide information for validating the
mechanical response of repository openings, while the heater experiments are
designed to provide information for validating models used in heat transfer
and thermomechanical calculations. The excavation investigations are
emphasized here, but descriptions and summaries of all the experiments are
available [2, 12]. A study plan that describes the first three experiments
described below will be available shortly [13].

Shaft Convergence Experiment

At three locations in one of the two vertical shafts, convergence anchors
and MPBXs will be installed in two parallel sets, about two meters apart
vertically. The locations for the experiments include both lithophysae-rich
and lithophysae-poor tuff. A plan view of a possible layout of instruments
is shown in Figure 6. The instruments will be installed following a drill
and blast round and as close as practical to the exposed face. The
installation of a concrete shaft liner will follow the excavation and
instrument installation, but access to the instruments will be maintained.
Pressure cells will be installed in the concrete liner.

Data from convergence measurements and the MPBXs will provide the first
opportunities to compare analytical predictions of mechanical response with
actual measurements from Yucca Mountain.

Demonstration Breakout Rooms

Two rooms with cross-sectional dimensions similar to the largest drifts in the repository will be excavated off of breakouts from one of the vertical shafts. The first will be mined in Unit TSw1, which contains significant amounts of lithophysal voids. The second will be constructed in Unit TSw2 at an elevation planned for the repository. Both rooms will be instrumented with convergence anchors, MPBXs, and rockbolt load cells. A possible layout for the instruments is shown in Figure 7. Again, the instruments will be installed as close to a newly mined face as practical. Measurements of accelerations during blasting and/or seismic measurements during artificial excitations may be used to gain insight into the damaged or relaxed zone around the excavations.

The construction of the demonstration breakout rooms will be the first confirmation that repository-size drifts can be constructed in welded tuff at Yucca Mountain using reasonably available technology (a regulatory requirement). Data from the rooms will also provide the first comparisons with analytical predictions for the mechanical behavior of drifts at Yucca Mountain. If the upper room (in tuff with a large amount of lithophysae) does not require unusually complex ground support, plans for repository construction may be relaxed with respect to this type of rock.

Sequential Drift Mining Experiment

A minimum of three drifts will be mined in sequence in a manner similar to the G-Tunnel mine-by experiment. At least two observation drifts will be mined first. MPBXs will be installed in holes that are drilled toward the volume around a future repository-size drift. Borehole stressmeters (BSM) and other instruments may also be installed in similar holes. The repository-size drift will be instrumented as excavation proceeds in a manner similar to the demonstration breakout rooms. Again, techniques for estimating the extent of damage or relaxed zones around the demonstration drift, such as seismic or permeability measurements, will be used. A possible layout for the drifts and the instruments is shown in Figure 8.

The data from instruments originating in the observation drifts will include measurements of the rock-mass response before, during, and after a drift is excavated. This information will provide the best opportunity for validation of the mechanical models. The ground-support systems will also be evaluated using convergence data, load measurements from the rockbolt load cells, and other measurements. Careful monitoring of the excavation process will provide input for the assessment of the excavation methods.

In Situ Design Verification

Three long drifts with a total length of about 1.5 km will be mined to characterize the nature of expected underground geologic structures. Sections of these drifts will be mined at full repository size and instrumented to provide an opportunity to observe the behavior of full-size drifts in a variety of ground conditions, including through fault zones.

The precise nature of instruments that will be installed in the long-drift sections is not known, but, at a minimum, the cross-drift convergence of the sections and the response of different ground-support systems will be monitored. Any local instabilities that occur will be documented. The construction of the drifts will be closely monitored to assess whether the techniques used are adequate and appropriate for continued use in the repository.

Other Experiments

As discussed previously, a number of heater experiments will be conducted. In addition, a large program of laboratory testing for mechanical and thermal properties of intact rock samples and the mechanical properties of jointed samples is planned. Finally, some direct measurements of in situ mechanical properties are anticipated. These include plate-loading tests for calculating rock-mass modulus and large-block shear tests for evaluating the large-scale strength of joints.

The DOE's plans for characterizing Yucca Mountain are presently the subject of consultation with the U.S. Nuclear Regulatory Commission, the State of Nevada, and others. The experiments described above may change as a result of these review activities. Also, more detailed descriptions of the experiments and the rationale for their selection will be presented in study plans that are presently being written and/or reviewed.

SUMMARY

A number of experiments that will provide insight into excavation effects in underground openings in tuff are planned for an exploratory shaft facility at Yucca Mountain, Nevada. These include shaft convergence measurements, demonstration breakout rooms, and a mine-by experiment. The planning for these experiments has been influenced by the successful conduct of a mine-by and other experiments in G-Tunnel, which has a thin bed of welded tuff that is similar to the much thicker welded tuff at Yucca Mountain. The experience to date suggests that openings in welded tuff can be constructed using reasonably available technology. Significant adverse effects from excavations have not been observed. The extrapolation of this experience to a repository will require an extensive data base from the proposed repository site. The planned excavation-effects experiments are intended to provide that data base. Additional experiments are planned to support and evaluate the impact of the addition of heat, which differentiates a repository from typical mined openings in similar geologic mediums.

REFERENCES

1. Nuclear Waste Policy Amendments Act of 1987: Title V-Energy and Environment Programs, Subtitle A-Nuclear Waste Amendments, Omnibus Budget Reconciliation Act of 1987, Conference Report to Accompany H. R. 3545, Report 100-495, December 21, 1987.

2. DOE (U.S. Department of Energy): "Site Characterization Plan, Consultation Draft, Yucca Mountain Site, Nevada Research and Development Area, Nevada," DOE/RW-0160, Office of Civilian Radioactive Waste Management, Washington D.C., January 1988.

3. Quiring, R. F.: "Precipitation Climatology of the Nevada Test Site,"
    WSNSO 351-88, U.S. Dept. of Commerce, National Oceanic and Atmospheric
    Administration, National Weather Service, Nuclear Support Office, Las
    Vegas, NV, 1983.

4. Montazer, P., and W. E. Wilson: "Conceptual Hydrologic Model of Flow in
    the Unsaturated Zone, Yucca Mountain, Nevada," Water-Resources
    Investigations Report 84-4345, U.S. Geological Survey, Lakewood, CO,
    1984.

5. Klavetter, E. A., and R. R. Peters: "Estimation of Hydrologic
    Properties of an Unsaturated, Fractured Rock Mass," SAND84-2642, Sandia
    National Laboratories, Albuquerque, NM, 1986.

6. Zimmerman, R. M., and R. E. Finley: "Summary of Geomechanical
    Measurements Taken in and Around the G-Tunnel Underground Facility,
    NTS," SAND86-1015, Sandia National Laboratories, Albuquerque, NM, 1987.

7. Nimick, F. B., L. E. Shephard, and T. E. Blejwas: "Preliminary
    Evaluation of the Exploratory Shaft Representativeness for the NNWSI
    Project," SAND87-1685, Sandia National Laboratories, Albuquerque, NM, in
    preparation.

8. Zimmerman, R. M., R. A. Bellman, Jr., and K. L. Mann: "Analysis of
    Drift Convergence Phenomena from G-Tunnel Welded Tuff Mining
    Evaluations," Rock Mechanics: Proc. of the 28th U.S. Symposium, I. W.
    Farmer,   J. J. K. Daemen, C. E. Blass, and S. P. Neuman, eds.,
    Tucson, AZ, 1987, 831-841.

9. Zimmerman, R. M., R. A. Bellman, Jr., K. L. Mann, and D. P. Zerga:
    "G-Tunnel Welded Tuff Mining Experiment Preparations," SAND88-0475,
    Sandia National Laboratories, Albuquerque, NM, in preparation.

10. Zimmerman, R. M., R. A. Bellman, Jr., K. L. Mann, D. P. Zerga, and
    M. Fowler: "G-Tunnel Welded Tuff Mining Experiment Data Summary,"
    SAND88-0474, Sandia National Laboratories, Albuquerque, NM, in
    preparation.

11. Bauer, S. J.: "Thermal/Mechanical Analyses of G-Tunnel Field
    Experiments at Rainier Mesa, Nevada," Proc. Workshop on Excavation
    Responses in Deep Radioactive Waste Repositories - Implications for
    Engineering Design and Safety Performance, OECD/Nuclear Energy Agency,
    Winnipeg, Canada, in preparation.

12. Blejwas, T. E.: "Planning a program in experimental rock mechanics for
    the Nevada Nuclear Waste Storage Investigations Project," Rock
    Mechanics: Proc. of the 28th U.S. Symposium, I. W. Farmer, J. J. K.
    Daemen, C. E. Glass, and S. P. Neuman, eds., Tucson, AZ, 1987, 1043-
    1051.

13. Shephard, L. E.: "Excavation Investigations Study Plan," SAND88-0201,
    Sandia National Laboratories, Albuquerque, NM, in preparation.

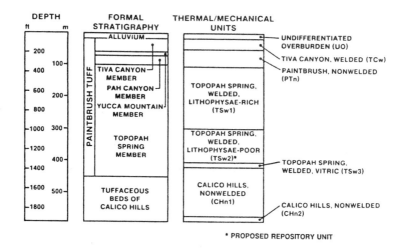

Figure 1.   Stratigraphic Units at Yucca Mountain

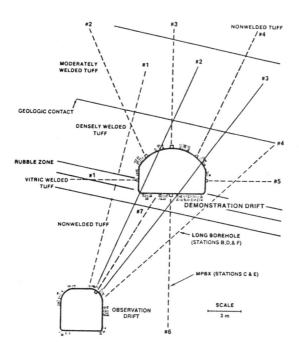

Figure 2.   Cross-Sectional View of G-Tunnel Mining Evaluation (Mine-By) [10]

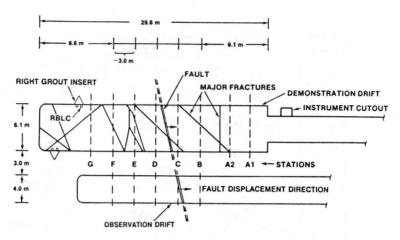

Figure 3.   Plan View of G-Tunnel Mining Evaluation (Mine-By) [10]

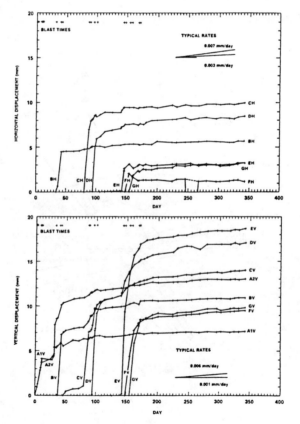

Figure 4.   Vertical and Horizontal Cross-Drift Convergence Measurements [10]

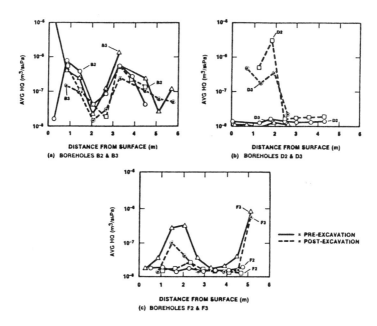

Figure 5.   Borehole Injection Measurements [10]

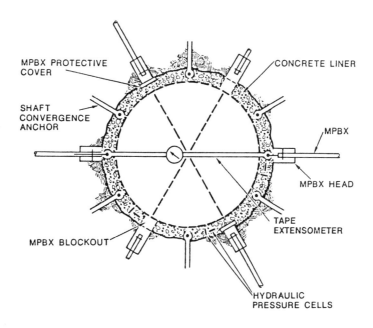

Figure 6.   Plan View for Shaft Convergence Measurements

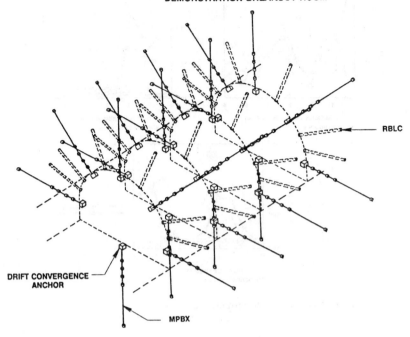

Figure 7.   Demonstration Breakout Rooms [12]

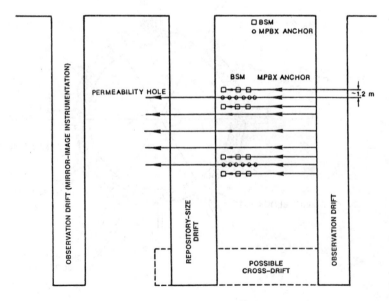

Figure 8.   Sequential Drift Mining Experiment [12]

THERMAL/MECHANICAL ANALYSES OF G-TUNNEL FIELD
EXPERIMENTS AT RAINIER MESA, NEVADA*

S.J. Bauer, L.S. Costin, E.P. Chen and J.R. Tillerson
Sandia National Laboratories
Albuquerque, New Mexico 87185 USA

ABSTRACT

Analysis methods (models) are currently being developed to support
thermal, mechanical, and thermomechanical aspects of repository design and
performance assessment of the candidate Yucca Mountain high-level nuclear
waste site. Credibility of these models, and therefore of design and
performance analyses, will in part be determined through comparison of
calculated and measured response (validation) for large-scale field
experiments. This paper discusses the models being developed, the rationale
behind the model development, and analyses of experiments performed at
G-Tunnel and planned as part of site characterisation at Yucca Mountain.

ANALYSES THERMOMECANIQUES D'EXPERIENCES IN SITU DANS
LE "G-TUNNEL" A RAINIER MESA, NEVADA

RESUME

On procède actuellement à la mise au point de méthodes d'analyse
(modèles) pour étayer les aspects thermiques, thermomécaniques et
thermochimiques de la conception et de l'évaluation des performances du dépôt
pour déchets nucléaires de haute activité susceptible d'être aménagé dans le
site de Yucca Mountain. La crédibilité de ces modèles, et donc celle de la
conception et de l'analyse des performances, dépendra partiellement de la
comparaison entre les réponses calculées et mesurées (validation) grâce à
l'exécution d'expériences in situ à grande échelle. Les auteurs examinent les
modèles en cours d'élaboration et les principes à la base de leur élaboration,
puis analysent les expériences réalisées dans le "G-Tunnel" ainsi que celles
prévues dans le cadre de la caractérisation du site de Yucca Mountain.

* This work was supported by the U.S. Department of Energy (US DOE) under
contract DE-AC04-76DP00789.

INTRODUCTION

Sandia National Laboratories, part of the Nevada Nuclear Waste Storage Investigations (NNWSI) Project administered by the Nevada Operations Office of the U.S. Department of Energy, is a participant in the process of characterizing a proposed site for geologic disposal of high-level nuclear wastes in the volcanic tuffs at Yucca Mountain, Nevada. In a repository, loads will be imposed on the rock mass as a result of excavation of the openings and heating of the rock by the nuclear waste. It is the effects of heating that are rather unique to the rock mechanics evaluations needed in support of repository design. In an attempt to gain a better understanding of the thermal, mechanical, and thermomechanical response of fractured tuff, a series of experiments have been performed and measurements have been taken in the welded and nonwelded tuffs at the G-Tunnel underground test facility at Rainier Mesa (Figure 1).

The experimental program has been complemented by an analysis program directed toward supporting experiment design, interpreting experimental results, developing and assessing numerical models, and obtaining a better understanding of rock mass response to thermal and mechanical loads. Finite element models of both linear and nonlinear material response are the primary models that have been applied to these field experiments. The experiments include two small-diameter heater experiments, a heated block experiment, a mine-by experiment, flatjack/slot deformation experiments, and in situ stress measurements. Beyond support of experiment design, analyses have focused on simulations of the above experiments by attempting to duplicate the initial, boundary, and loading conditions of the experiments. Prediction of temperatures, stresses, and displacements are the products of the analyses performed; measurements of these same parameters are made in the experiments. The ability to compare the calculated values with the measured values has assisted in the model development program. The good comparisons observed between measured and calculated responses has provided valuable insight into interpretation of experimental results and has allowed development of an initial understanding of response of a fractured rock mass to thermal and mechanical loads.

GEOMECHANICAL FRAMEWORK FOR MODEL DEVELOPMENT

Analysis of the regional geologic structures indicates that the stress state within the Basin and Range tectonic province, which includes the Nevada Test Site (NTS), has changed from crustal shortening to crustal extension within the last 20 to 30 million years [1]. The regional extension has resulted in a series of faults and joints that strike approximately north northeast (Figure 1). The direction of maximum extension is approximately N50W, based on regional structures [2], fracture orientations in the playa of Yucca Flat [2], seismicity studies [3,4], analyses of borehole ellipticities [2,5], stress measurements [e.g., 6,7], and strain measurements [8]. These studies, together with numerical analyses of the minimum in situ stress, imply that the maximum principal stress within the area of study is near vertical (and may be determined by the overburden) and the minimum horizontal stress is approximately 30 to 80% that of the overburden (with local modifications resulting from topographic effects and effects of mechanical stratigraphy) [2].

Further studies of the structural geology of Yucca Mountain [9,10,11] indicate that the mountain is composed of a sequence of volcanic tuff layers. Each thermal/mechanical tuff unit has its own set of thermal and mechanical properties and fracture characteristics. The north-northeast trending fractures tend to have near-vertical to steep westward dips. Fracture densities are greater in welded than in nonwelded tuffs [9].

It has been argued by Swolfs and Savage that the fractures impart a mechanical anisotropy to the rocks [12]. Scott and Castellanos [9] conclude that the fracture anisotropy caused the observed deviations in the direction of deep drillholes. Complications in understanding the behavior of fractured rock are implied by the observed nonlinear and stress-dependent mechanical response of individual fractures [cf. 13].

DESIGN AND PERFORMANCE FRAMEWORK FOR MODEL DEVELOPMENT

Geomechanical analyses consider mechanical, thermal, and thermomechanical response of the rock mass through the emplacement, retrieval, and postclosure performance periods. The geomechanical analyses planned for design and performance assessment indicate the need to understand the response of the host rock to excavation and thermally induced stresses. Excavation-induced stresses depend on the in situ stress field and the size and geometry of the opening relative to the local structure of the rock mass. Thermal stresses developed in the rock as a result of heating depend primarily on the magnitude of individual or combined heat sources, thermal properties of the rocks, and the heat-transfer mechanism; thus, thermal stresses vary with time and position. The thermally induced stresses are also concentrated near the opening. The effects of excavation-induced stresses are evaluated and then the effects of thermal stresses are calculated; thus, the thermomechanical response also varies with time and position.

Currently, multiple models are being used to assist in developing an understanding of the mechanical response of tuff. The fractured rock mass is being treated as an elastic, elastic-plastic, or a continuum composed of intact rock and fractures. In applications for which the physical situation may dictate, the modeling of discrete faults or fractures is performed. The continuum approach to modeling intact rock and fractures is being addressed through a compliant-joint model developed by Thomas [14]. This approach is being updated through two separate efforts, called the Chen model [15] and the JEM model [16]. Both models include two parts: (1) a continuum approximation based on average discontinuous displacements across joint planes within a representative elementary volume and (2) a material constitutive description based on linear-elastic-matrix material behavior and nonlinear normal and shear joint behavior. For both models the deformation response normal to the joint is assumed to be nonlinear elastic (Figure 2).

The Chen model is derived from laboratory testing of joint samples (Figures 2 and 3). For the Chen model, the joint shear response is treated as linear elastic in the shear stress versus shear displacement (slip) relationship before attaining a stress level governed by the Coulomb friction criterion (Figure 3). Beyond this critical stress value, a line with finite slope, analogous to strain-hardening plasticity, governs the rest of the slip relationship. Full coupling of normal and shear response is being implemented. For the JEM model [16] (Figure 4) in which shear and normal response is coupled, the joint shear response is identical to that determined empirically by Barton [17].

These approaches to modeling the mechanical response have been based on three integrated sources of data: (1) field studies of the local and regional geology of the site, (2) laboratory and field data indicative of the mechanical and thermal response of the rock from Yucca Mountain, and (3) data from in situ field experiments at Rainier Mesa and from laboratory experiments on rock samples from Rainier Mesa. Boundary element and finite element codes are being used as analysis tools to supplement tunnel index methods in determining underground support requirements.

Currently, the dominant heat-transfer mechanism in the unsaturated tuff is understood to be temperature-dependent conduction. Thermal properties are allowed to vary as a function of temperature (Figure 5). Previous laboratory and field tests and analyses, including that at G-Tunnel, indicate that the ability to accurately predict the heat flow and temperatures developed from the emplacement of a decaying heat source in a geologic media is possible with available codes, provided that the primary heat transfer mechanism is conduction [18].

A thorough understanding of the temperature field resulting from waste emplacement is important because (1) the thermal load drives the deformation of the rock mass after excavation and (2) analyses thus far indicate that the thermally induced loads act to rotate the stress field so that the maximum principal stress is near horizontal [18] (Figure 6). The rock mass is thus predicted to be subject to a stress field that it has probably not experienced.

ANALYSES OF EXPERIMENTS FIELDED AT G-TUNNEL

In the following, a summary of the results of the analyses of experiments fielded at the G-Tunnel underground test facilities is presented. Initial analyses were performed to support experiment design; later analyses were intended to assist in interpreting the rock mass response and to evaluate the modeling techniques that were applied.

IN SITU STRESS ANALYSIS

In this study, finite element calculations based on simple assumptions about the stress field and material properties were used to estimate the in situ minimum horizontal stress at Rainier Mesa.

Gravitational stress was the only loading applied. Other assumptions used in the calculations were plane-strain conditions and linear-elastic material responses. The mechanical effects of pore pressure were not included except as pore water modifies the effective mass that induces gravitational loads. The validity of these approximations was examined by comparing the calculated stresses with measured stresses at Rainier Mesa [20].

Two potential effects upon the gravitationally induced components of minimum horizontal stress were studied: topography and mechanical stratigraphy. These separate effects were analyzed by comparing calculations that assigned one set of homogeneous elastic properties to the entire modeled region with calculations that included the vertical variation of mechanical properties for Rainier Mesa.

The calculations that used homogeneous elastic properties resulted in stress magnitudes that varied smoothly with depth (Figure 7). The set of calculations that included appropriate elastic properties for selected mechanical units predicts a vertical variation in horizontal stress. The variation in calculated stress is consistent with trends in measurements of in situ stress (Figure 8). The stress magnitudes are generally higher in the units characterized by high values of Poisson's ratio and low values of Young's modulus. The success of this approach in developing an understanding of the vertical variation in horizontal stress at Rainier Mesa has prompted similar analyses for Yucca Mountain to be performed [21].

SMALL-DIAMETER HEATER EXPERIMENTS

Small-diameter heater experiments were performed in welded tuff and nonwelded tuff [18]. Here, analysis of the second experiment performed in welded tuff in which the heater was placed in a horizontal borehole and operated in four steps of increasing power levels: 400, 800, and 1200 W for 8 days each, and 1200 W for 11 days is reported [22].

The experiment geometry (Figure 9) has been modeled axisymmetrically as shown in Figure 10. Inside the stainless steel heater shell, the heated section consists of a central heater rod surrounded by air space. Heat for the problem originates in this central rod, is radiated to the heater shell, and then reradiated to the borehole wall. Above the heater rod is a length of conduit with thermal properties identical to the heater rod, but without heat generation. The conduit is surrounded by bubbled alumina in the insulated section, and epoxy in the terminal section. Low-resistance wires are connected to the heater rods in the terminal section. These wires, as well as the heater handling pipe, are represented by a continuation of the conduit rod running up the center of the assembly. The heater handling pipe is surrounded by aluminum honeycomb insulation. A reradiating surface was included at the top of the borehole to simulate the effect of the heater pressure unit covering the experiment.

The tuff surrounding the heater is modeled using a special algorithm to account for the phase change in the groundwater during heating [23]. Initially, the tuff is assumed to be saturated with groundwater. As the temperature passes the water's boiling point, the groundwater absorbs energy equal to its heat of vaporization; this is modeled as a temporary increase in the tuff's heat capacity (Figure 5). After the groundwater has vaporized entirely, thermal properties of dry tuff are used.

The mesh used in this analysis extends 14 m above and 16 m below the floor of the alcove and radially 12 m from the heater. It contains 387 eight-node quadrilateral elements, with a total of 1222 nodes. Heat flux on the boundaries was prescribed to be zero. Radiative heat transfer across air gaps in and around the heater is significant in this problem and is discussed in detail in Reference 22.

Calculated thermal results are presented for two assumed values of initial saturation 100 and 60% and are shown in Figure 11 with the measured temperatures also plotted. The effect of lowering the saturation is to raise the calculated temperature. This is believed to be reasonable, because less heat is required to vaporize the reduced amount of groundwater. In general, temperatures calculated are close to the average measured temperatures.

As a supplement to these analyses, a thermomechanical analysis was performed in which the calculated temperatures were used as input to drive a linear-elastic mechanical model. The calculated results (Figure 12) are shown for two times and are compared with the measured displacements. The comparison is best at points away from major fractures and the free surface.

G-TUNNEL MINING EVALUATION

The G-Tunnel mining evaluations experiment was a mine-by experiment intended to evaluate excavation methods and provide an empirical understanding for the excavation response of welded tuff for a repository-sized and shaped opening using monitored excavation methods. The analyses, which are described in the following paragraphs, are preliminary linear-elastic calculations that were performed contemporaneously with the experiment. They are being upgraded as more information about the structural geology becomes available and model development progresses.

The geometry of the experiment, as fielded, is shown in Figure 13. A two-dimensional cross-section of the modeled region showing loading and boundary conditions and material properties is shown in Figure 14. At Time 0, the problem was initialized and equilibrium with in situ stresses took place; at Time 1, Drift 12 was excavated; and at Time 3, the Demonstration Room was excavated.

In Figure 15, comparisons between the measured and calculated displacements for multi-point borehole extensometer (MPBX) measurement locations after excavation of the Demonstration Room are shown. In general, the measurements and calculations are of the same magnitude and trend. Both are small. It appears that the greatest differences are observed for points very near the room opening (a free surface), where local block movements could have occurred.

PRELIMINARY ANALYSES OF EXPLORATORY SHAFT EXPERIMENTS

The rock mechanics experiments planned for the Exploratory Shaft Facility (ESF) at Yucca Mountain are in part intended to supply information to assist in model validation. The analyses that were performed thus far were meant to assist in the design of the experiments.

SHAFT CONVERGENCE

The Shaft Convergence Test (SCT) [24] is designed to measure the response of several rock mass units through which the exploratory shaft (ES) will pass. The ES, in current conceptual design, will be excavated using drill and blast techniques. Approximately 2 m of material will be removed during each round of drilling, blasting, and mucking. A 0.3-m unreinforced concrete liner will be poured in sections as the shaft is deepened. The liner will be continually advanced to within 5 to 10 m of the shaft bottom as shaft sinking proceeds.

The SCT includes measurements of in situ stresses, rock mass deformation, and liner stresses at three nominal depths: 80, 200, and 320 m. Instrumentation will be installed approximately 1 m above the floor of the shaft at each station to measure changes in the diameter of the shaft. Mining will then proceed. The shaft diameter will be measured after each round of drilling, blasting, and mucking until the installation of the shaft liner prohibits further data collection.

Analyses were conducted to determine the extent of the rock/liner interaction and its implications for the measurements to be made in the SCT. The specific objectives were to
(1)   estimate the stress state in the region of proposed stress measurements,
(2)   characterize the stress redistribution as a result of excavation, including that caused by fracturing (as defined by a Drucker-Prager failure criterion),
(3)   estimate the closure of the ES,
(4)   estimate the influence of the shaft liner on closure of the shaft, and
(5)   estimate the stress magnitudes that shaft convergence may cause in the concrete liner.

The analysis approach has two important features: the size of the problem domain and the manner in which excavation is simulated.

The excavation-related displacements and stresses at each of the measurement stations are not expected to affect those at the other stations because the measurement stations are separated by large distances (120 m). Therefore, the approach of analyzing three disk-shaped regions encompassing each of the three measurement stations (Figure 16) was adopted. The radial and vertical dimensions of the modeled region were determined on the basis of elastic analyses using a coarse mesh.

Shaft excavation and lining are processes in which material is removed and added, respectively. These processes were simulated numerically by changing material properties within the appropriate regions of the domain during successive time steps of the analysis.

A finite element analysis of a cylindrical region was performed. The medium was modeled in an axisymmetric geometry and was represented as an elastic-plastic material. Identical finite element idealizations (meshes) were used to analyze each modeled region. The material properties, boundary conditions, and initial conditions were changed to simulate the conditions at each measurement depth.

The ES was modeled as a 2.14-m radius cylinder with a 0.3-m thick unreinforced concrete liner. Each round of excavation and lining took place in 2.5-m increments. The assumption of a blast-damaged zone (1 m below the floor and into the walls) surrounding the shaft was incorporated into some of the analyses because it is a region of relatively high compliance and would increase the shaft convergence. The blast-damaged zone was modeled by elements whose elastic modulus and cohesion were reduced by 50% from the corresponding values assigned to the rest of the region. The mesh was formed such that each excavation round defined a repeating module that represented 2.5 m of shaft height and extended 30 m radially. Ten excavation modules were stacked to model the extent of excavation required to capture the history of motion at the measurement station. Nine separate analyses were performed to simulate the three depths and to examine the effect of blast damage and liner placement on shaft convergence.

The following observations and conclusions may be drawn from the analyses performed:

(1) For the analyses performed, the effects of geologic anomalies or the potential effects of joints have not been considered. There will probably also be uncertainties in the stress measurements. From the relatively simplified analyses performed, it is predicted that the horizontal stresses in a region between 5 and 10 m below the shaft floor (the region where stresses are to be measured) are not significantly different from the in situ stress (Figure 17).

(2) The calculated stress state around the shaft is within the elastic range of the tuff assumed for these calculations. The only yielding that was observed was in a very restricted region near the stress singularity at the shaft wall/floor intersection for each excavation round.

(3) The maximum shaft convergence varied from about 0.5 mm for the calculation in Unit TSwl without the damage zone to about 2.5 mm for the calculation in Unit PTn with the damage zone. All other estimates of convergence fall within this range (Figure 18).

(4) Because of motion of the measurement station before excavation and gauge installation and motion of the proposed gauge anchor position, the predicted measurable convergence is only 25 to 35% of the calculated maximum convergence.

(5) Calculated stresses in the concrete liner are very low (tangential stresses are less than 5% of the unconfined compressive strength) for the assumed boundary conditions, which are believed to be on the conservative side of overestimating the stress.

(6) Most of the shaft convergence is predicted to occur before liner installation, even when the liner is installed 5 m from the shaft face. Continued excavation causes minor convergence of the shaft, and therefore the liner is predicted to have negligible effects on shaft convergence magnitude.

DEMONSTRATION BREAKOUT ROOMS

The Demonstration Breakout Rooms (DBRs) are intended to be approximately the same shape and dimensions as the largest of emplacement drifts. The drifts will be excavated laterally from breaks in the ES lining at two depths, 160 and 366 m, using controlled blasting techniques. It is proposed that excavation will proceed to the first measurement station and stop while instrumentation is installed, and then this process will be repeated for each of four more measurement stations.

The purpose of analyses of the DBRs was to estimate the magnitudes of drift convergence, potential stresses in the rockbolts, and the extent of the zone of relaxation near the DBRs [24].

The DBRs will be new excavations in tuff starting more than four shaft diameters from the shaft wall. Consequently, the perturbed stress state near the shaft does not have to be considered in the analysis, and the expected initial stress state for the DBR is the assumed in situ stress state in the rock far removed from the ES.

The approach was to analyze a two-dimensional cross section perpendicular to the axis of each DBR. Three finite element analyses were conducted, two for the DBR at the 160-m depth and one for the DBR at the 366-m depth. The DBR at the 160-m depth was modeled with and without a damaged zone. Calculations made without simulating a damage zone are elastic; those with a damaged zone are elastic-plastic (using Drucker-Prager yield criterion). The geometry and boundary conditions are shown in Figure 19.

The following calculational sequence was used:

(1) The in situ stress state was imposed on the problem domain.

(2) The properties of a section of the modeled region were changed to simulate excavation of the DBR and, for the calculations that included a damage zone, blast damage to the rock (in a manner similar to the SCR

analyses). The blast-damaged region was simulated as extending 1 m into the rock wall.
(3) The equilibrium stresses and displacements were calculated throughout the modeled region.

The calculated room convergence can be taken as an estimate of the maximum convergence that would occur from the progressive excavation of the room in a linear-elastic material.

The results, as summarized by example with Figures 20 and 21, suggest that introducing a damage zone tends to increase the displacements, to lower the stress at the excavation surface, and to produce a secondary peak in stresses in the undamaged rock mass. No yield of the rock mass was predicted in these calculations.

REFERENCES

1. Eaton, G. P.: "A Plate-Tectonic Model for Late Cenozoic Spreading in the Western United States," in Rio Grande Rift: Tectonics and Magmatism, R. E. Riecker, ed., 1979, pp. 7-32.
2. Carr, W. J.: "Summary of Tectonic and Structural Evidence for Stress Orientation at the Nevada Test Site," OFR 74-176, United States Department of the Interior, Geological Survey, Denver, CO, 1974.
3. Hamilton, R. M., and J. H. Healy: "Aftershocks of the BENHAM Nuclear Explosion," Bull. Seis. Soc. Am., V. 59, 1969, pp. 2271-2281.
4. Fischer, F. G., P. J. Papanek, and R. M. Hamilton, "The Massachusetts Mountain Earthquake of 5 August 1971 and Its Aftershocks, Nevada Test Site," Report USGS-474-149, U.S. Geol. Survey, Denver, CO, 1972 16 pp.
5. Stock, J. M., J. H. Healy, S. H. Hickman, and M. D. Zoback: "Hydraulic Fracturing Stress Measurements at Yucca Mountain, Nevada, and Relationship to the Regional Stress Field," Journal of Geophysical Research, V. 90, No. B10, 1985, pp. 8691-8706.
6. Haimson, B. C., J. Lacomb, A. H. Jones, and S. J. Green: "Deep Stress Measurements in Tuff at the Nevada Test Site," Advances in Rock Mechanics, National Academy of Science, Washington, DC, 1974, pp. 557-561.
7. Obert, L.: "In Situ Stresses in Rock, Rainier Mesa, Nevada Test Site," Report WT-1869, Applied Physics Laboratory, U.S. Bureau of Mines, 1964, 27 pp.
8. Smith, S. W., and R. Kind: "Observations of Regional Strain Variations," Journal of Geophysical Research, V. 77, No. 26, 1972, p. 4976.
9. Scott, R. B., and M. Castellanos: "Stratigraphic and Structural Relations of Volcanic Rocks in Drill Holes USW GU-3 and USW G-3, Yucca Mountain, Nye County, Nevada," USGS-OFR-84-491, Denver, CO, 1984.
10. Spengler, R. W., F. M. Byers, and J. B. Warner: "Stratigraphy and Structure of Volcanic Rocks in Drill Hole USW G-1, Yucca Mountain, Nye County, Nevada," USGS-OFR-81-1349, Denver, CO, 1981.
11. Spengler, R. W., M. P. Chornack, D. C. Muller, and J. E. Kibler: "Stratigraphic and Structural Relations in Volcanic Rocks in Drill Hole USW G-4, Yucca Mountain, Nye County, Nevada, "USGS-OFR-84-789, Denver, CO, 1984.
12. Swolfs, H. S., and W. Z. Savage: "Topography, Stresses, and Stability at Yucca Mountain, Nevada," In Proc. 26th U.S. Symp. on Rock Mech., E. Ashworth (ed.), Rapid City, SD, 1985, pp. 1121-1129.
13. Goodman, R. E.: Methods of Geological Engineering in Discontinuous Rocks, West Publishing Co., St. Paul, MN, 1976.
14. Thomas, R. K.: "A Continuum Description for Jointed Media," SAND81-2615, Sandia National Laboratories, Albuquerque, NM, 1982.
15. Chen, E. P.: "Two-dimensional Continuum Model for Jointed Media with Orthogonal Sets of Joints," In Proc. 27th U.S. Symp. on Rock Mech. Key to Energy Production, H. L. Hartman, ed., Univ. of Alabama, Dept. of Min. Eng. Tuscaloosa, AL, 1986, pp. 862-867.

16. Blanford, M., and S. Key: "The Joint Empirical Model--An Equivalent Continuum Model for Jointed Rock Masses," SAND87-7072, Sandia National Laboratories, Albuquerque, NM, in preparation.
17. Barton, N.: "Modeling Rock Joint Behavior From In Situ Block Tests: Implications for Nuclear Waste Repository Design," ONWI-308, Office of Nuclear Waste Isolation, Batelle Memorial Institute, Columbus, OH, 1982
18. Zimmerman, R. M., M. L. Blanford, J. F. Holland, R. L. Schuch, and W. H. Barrett: "Final Report: G-Tunnel Small-Diameter Heater Experiments," SAND84-2621, Sandia National Laboratories, Albuquerque, NM, 1986.
19. St. John, C. M.: "Reference Thermal and Thermal/Mechanical Analyses of Drifts for Vertical and Horizontal Emplacement of Nuclear Waste in a Repository in Tuff," SAND86-7005, Sandia National Laboratories, Albuquerque, NM, 1987.
20. Warpinski, N. R., D. A. Northrup, R. A. Schmidt, W. C. Vollendorf, and S. J. Finley: "The Formation Interface Fracturing Experiment: An In Situ Investigation of Hydraulic Fracture Behavior near a Material Property Interface," SAND81-0938, Sandia National Laboratories, Albuquerque, NM, 1981.
21. Bauer, S. J., and J. F. Holland: "Analysis of In Situ Stress at Yucca Mountain," In Proc. 28th U.S. Symp. on Rock Mech., Tuscon, AZ, June 29-July 1, 1987, pp. 707-713.
22. Blanford, M. L. and J. D. Osnes: "Numerical Analyses of the G-Tunnel Small-Diameter Heater Experiments," SAND85-7115, Sandia National Laboratories, Albuquerque, NM, 1987.
23. Morgan, K., R. W. Lewis, and O. C. Zienkiewicz: "An Improved Algorithm for Heat Conduction Problems with Phase Change," Int. J. Num. Meth. Engng., Vol. 12, No.7, 1978, pp. 1191-1195.
24. Costin, L. S., and S. J. Bauer: "Preliminary Analyses of the Excavation Investigation Experiments for the Exploratory Shaft at Yucca Mountain, Nevada Test Site," SAND87-1575, Sandia National Laboratories, Albuquerque NM, in preparation.
25. Ekren, E. B.: "Geologic Setting of Nevada Test Site and Nellis Air Force Range," in Nevada Test Site, GSA Mem. 110, Geological Society of America, Boulder, CO, 1968.

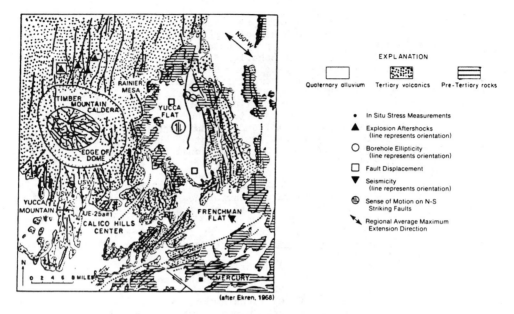

Figure 1. Map of Nevada Test Site and Vicinity Showing Basin and Range Faults, and Indicators of Maximum Extension Direction

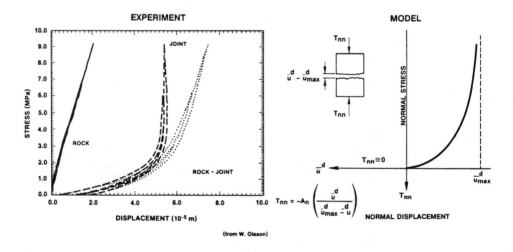

Figure 2. Nonlinear Elastic Normal Joint Behavior

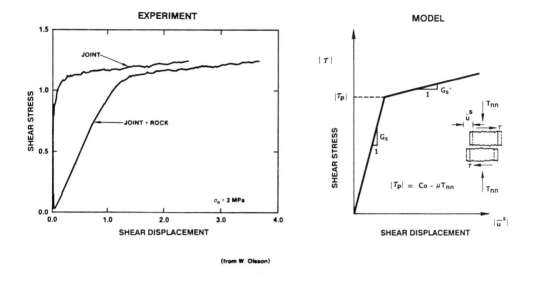

Figure 3. Nonlinear Shear Behavior of Joints (Compliant Joint Model)

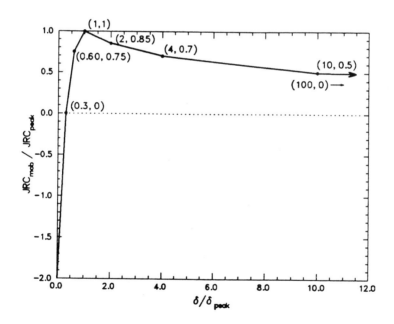

Figure 4. Nonlinear Shear Behavior or Joints (Joint Empirical Model)

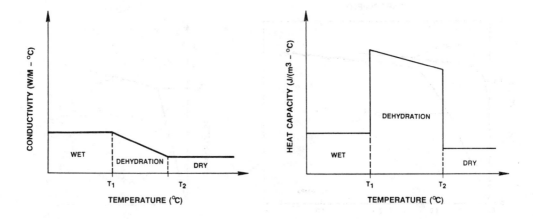

Figure 5.   Schematic of Temperature Dependence of Thermal Properties

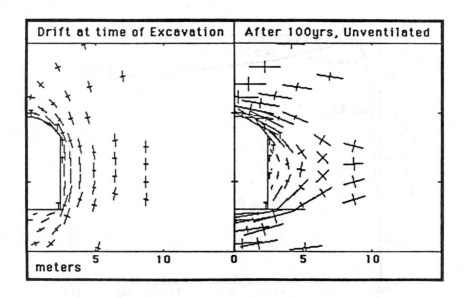

Figure 6.   Predicted Principal Stresses Through Time for the Vertical
Emplacement Scheme

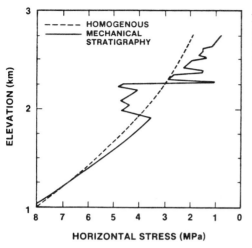

RAINIER MESA BASELINE = 1.83 km
HORIZONTAL STRESS PROFILES
CONTRAST

Figure 7. Variation of Calculated
Horizontal Stress with Depth

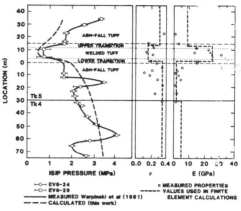

Figure 8. Comparison of Calculated
Horizontal Stresses with
Instantaneous Shut In Pressures
(ISIP) Measured at Rainier Mesa.
(Variations of elastic properties
are shown.)

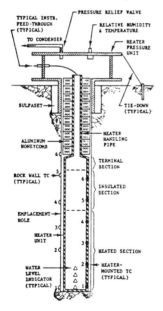

Figure 9. Small-Diameter Heater
Section (Actual orientation of
this experiment was horizontal.)

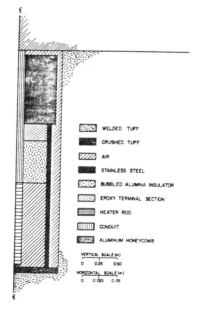

Figure 10. Model Abstraction and
Material Identification for the
Welded Tuff Small-Diameter Heater
Experiment

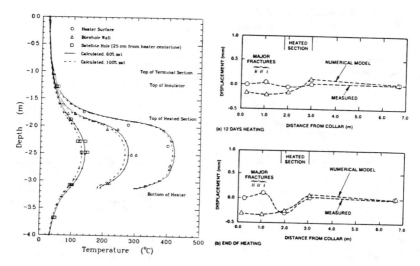

Figure 11. Comparison of Measured and Calculated Temperature Profiles for the Welded Tuff Small-Diameter Heater Experiment

Figure 12. Comparison of Measured and Calculated Displacements for the Welded Tuff Small-Diameter Heater Experiment

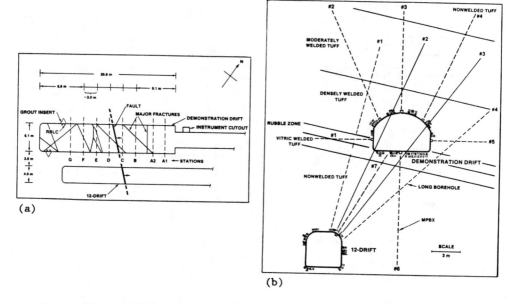

Figure 13. Plan View (a) and Cross Section (b) of the G-Tunnel Mining Evaluations Experiment

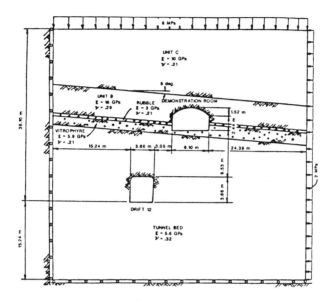

Figure 14.   Cross Section of the Meshed Region for the G-Tunnel Mining
Evaluations Experiment

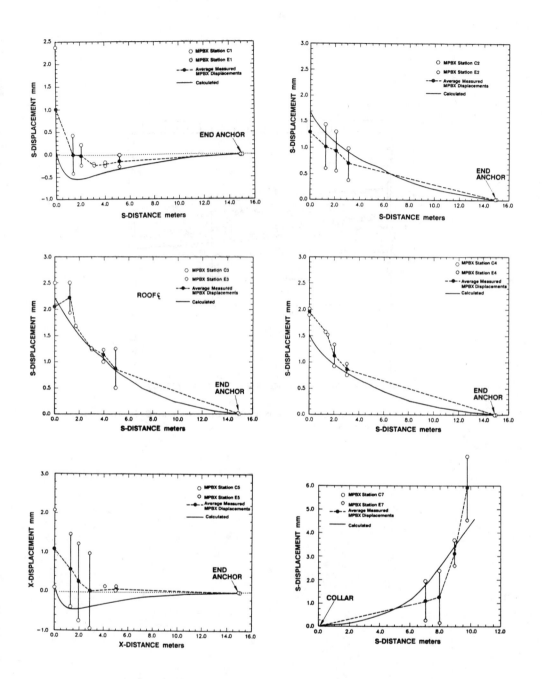

Figure 15. Comparisons Between Measured and Calculated Displacements for Multipoint Borehole Extensometer Measurement Locators

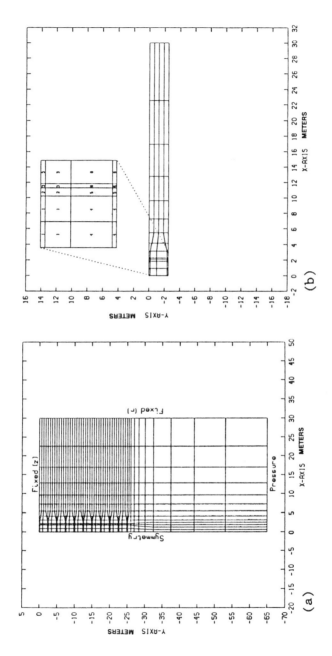

Figure 16. Finite Element Mesh Showing (a) Geometry and Boundary Condition and (b) an Excavation Module

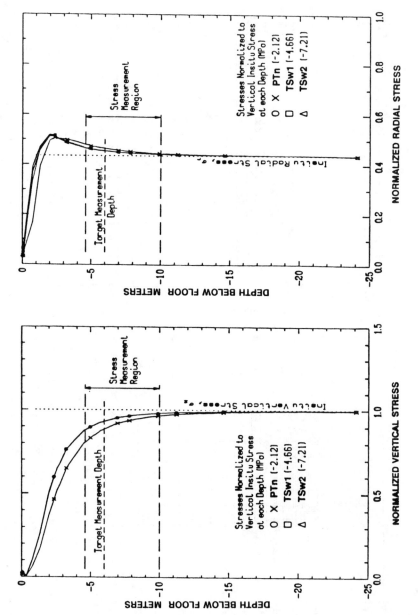

Figure 17. Normalized Vertical and Radial Stress Distribution in the Shaft Centerline Below the Floor

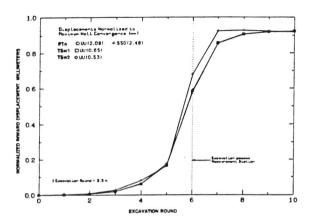

Figure 18. Normalized Relative Displacement History of Shaft Wall at
Measurement Station Depth for Four Simulations of Shaft Excavation

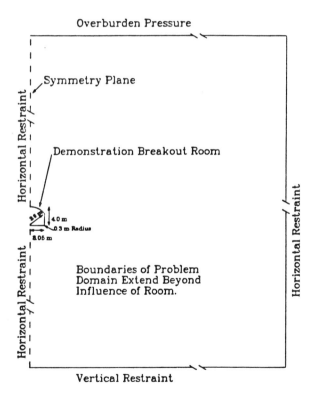

Figure 19. Problem Domain for Plane Strain Analysis of Demonstration Breakout
Rooms

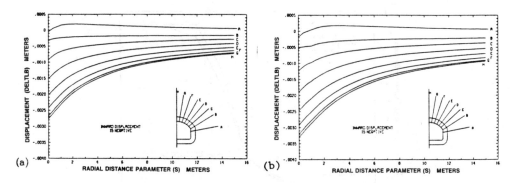

Figure 20.   Comparison of Displacements for Demonstration Breakout Room
             Experiment With (a) and Without (b) a Damage Zone

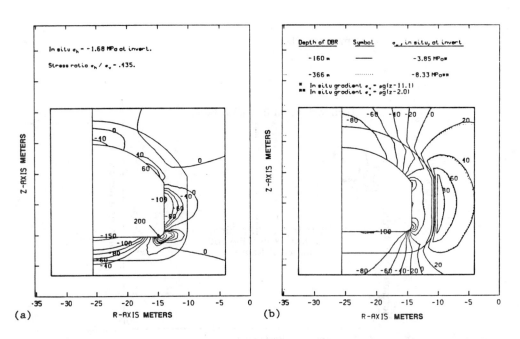

Figure 21.   Comparison of Horizontal Stresses (in Percent Change From the In
             Situ) for the DBR With (a) and Without (b) a Damage Zone

ANALYSIS REQUIREMENTS FOR EVALUATING
EXCAVATION RESPONSE IN HARD ROCK

E.N. Lindner
Battelle Memorial Institute
United States

ABSTRACT

An overview of analysis requirements for an Excavation Response
Experiment is presented as it addresses the concerns of a nuclear waste
repository program. The tools, procedures and requirements to perform such
analyses are examined, and the implications of these efforts are noted. The
analytical effort is subdivided into specific categories and examples of
modelling concerns for each category are presented. Also, the various
computer codes available for mechanical analyses are reviewed, together with
the input requirements, and constitutive capabilities.

EVALUATION DES EFFETS DES TRAVAUX D'EXCAVATION
DANS LES FORMATIONS DURES : BESOINS EN MATIERE D'ANALYSES

RESUME

L'auteur recense les besoins en matière d'analyses en vue d'une
expérience sur les effets des travaux d'excavation s'inscrivant dans un
programme d'aménagement d'un dépôt destiné à recevoir des déchets nucléaires.
Il examine les instruments, les procédures et les conditions nécessaires pour
réaliser ces analyses et évoque ce qu'impliquent ces travaux. Ces travaux
d'analyse sont subdivisés en catégories spécifiques et des exemples des
problèmes de modélisation qui se posent sont présentés pour chaque catégorie.
L'auteur passe également en revue les divers programmes de calcul disponibles
pour les analyses mécaniques, assortis, des besoins correspondants en matière
de données d'entrée, et leurs potentialités.

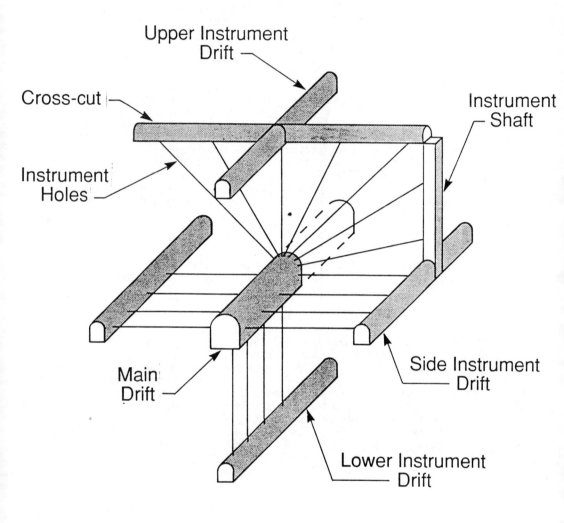

Upper Instrument Drift

Cross-cut

Instrument Shaft

Instrument Holes

Main Drift

Side Instrument Drift

Lower Instrument Drift

Figure 1. Idealized Excavation Response Experiment
Configuration with Monitored Cross-Section

1. INTRODUCTION

In the regulatory environment, one fundamental purpose of in situ testing is to demonstrate the ability to predict rock mass response. These predictions must incorporate the state-of-the-art theories and procedures, to simulate the response of a relatively complex material - a rock mass. In addition, these predictions (and, in a larger context, all analytical efforts) are under intensive review when performed for a potential repository site. As such, the analytical tools, i.e., computer codes, must have demonstrated ability to predict rock mass behavior to assure a good engineering design and meet licensing requirements.

To perform such predictions, extensive preparation and effort is required. The computer codes to be used are not standard off-the-shelf items, but rather must be modified for the specific in situ test. The codes must be tested against actual case studies prior to use in predictions. Moreover, the development of specialized constitutive models, and of pre- and post-processors to handle data input/output are dependent on various factors such as rock type and involve long lead times. Code development activities are also concurrent with other analysis requirements for test planning, such as scoping and design computations which are used to identify the test configuration and instrumentation.

Hence, in the planning and scheduling of an in situ testing program, a definitive analysis program for each test should be identified at the onset. This is to insure that all required steps are performed when necessary, and that the predictions will actually demonstrate the state-of-the-art capability.

The following paper examines some of the analysis requirements for one particular underground test, an Excavation Response Experiment. The requirements are addressed in a general manner, but specific examples are presented as applied to hard rock repositories. Many of examples and concepts are based on the ongoing deliberations of Joint ERE Planning Committee for the Underground Research Laboratory, as part of a cooperative agreement between the U.S. Department of Energy and Atomic Energy of Canada Limited.

2. TEST DESCRIPTION

An Excavation Response Experiment (ERE) is an in situ test to examine the response of a rock mass to the creation of an underground opening. The experiment, which is also referred to as a Mine-by Test [1] or a Sequential Drift Mining Test [2], involves the controlled excavation of a main drift through an instrumented body of rock. The main drift is to be representative of a repository storage room or haulage. Instrumentation is placed prior to excavation from adjacent instrument drifts, shafts and/or cross-cuts to monitor changes at varying distances from the excavation. One possible test configuration for an ERE is shown in Figure 1.

The scope of the experiment is dependent on the site characterization or research program undertaken. An ERE can examine solely the mechanical response of the rock, i.e., the displacements and stress

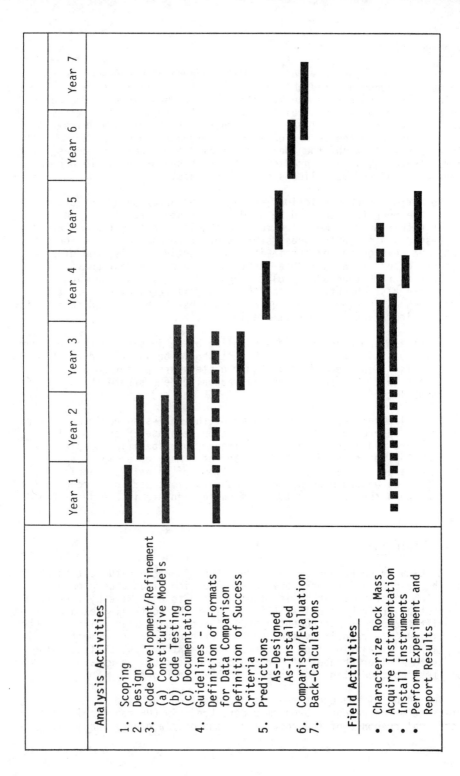

Figure 2. Idealized Schedule for Excavation Response Experiment

changes that are induced in the rock mass by a specified excavation process and sequence. The scope can be broadened to examine different excavation methods (e.g., drill and blast versus mechanical methods) and/or different procedures (e.g., full-face versus a pilot-and-slash approach) for various segments of the main drift.

The test can also examine the hydrological response of the rock mass and how it interacts with (or is "coupled to") the mechanical response. This coupling is of obvious importance to a repository below the water table. Repository construction will disturb the rock mass and ground-water flow around the repository and the extent of this disturbance can be evaluated with an ERE. However, this increased scope adds to the complexity of required constitutive models, to the complexity of computer codes and to analysis requirements.

The scope of an ERE has a major impact on the type and extent of analysis efforts that are to be required. For the present paper, it will be assumed that the focus of the experiment is on the mechanical processes alone, although occasional references will be made to hydrological concerns.

3. TYPE OF ANALYSIS ACTIVITIES

3.1 General

The analysis effort associated with any underground test is a major factor in the test schedule, and requires significant resources. The analysis effort can be subdivided into arbitrary categories as follows:

- Scoping
- Design
- Code Development and Refinement
- Back-Calculations
- Demonstration Predictions
- Comparison and Evaluation
- Guideline Definition.

Many of these activities are concurrent with each other, but certain inter-dependencies exist. Assuming that the experiment will involve two analysis groups (for required redundancy) of three to five professionals each, a schedule for a "typical" ERE experiment is presented in Figure 2. As shown, most inter-dependencies are obvious; it is necessary that scoping activities precede design activities, which in turn precede the predictions for the experiment as well as the acquisition of instrumentation. Perhaps not as readily evident is that the final design cannot be performed without some characterization data, and that the definition of analysis results (e.g., how final predictions will be presented) must also be performed prior to the finalization of code output format and/or post-processors which must be included in final documentation.

The duration of each of these activities is dependent on available resources, but there exist certain minimum durations. The durations shown in Figure 2 represents a best guess, and assume significant resources are readily available. These activities are discussed in more detail in the following sections.

3.2 Scoping Computations

Scoping activities involve an examination of the various alternatives and concerns to formulate a general test configuration and approach for the ERE. Scoping computations are generally performed prior to characterization of the rock mass and, hence, computations must consider several possible geologies as well as several possible boundary conditions (e.g., several possible stress fields). Further, the ERE configuration should be oriented where it will induce the most rock mass response, not influence other test programs, and provide the most benefit for the test program.

For example, if the experiment is to examine response of a discrete fracture zone, the placement of main drift relative to the zone should be at some oblique angle to produce both normal and shear displacement of the zone; approaching the zone at right angles produces little effect on the zone.

During scoping, the inter-relationship of the ERE with other ongoing or planned underground tests must also be examined. Obviously, the performance of the Excavation Response Experiment should not influence the stress or displacement conditions of other underground tests being performed at the same time and visa versa. Assuming an elastic rock mass and a 5% stress change criterion, an excavation should be spaced approximately five room diameters from another drift. The actual spacing is dependent on range of rock mass parameters and the stress criterion adopted.

Scoping computations must also consider the testing program as a whole to integrate test concerns and minimize unnecessary cost. It should be noted that the emplaced instruments and/or excavations of the ERE may be incorporated into some future experiment. Specifically, the instrumented rock of the ERE may be used for a heater experiment or the instrument shafts may be used for sealing experiments. This reuse of the ERE resources can also provide a more comprehensive understanding of the rock mass of the site, thereby improving back-calculations of the ERE itself.

3.3 Design Computations

Design computations are required to examine the details of the experiment and provide a basis for other tasks such as instrument acquisition. The computations must determine the optimum monitoring positions for the test configuration, the type, number and accuracy of required instruments, the length of instrument holes, and the required ventilation and support requirements. These computations require some direct geologic knowledge of the ERE location; ideally this knowledge would be from boreholes located along the proposed main drift axis or along the instrument drift locations.

An example of one design concern is how to place the instrumentation around the main drift for the best results. At some point along the axis of the main drift, a fully instrumented vertical cross-section is considered mandatory. This cross-section is to provide a comprehensive understanding of the deformation/stress field around the excavation, more so than one or two boreholes from side/upper instrument drifts could provide (see Figure 1). This can be achieved in several ways but each has its limitations.

One method is to place a fan of boreholes from side instrument drifts, as per the Spent Fuel Test at the Climax Facility [3]. This approach is limited by the extensometers which record displacement only axially along the instrument holes, and a major component of deformation (i.e., radial to main drift) is not monitored. Another approach is to place a fan of instrument holes from the main drift as close as possible to the excavation face as per the Room 209 Experiment at the Canadian Underground Research Facility [4]. This approach is limited by the loss of part of the monitoring record prior to installation of the instruments.

A third approach is to excavate a series of cross-cuts and shafts, and place instrumentation radially around the main drift prior to its excavation. As illustrated in Figure 1, the cross-cut and shafts are both used to place instrumentation both above and below the main drift. While this approach addresses some of the problems of the two aforementioned approaches, it is an expensive solution, and must be evaluated within economic constraints and against data needs for model validation.

3.4 Code Development and Refinement

Code development encompasses a variety of tasks necessary to prepare a computer code for use in test predictions. A starting point for the effort is the determination of which constitutive models will be required for a particular ERE and which code will be used for predictions. As no single code contains all possible or needed constitutive models, efforts are directed into modifying the selected code for the desired ends. Thereafter, the code must be extensively documented and tested prior to use in predictions.

ERE modelling requirements in hard rock include a range of both response and failure models. The selection of the constitutive models is to a large extent dictated by the rock type and in situ conditions at the test site. However, a rock mass will seldom behave as a simple isotropic/elastic material even in the most massive formations. Non-linear elasticity and anisotropy capabilities can be stated as minimum requirements. For fractured rock, modelling concerns escalate to choices between continuum and discontinuum based codes and discussions arise as to what is a significant fracture. For failure modelling, ERE predictions must consider excavation damage, as well as the potential for failure in the mass, especially tensile failure in the excavation floors and walls.

From previous experiments, there are several possible areas for future ERE code development. Evidence suggests that non-linear elasticity is a significant factor together with the excavation damage zone in massive rock. To consider filled fractures or fracture zones, it may be necessary to modify existing fracture models to initially close due to shear loading, and subsequently dilate under increased displacement as suggested by recent field results. The potential for movement of discrete blocks towards the excavation in a inelastic fashion must also be evaluated. Modelling must also incorporate the excavation sequence and consider the loading of the excavation process as well as the subsequent unloading from excavation.

3.5 Guideline Definition

As part of the analysis effort, it is necessary to develop various guidelines for handling the enormous amount of information developed during the ERE. Aside from the specification of report content, tape formats, etc., there are the tasks of defining the output format, and establishing success criteria.

The definition of output formats at first glance appears to be insignificant and of little value. However, during the comparison and evaluation stage of the test, the simple request to plot the predictive data in another manner can require significant effort and lost time due to handling of large volumes of data, recovering data from various storage locations, writing new programs to process it, documenting these programs, and finally producing the plot. Investing in defining an output format can pay large dividends at a crucial step in the experiment.

A similar item is the definition of success criteria, or in more simple terms, what is considered a "good" comparison between predicted and field results? On this simple question rides the judgement of the experiment as a success or failure. In the past, the definition of good has been subjective and hence what is seen as a success to some is a monumental failure to others. However, the definition of good or bad should reside on a reasoned logic. This logic should be considered the significance of the parameter to specific repository goals such as safety.

A simple example of such logic would be the consideration of a stress change value in view of the goal of preclosure safety. A safe opening is based on several considerations of support and stability. Within these considerations, the stability computation must be evaluated as to what variation is acceptable and still maintain the assurance of a safe opening. A range of stress can then be determined which will maintain the stability computation within this acceptable variation. This range provides a yardstick for what can be termed a "good" comparison. The actual process of defining these success criteria is beyond the scope of this overview.

3.6 Demonstration Predictions

As noted earlier, the use of the codes in predicting the rock mass response will be under intense review from regulators and the public. Predictive efforts must not only involve the clear documentation of the code used, but must also detail the method of selecting input parameter values. Careful cross-checking will be required to insure the accuracy of input values and code setup. An comprehensive report is required on the entire process.

One point of discussion for ERE predictions involves the selection of the conceptual basis for modelling. The selection must balance the desire to reduce the uncertainty in modelling versus the need to demonstrate the ability to predict. Modelling can be based solely on the design of experiment, so-called "As-Designed" predictions. Uncertainty arises in that instruments are not always installed at the ideal locations (for whatever reasons) and that geometry of the main drift and rock conditions will vary during the actual excavation. To reduce such uncertainties, predictions can also be based on the test configuration after instruments have been installed but prior to excavation, ("As-Installed"), or after the main drift has been excavated ("As-Constructed").

However, to demonstrate the ability to predict, the usefulness of the As-Constructed design must be questioned. Although the analyst does not have access to the field data, his predictions are open to the speculation that they have somehow been adjusted to fit the results, and are not valid. The As-Installed predictions are not open to this criticism, but require a significant period of time (approximately one year) between the installation of instruments and the actual ERE excavation. This impact on the schedule is often times not acceptable, and As-Designed predictions must suffice.

Another aspect of uncertainty arises from the variation of material properties; rather than one value for an elastic modulus, laboratory test data will provide a range of possible values. On a larger scale, the rock mass is seldom homogeneous with regard to any parameter, and this uncertainty must also be addressed. To correctly capture this variation, it is necessary to perform not a single prediction but rather a series of simulations or computer runs to identify a range of possible outcomes. The final result is an expected value together with a probable range for each parameter. It is suggested that predictions for a ERE should not be single value or curve for each output parameter (e.g., stress change at some point), but rather must be presented as a expected value together with a possible range of results based on stochastic analysis of input values.

3.7 Comparison and Evaluation

Upon excavation of the main drift of the ERE, the recorded instrumentation data is compared to the model predictions. This involves more than the tabulation of monitored and predicted data together to provide a subjective basis of comparison. The divergence of the predictions from the monitored results must be assessed both graphically and mathematically to identify relationships. Examples of such comparison techniques are time-series plots, spacial variation and contouring at various time points, pair-wise comparisons of measured/predicted data, and comparison of regression parameters for measured/predicted data. These assessments are then used together with the success criteria (see Section 3.5) to judge if the ERE has achieved its goal.

One item that arises in this evaluation is the significance of data points that do not fit the expected trend or do not agree with predictions. It is desirable to eliminate spurious data that can confuse the comparison, but the elimination of the data point must be fully justified by the evidence at hand. As such, the analyst cannot be permitted to judge the quality of monitored data; the evidence that a point seems spurious based on analytical results is not sufficient. It can be suggested by others that he is adjusting the field data to agree with the computer results. Rather, the personnel responsible for the field measurement must establish the accuracy of monitored data in isolation from the analyst. The determination of this accuracy is more than just a assessment of whether the field data is good or bad, it must include a possible range of values (e.g., an error bar) based on the accuracy of the individual instrument together with the accuracy of the monitoring system.

3.8 Back-Calculations

Upon the presentation of the data comparison, it may be thought that the analysis effort is completed. However, even with the most comprehensive and accurate computer code, there will exist differences between monitored and predicted results which must be examined.

As noted, uncertainty can arise from several sources such as variation in actual test configuration from that assumed by the model, together with possible unexpected geological structures. Back-calculations are performed to consider these variables to demonstrate whether the observed differences can be explained by more closely capturing the actual test.

If they cannot, it is probable that these differences represent a lack of knowledge as encompassed in the computer code. This may indicate that the rock mass behavior was more complex than considered by the analyst, or that the state-of-the-art constitutive models were insufficient to describe rock mass response. In any event, the unexplained differences must be clearly identified to assist future analysis efforts.

Therefore, back-calculations should be performed as part of the analytical effort for an ERE, examining the various differences between the modelled and the as-is test. These efforts should involve the examination of a limited "what-if" concerns to identify the cause of variances noted that can be explained from a change in test configuration.

4. SIMULATION REQUIREMENTS

4.1 Simulation Technique

The underlying simulation techniques for the variety of available codes can be subdivided into two general categories: 1) continuum techniques and 2) discontinuum techniques. Continuum techniques represent the rock mass as a continuous body, and include numerical methods such as finite element, finite difference and boundary integral approaches. On the other hand, discontinuum techniques represent the rock mass as a collection of discrete bodies or blocks, and include methods such as discrete element and discontinuous analysis approaches.

It should be noted that continuum methods can be employed in fractured rock by the use of special elements. Discrete-joint or ubiquitous-joint elements are used to describe the behavior of discrete fractures or closely-spaced fracturing, respectively. These elements cannot address large rotations of individual blocks which require a discontinuum approach. However, discontinuum codes require more extensive input definition such as the aperture geometry and orientation of individual fractures. These data may not be available to the analyst for the ERE.

The limitations and advantages of various methods have led to the development of another category of codes known as hybrid methods. A hybrid code melds the capabilities of two methods to obtain a code that can model the rock mass more accurately and effectively. Combination of discrete element and boundary integral methods can reduce the effort required to define the rock mass while establishing more realistic boundary conditions [5]. Other combinations are possible as well. Regrettably, the development of these hybrid codes in the public domain has been limited.

4.2 Available Codes And Capabilities

There is an extensive set of available computer codes to model the mechanical response of rock. Each of these codes has been developed with a prescribed problem (or set of problems) in mind and hence the available constitutive models vary dramatically from code to code. Some of the codes used in the U.S. and Canadian research programs for nuclear waste disposal are shown in Table I.

Examining the capabilities of these codes in Table 1, it is evident that no single code can do everything. Even if a capability has been noted in this table (as designated by a "X") it may be inappropriate to the rock type under consideration. Of course, not all capabilities are required in a particular rock mass, but it is highly probable that any selected code will require development and refinement as described in Section 3.4. Some capabilities are not incorporated in any of these codes such as the ability to address variability in input parameters ("Stochastic Processes") or incorporate the loading of the rock mass from the excavation process ("Explosive loading").

As to minimum requirements, the prediction and design activities of the ERE requires a 3-dimensional capability to simulate the rock response to the progressive excavation of the main drift. This would remain a requirement even if the rock mass is linear-elastic and isotropic in nature.

However, for the scoping of the ERE, a 2-dimensional code may suffice and would be more efficient to study the range of possibilities. Back-calculations may also employ 2-dimensional codes again to effectively examine a set of possible situations or a change in constitutive modelling.

5. SUMMARY

In summary, the following broad points have been stated:

1. An comprehensive set of analytical activities have been presented for a Excavation Response Experiment. These activities can be subdivided into six categories: scoping, design, code development/refinement, predictions, comparison/evaluation, back-calculation, and guideline definition.

2. Detailed planning is necessary to accomplish these activities and various analytical concerns must be addressed during this process.

Table I. Features of Some Available Mechanical-Analysis Computer Codes

Code Name	Reference	Geometry — Plane stress/strain	Axisymmetric dimensionality	Rectangular dimensionality	Large rotations	Linear elastic	Nonlinear elastic	Elasto-plastic	Visco-elastic/plastic	Joint/slip model	Crack propagation	Anisotropy	Dilatancy/consolidation	Thermal dependence/coupling	Damage model	Ubiquitous jointing	Work softening	Large displacements	Large strain	Thermal spall	Stochastic processes	Initial stress/strain	Structural components	Material removal/addition	Hydr./chem./radiation coupling	Explosive loading	Pre-/post-processors	Finite difference	Finite element	Boundary element	Distinct element	Explicit	Implicit
ABAQUS	(f)	x	x	3		x	x	x	x	x	x	x	x	xg	x			x	x			x	x	x	x		x		x				x
ADINA	SAI 1981	x	x	3		x	x	x	x	x	x	x		x	x		x	x	x			x	x	x			o		x				x
ANSYS	Curtis 1983	x	x	3		x	x	x	x	x	x	x		x				x				x	x	x			x		x			x	x
BMINES	Van Dillen 1981	x	x	3		x	x	x	x	x	x	x	x	x				x				x		x			xa		x				x
FLAC	(f)	x		2	x	x	x		xc	x			x					x				x						x				x	
JAC	Biffle 1984	x	x	3		x	x	x	x	xb	xb	x						x				x							x			x	
MARC	SAI 1981	x	x	3		x	x	x	x	x	x	x	x	x				x	x			x	x	x			x		x			x	x
MINAQ	(e)	x		2		x				x			x	x								x							x				x
REPOS	(e)	x	x	3		x	x		x	xb	x	x	x	x		o						x		x					x				x
ROCKMAS	(f)	x	x	2		x								x								x			x					x			x
SAFE	(f)	x	x	3		x	x				o	x		x										x			xd		x	x			
SANCHO	Stone 1985	x	x	2		x	x	x	x			x					x	x	x					x				x				x	
SPECTROM 32	(e)	x	x	2		x	x	x	x	x		x		x					x			x							x			x	
STEALTH	SAI 1981	x	x	3		x	x	x	x	xb	x	x		x		x		x	x			x		x	x		xa					x	
3DEC	(f)	x		3	x	x	x	x	x	x	x	x				x		x	x			x		x	x		x				x	x	
UDEC/MUDEC	(f)	x	x	2		x	x	x	x	x	x	x				x		x	x					x	x		x				x	x	
VISCOT	Intera 1983	x	x	2		x	x	x	x					x				x	x			x			x				x			x	

x Existing code feature
o Code feature being developed

(a) Mesh generator
(b) Frictional interface only
(c) Only along excavation horizon

(d) Post-processor available
(e) Documentation in draft stage
(f) Software supplier's literature
(g) Fully-coupled, not just forward-coupled

Modified from St. John, 1983

3. Uncertainty must be considered when making predictions as well as during the evaluation of monitored data. Both predictions and data must be presented within a range which displays this uncertainty.

4. The selection of the code for ERE predictions must be made from a variety of codes and constitutive models; no one code will fill all the analyst needs and must be modified and refined for the experiment.

In closing, it is the contention of the author that while a variety of capabilities exist, they have as yet not been fully or effectively utilized in the analysis of an ERE-like experiment. Further, based on available literature and a subjective assessment of previous experiments, it is concluded that there have been no successful predictions made for any Excavation Response Experiment to date.

6. REFERENCES

(1) Roberds, W., F. Bauhof, L. Gonano, E. Wildanger, W. Dershowitz, F. Marinelli, K. Jones, C. Weldon, B. Nebbitt, M. Stewart, R. Gates, D. Pentz, J. Byrne: In Situ Test Programs Related to Design and Construction of High-Level Nuclear Waste (HWL) Deep Geologic Repositories, NUREG/CR-3065, 813-1162, Vol. 2, prepared by Golder Associates for the U.S. Nuclear Regulatory Commission, pp. A. 4-1 to A. 4-44., 1982.

(2) U.S. Department of Energy (DOE) 1987: Thermal and Mechanical Properties, Chapter 8, Section 8.3.1.15, Draft Site Characterization Plan, Yucca Mountain Site, Nevada Research and Development Area, Nevada, prepared by the U.S. Department of Energy, Washington, DC, pp. 8.3.1.15-46 to 8.3.1.15-49., 1985.

(3) Wilder, D. G., and J. L. Yow, Jr.: Geomechanics of the Spent Fuel Test - Climax, UCRL-53767, prepared by Lawrence Livermore National Laboratory University of California, Livermore, CA., p. 84., 1987.

(4) Wright, E. D. (Editior): Semi-Annual Status Report of the Canadian Nuclear Fuel Waste Management Program, 1986, April 1 - September 30, Technical Record TR-425-1, Atomic Energy of Canada, Ltd., pp. 106, 116-117., 1987.

(5) Lorig, L. J. and B. G. H. Brady: A Hybrid Computational Scheme for Excavation and Support Design in Jointed Rock Media ISRM Symposium, Design and Performance of Underground Excavations, International Society for Rock Mechanics, pp. 105-112., 1984.

(6) Curtis, R. H., R. J. Wart and E. L. Skiba: A Summary of Repository Design Models, NUREG/CR-3450, prepared for the Division of Waste Management, Office of Nuclear Material Safety and Safeguards, U.S. Nuclear Regulatory Commission, Washington, DC., 1983.

(7)   Intera Environmental Consultants: VISCOT: A Two-Dimensional and Axisymmetric Nonlinear Transient Thermoviscoelastic and Thermoviscoplastic Finite Element Code for Modelling Time-Dependent Viscous Mechanical Behavior of a Rock Mass, ONWI-437, prepared for the Office of Nuclear Waste Isolation, Battelle Memorial Institute, Columbus, OH., 1983.

(8)   Science Applications Inc. (SAI): Tabulation of Waste Isolation Computer Models, ONWI-78, prepared for the Office of Nuclear Waste Isolation, Battelle Memorial Institute, Columbus, OH., 1981.

(9)   St. John, C. M.: Fractured Rock Modelling in the National Waste Terminal Storage Program: A Review of Requirements and Status, ONI-5, prepared by J.F.T. Agapito and Associates, Inc. for the Office of NWTS Integration, Battelle Memorial Institute, Columbus, OH., 1983.

(10) Stone, C. M., R. D. Krieg and Z. E. Beisinger: SANCHO, A Finite Element Program for the Quasistatic Large Deformation Inelastic Response of Two-Dimensional Solids, SAND84-2618, Sandia National Laboratories, Albuquerque, New Mexico., 1985.

(11) Biffle, J. H.: JAC: A Two Dimensional Finite Element Computer Program for the Non-Linear Quasistatic Response of Solids with the Conjugate Gradient Method, SAND81-0998, prepared by Sandia National Laboratories, Albuquerque, New Mexico, 1984.

```
* * * * * * * * * * * * * * * * * * * * * * * * * * * * * *
```

Session II(a) - Séance II(a)

CHAIRMAN'S SUMMARY - RAPPORT DE SYNTHESE DU PRESIDENT

R.A. ROBINSON (Battelle, United States)

```
* * * * * * * * * * * * * * * * * * * * * * * * * * * * * *
```


The following discussion provides a summary of the key issues, conclusions and recommendations that were identified in the nine presentations of this part of Session II. The issues are separated into two time periods covering excavation response during: a) construction and operation of the repository facilities; and b) long-term isolation after closure of the facilities.

I.   KEY ISSUES

A. EXCAVATION-INDUCED RESPONSE DURING CONSTRUCTION AND OPERATION
    (t $\leq$ 80 yr)

-   Mechanical: stability of shafts and tunnels

-   Hydrologic: induced inflow to shafts and underground excavations

-   Thermo-mechanical: induced rock instability and water inflow.

B. EXCAVATION-INDUCED RESPONSE DURING LONG-TERM ISOLATION
    (t up to 10,000 yr)

-   Mechanical: stability of engineered barriers (e.g., shaft and tunnel seals and backfill)

-   Hydrologic: sealing and isolation performance of backfilled shafts and tunnels

-   Thermo-mechanical: induced effects on the hydrology of the site (i.e., interactions that affect the transport of long-lived radioactive isotopes of I, Cs, Tc and Pu).

II. CONCLUSIONS

The following statements summarise the main conclusions that each of the first nine speakers contributed to this session. R. Lieb stated that analysis and evaluation of hydrologic and long-term rock mechanical behaviour are crucial to determining the safety of waste isolation in crystalline rock. L. Liedtke indicated that the validity of numerical flow models must be checked with in-situ field tests. Based on ventilation and heater tests conducted in the Grimsel Test Facility, significant stress- and temperature-induced effects on the permeability of the rock around a tunnel (excavated by tunnel boring) could not be measured according to T. Brasser. Based on his hydrologic analyses, J.M. Hoorelbeke concluded that designers should opt for "smooth"-boring methods (i.e., controlled blasting or boring machines) and careful sealing of storage cavities. This will result in a reasonable limitation of induced porosity using conventional mining methods. Also, underground design should aim at providing minimum mechanical and hydrologic interaction between the storage cavities. R. Pusch stated that conventional drill and blast methods used in excavation of shafts and drifts will produce a disturbed zone around the excavation that has a significantly enhanced permeability. However, based on recent shaft and drift sealing tests conducted in the OECD/NEA Stripa Project, there are proven sealing techniques available to isolate and seal off these excavation-induced, high-permeable zones around shafts and tunnels. A. Jakubick and D. Keil stated that the hydrologic conditions in the excavation response zone cannot be easily controlled by engineering design. Careful documentation of the hydrologic properties of this zone is essential! S. Bauer stated that his analyses of excavation response tests in G-tunnel tuffs showed that the damage zone extends about one meter into the wall of the drift, and he was able to successfully model the excavation response of the experiment. T. Blejwas concluded that excavations will alter the stress regime and modify physical characteristics of the rock around the opening. Also, heat will induce additional stresses, change rock properties and, perhaps, affect water movement. The final speaker, E. Lindner, indicated that careful planning and co-ordination of analysis efforts are mandatory for conducting successful in-situ tests. This is especially important since in-situ testing is usually quite expensive and time consuming.

III.   RECOMMENDATIONS

The major recommendations that the speakers proposed in their presentations in this Session included the following. R. Lieb recommended that the focus of future excavation response testing and analysis should be on identifying governing parameters that will affect waste isolation. E. Lindner and L. Liedtke stated that the validity of numerical models needs to be established by in-situ field tests prior to using the models to predict the safety of the repository site for waste isolation. T. Blejwas stated that the recommended excavation-effects experiments for the proposed U.S. repository in tuff at Yucca Mountain include shaft convergence, demonstration break-out rooms, sequential drift mining and in-situ design verification. Finally, to reinforce the earlier recommendations of K. Dormuth and R. Lieb, I recommend that the total system performance assessment modellers be more directly involved in the initial planning of excavation-effects tests. This is critical to the identification of the important hydrologic, mechanical and thermo-mechanical parameters, and the determination of the magnitudes where these governing parameters become significant to waste isolation.

SESSION II(b)

EXCAVATION RESPONSE IN CRYSTALLINE ROCK
- SPECIFIC GEOTECHNICAL STUDIES

SEANCE II(b)

EFFETS DE L'EXCAVATION SUR LES ROCHES CRISTALLINES
- ETUDES GEOTECHNIQUES SPECIFIQUES

Chairman - Président

A. BARBREAU
(DPT/CEA, France)

MODELLING EXCAVATION RESPONSES IN JOINTED ROCK

N. Barton and A. Makurat
Norwegian Geotechnical Institute
Taasen, Norway

ABSTRACT

Tunnel excavation in jointed rock causes complex changes in the surrounding rockmass that will generally result in increased permeability. New methods of describing the relevant rock joint behaviour, allow the physical and conducting joint apertures to be tracked in the course of complex tunnel-related loading histories, i.e. during joint closure, shear, dilation, shear reversal, cyclic normal loading, with asperity wear etc. This joint behaviour model has been incorporated as a sub-routine in the universal distinct element code (UDEC) for studying the changes in joint geometries that occur in the rock mass surrounding tunnel excavations, with and without internal or external fluid pressure. The graphical presentation of behaviour in the disturbed zone includes plots showing stresses and displacements, the magnitude and location of joint shearing, and the corresponding magnitudes of mechanical and conducting apertures of joints. A special tunneling problem in Oslo include analyses of joint conductivity changes caused by initial excavation, leakage control by subsequent systematic bolting, followed by final concrete lining with build-up of full external ground-water pressure.

MODELISATION DES ESSAIS DES TRAVAUX D'EXCAVATION DANS UNE FORMATION DIACLASEE

RESUME

L'excavation d'une galerie dans une formation diaclasée entraîne des modifications complexes dans la masse rocheuse encaissante qui se traduiront généralement par une augmentation de la perméabilité. De nouvelles méthodes décrivant le comportement des diaclases de la roche permettent de suivre l'évolution physique et hydraulique de l'ouverture des diaclases pendant le déroulement des phénomènes dynamiques complexes liés aux travaux d'excavation, c'est-à-dire fermeture des diaclases, cisaillement, dilatation, inversion du cisaillement, phénomènes dynamiques cycliques normaux, s'accompagnant d'une usure des aspérités etc. Ce modèle simulant le comportement des diaclases a été incorporé en tant que sous programme dans le programme universel fondé sur la méthode des éléments finis (UDEC) permettant d'étudier les modifications de la configuration des diaclases qui se produisent dans la masse rocheuse où sont pratiquées les excavations correspondant à la galerie, avec et sans pression hydraulique externe. La présentation graphique du comportement dans la zone perturbée comprend des courbes faisant apparaitre les contraintes et les déplacements, l'ampleur et l'emplacement des phénomènes de cisaillement des diaclases et l'ampleur correspondante des ouvertures mécaniques et hydrauliques. Dans le cadre de ces travaux de creusement, les chercheurs norvégiens ont analysé certains aspects particuliers comme les modifications de la conductivité des diaclases provoquées par l'excavation initiale, la limitation ultérieure des fuites par un boulonnage systématique suivie par la pose d'un revêtement final de béton accompagné d'un rétablissement complet de la pression de l'eau souterraine à l'extérieur de la zone.

1. DISTURBED ZONE IN DEEP SHAFT

Permeability enhancement due to excavation can be measured directly by increased inflow, or inferred from deformation measurements. As an example of disturbed zone effects, the case of a deep shaft excavated in jointed quartzites can be utilized. Elastic continuum analyses were performed to compare theoretical solutions with the measured distribution of deformation to a depth of 10 m surrounding the 6 m diameter, 1600 m deep shaft. The deformations were measured with multiple position borehole extensometers (MPBX) at three locations around the shaft. A sketch of the set-up and some results are shown in Figure 1.

The five measured points in the upper diagram are derived from the MPBX measurements. One set of the latter are shown in the lower diagram, for different depths in the rock mass. Non-linear anisotropic behaviour was exhibited, as one would expect for a shaft excavated in a steeply dipping structure at a depth of nearly 1600 m. Maximum (surface) deformations of 15,5, 4,8 and 3,3 mm were recorded from the MPBX installed respectively perpendicular, parallel and at 45° to the dominant bedding joints.

Rock mass deformation moduli estimated from small-scale in situ tests (CSM/Borehole jack) varied from 48 to 76 GPa. Values back-calculated from the measured depth dependent behaviour (Figure 1) varied from 3 to 14 GPa for one side of the shaft, and from 21 and 28 GPa for the adjacent side. The need to utilize such a wide range of E moduli (3 to 28 GPa) to "explain" measured behaviour is a simple illustration of what a disturbed zone means in terms of continiuum response. Improved models incorporating joints are clearly needed in order to quantify what is really occurring in the surrounding rock mass. Surprising and quite interesting results can be demonstrated.

2. JOINT DEFORMATION COMPONENTS

A simplified picture of the two fundamental joint deformation components; closure and shear, are illustrated in Figure 2. The obviously non-linear behaviour results in concave-shaped curves under closure, and convex-shaped curves under shear. The latter may depend on the size of block sheared; more complex peak-residual behaviour may be exhibited by smaller block sizes.
When these two components are coupled during the deformation of a rock mass, the resulting load-deformation behaviour may exhibit curves that are concave, linear or convex. This will be determined by the dominance of one of the components (N) or (S) as illustrated in Figure 2 (lower diagram). The increasing degree of hysteresis exhibited by rock masses A, B and C is a function of the irreversible shear components (S) and of accompanying dilation (if the joints are non-planar and stresses are moderate).

3. JOINT APERTURES

One of the objectives of disturbed zone modelling is to track potential conductivity changes in the rock mass. The preceeding observations on joint closure, shear and dilation need therefore to be
interpreted as joint aperture changes. At the same time, consideration must be given to the fact that joints under stress have
varying degrees of contact. Any fluid flow will tend to be channeled
around these contacting areas, although fluid pressures may be transferred throughout the whole joint plane.

A joint of a given physical aperture (E) will have a corresponding
theoretical smooth-wall (parallel plate type) conducting aperture (e).
The conductivity (K_j) of the joint will be given by:

$$K_j = e^2/12 \qquad (1)$$

where $e \leqslant E$, and where $e \ll E$ at high stress, due to large areas in
contact.

An empirical model of the relationship between (e) and (E) and the
influence of joint roughness (JRC) is illustrated in Figure 3. Rough
joints that are under very high stress, and therefore have extremely
small apertures, can show (E/e) ratios greater than 10 as indicated in
current Stripa waste isolation studies in Sweden.

In field applications, an initial value of (e) and its statistical
variation, is obtained from analysis of permeability tests using for
example Snow's (1968) statistical method. This initial value of e_o
which is depth dependent) is converted to a physical aperture E_o
(using Figure 3) for use in subsequent modelling of any changes (ΔE)
caused by shear, dilation closure, or shear reversal, where:

$$E = E_o \pm \Delta E \qquad (2)$$

4. CONSTITUTIVE JOINT MODEL

A brief summary of the constitutive joint sub-routine that is now in
operation in the special version of μDEC-BB operated by NGI will be
given here, with appropriate references to the fuller treatment to be
found elsewhere.

The shear strength-displacement-dilation behaviour are described in
principle by the following two generalized equations:

$$\phi_{mob} = JRC_n \text{ (mob) } \log \left(\frac{\sigma_1 - \sigma_3}{\sigma_n'} \right) + \phi_r \qquad (3)$$

$$d_n(mob) = 1/2 \text{ JRC (mob) } \log \left(\frac{\sigma_1 - \sigma_3}{\sigma_n'} \right) \qquad (4)$$

where ϕ_{mob} = the mobilized friction angle at any given
                     displacement

          $JRC_n(mob)$ = the full-scale mobilized joint roughness
                     coefficient at any given displacement

$(\sigma_1 - \sigma_3)$ = the confined asperity strength, simplifying to JCS_n (full-scale joint wall compression strength) at low stress (Barton 1977)

ϕ_r = residual friction angle

σ_n' = effective normal stress

$d_n(mob)$ = full-scale dilation angle mobilized at any given displacement

   Fuller details of this joint constitutive model are given by Barton et.al. (1985).

The normal stress-closure behaviour is also based on the index parameters (JRC) and (JCS) using the Bandis (1980) hyperbolic formulation:

$$\frac{\Delta V_j}{\sigma_n'} = a - b\Delta V_j \tag{5}$$

where ΔV_j = joint closure

   a/b = asymtote to hyperbola = V_m (maximum joint closure)

   a = reciprocal of initial gradient = K_{ni} (initial normal stiffness)

Empirical exressions for K_{ni} and V_m based on JRC and JCS complete the modelling of normal closure (see Bandis et.al. 1983).

5.   INPUT PARAMETERS

The basic index parameters (JRC) and (JCS) are obtained from tilt tests (Figure 4) and Schmidt hammer tests conducted on the jointed core. Such tests might appear absurdly simple in comparison to the sophisticated use to which the parameters are put. This, however, is one of the strengths of the method; data can be obtained from large numbers of inexpensive index tests which gives a sound statistical base to the final values used as input.
Relevant stress levels needed to define the modelled boundary conditions are either measured (using the hydraulic fracturing technique) or are given by the client.

6.   JOINT SUB-ROUTINE OUTPUT

Typical sets of stress-closure-conductivity and shear-dilation-conductivity curves will be presented here, so that the reader becomes familiar with the physical appearance of the constitutive model described by equations 1 to 5.
Figures 5 and 6 illustrate appropriate sets of curves that were generated on an HP 41 CV programmable calculator and peripherals by

Bakhtar, using the program described by Barton and Bakhtar (1987). The particular input data was obtained from tilt test and Schmidt hammer characterization of the welded tuff in G-Tunnel, Nevada Test Site, in a contract for Sandia National Laboratories.

The calculator program has since been expanded in FORTRAN by Christianson and Hårvik (1985) for use as a sub-routine in UDEC, or in the micro version µDEC-BB. A spread sheet LOTUS version helps the user select appropriate input data to match available pre-excavation joint permeability data.

7. EXAMPLES OF DISTURBED ZONE ANALYSES

a) TBM-Tunnels in jointed rock

A useful starting point for the evaluation of disturbed zones around tunnels are the UDEC studies of highly stressed, circular sub-sea tunnels performed by Christianson and Hårvik (1985). In these preliminary studies, linear joint properties were used i.e. single values of c, ϕ, Kn (normal stiffness) and K_s (shear stiffness). Figure 7 illustrates the zones of joint shear caused by excavation of tunnels in isotropic and anisotropic stress fields. Note that joints that are BETWEEN the sectors 3, 6, 9 and 12 o'clock suffer shear, while those that FOLLOW the theoretical lines of zero shear (at 3, 6, 9 and 12) do indeed show zero shear. As one might expect, the joints in the sectors 2 to 4 o'clock and 8 to 10 o'clock tend to be closed rather than sheared by the higher vertical stress in the second example, with σ_v = 1.5 σ_h, where σ_v = 10 MPa.
The zones of joint shear will usually represent potential increases in conductivity, unless the rocks are soft and joint gouge is produced (Makurat and Barton, 1987). The magnitudes of these increases depends on the ratio (JCS/σ_n') and on the joint roughnes (JRC), since these determine the initial joint apertures and the magnitude of potential dilation if shearing occurs. There is a certain balance between joints that are too rough to shear, and joints that can shear, but dilate less.

b) Drill and blasted tunnels in jointed rock

The previous example was a hypothetical study using assumed rockmass paramenters, and simplified linear descriptions of the joint behaviour. The example to be described here is a more sophisticated non-linear treatment of disturbed zone modelling using the µDEC-BB code.

The numerical modelling philosophy we have adopted to simulate real conditions as closely as possible has the following basic steps:

1. Specify jointed rock mass and input parameters
2. Consolidate modelled rock mass to in situ effective stresses
3. Excavate desired tunnel cross-section
4. Run program to a limited maximum displacement
5. Apply temporary support (systematic bolting)

6. Run program to equilibrium
7. Apply permanent support (cast concrete, concrete elements etc.)
8. Build up water pressure in joints and matrix
9. Run program to final equilibrium

(Step 6 may be replaced by a limited displacement command, if lining placement is "close" to the face, i.e. if rock displacements are incomplete).

The tunnels in question are twin motorway tunnels currently under construction under Oslo. Special concern for the effect of excavation on the pre-injected ring ahead of the tunnel face led to the use of μDEC-BB. Strict control of water inflows to limit settlements in overlying clays has resulted in special recommendations to the owners. The example chosen is a relatively shallow tunnel with moderate stress levels. Joint response in the weak sedimentary rocks may be equivalent to joint response in harder rocks such as granite in tunnels at greater depth.

In order to obtain necessary input data, NGI's engineering geologists and rock mechanics engineers have mapped and characterized the jointing, analysed the water pumping tests using Snow's (1968) method, conducted tilt tests on jointed core, measured stresses using the minifrac technique, and set up the necessary joint input and output for the UDEC sub-routine. The latter involves generation of joint performance curves such as those illustrated in Figures 5 and 6. An important detail is the matching of the depth dependent joint apertures with those derived from water pumping tests.
As illustrated in Figure 8 (upper diagram) the dipping nodular limestones and shales are modelled with two different joint properties to represent the foliation (1) and cross-joints (2), and joints in the igneous dyke have a third character (3). Along dyke boundaries low friction surfaces are modelled (4).

Figure 9 indicates the redistribution of prinipal stress caused by excavation under the assured stress field (with $K_o = \sigma_h/\sigma_v = 0,5$). A maximum vertical tangential stress of 6,5 MPa is indicated compared to the maximum pre-excavation vertical stress of 2,5 MPa at the base of the model.

Figure 10 illustrates the deformation resulting from excavation. Of special note is the wide zone of influence of these closely spaced tunnels, with deformation up to the surface equal to about 40 % of the deformation in any one tunnel. The distribution of joint shearing (Figure 11) is also surprising, with maximum values of 2,2, mm, and values of at least 1 mm extending some 10 to 15 m from the opening. Shearing disturbances can be noted some 30 m away from the tunnels.

Figures 12 and 13 illustrate the behaviour of the right hand tunnel in more detail. The influence of the stiffer dyke material is clearly seen in the deformation vectors. Some interesting vector direction changes are seen in the floor and roof of the tunnel, which result in marked joint shearing (Figure 13).

Figure 14 and 15 illustrate details of behaviour around the left hand tunnel, indicating the special effects that jointed media have on stress distribution. The zone of tension (arrows) seen in the floor and right springline are reflected in some large conducting apertures, with a maximum value of 1,6 mm. The local change of aperture from 20 to 60 μm prior to excavation, up to mm size after excavation, can obviously have an enormous influence on local inflows.

Figure 16 shows the potential positive influence of early rock bolt installation (close to the face) in controlling joint apertures. More extensive bolting of the right springline would have been required to control the zone of large joint apertures seen in this location. The "lightness" of the bolting in this zone is reflected in the high values of bolt tension.

CONCLUSIONS

1. Disturbed zones around excavations in jointed rock are reflected in subtle changes in the rock mass which can be only crudely recorded by conventional deformation measurement. Axial type MPBX measurements in boreholes indicate increasing deformation gradients close to the excavations. When attempting to model such behaviour with elastic continuum analyses, it is necessary to utilize successively lower deformation moduli as the excavation surface is approached. Reductions as large as 1:25 are indicated.

2. The modelling of conductivity changes in disturbed zones in real jointed media requires continuous tracking of the physical aperture changes caused by joint closure, shear, dilation, or tensile opening. Physical apertures are then converted to conducting apertures.

3. Numerical discontinuum analyses using the μDEC-BB code indicate significant zones of joint shear and joint aperture increase surrounding tunnels that are excavated in jointed media. Tunnels excavated at great depth in soft media such as shale could be expected to show large shear magnitudes. The soft material may on the other hand cause gouge production which helps to seal joints.

4. Joint shear and resulting conductivity increases can occur around stable circular tunnels excavated in isotropic stress fields.

5. Some joint shear can probably be accommodated without significant increases in conductivity if the rock mass is massive due to large block sizes. Rough joints which would dilate most and cause greatest permeability changes, will also be less likely to shear when tunnel excavation occurs. Consequently the smoother joints may cause greatest changes.

ACKNOWLEDGEMENTS

Our capability of modelling disturbed zones can be traced to major
contributions by Peter Cundall, Stavros Bandis, Mark Christianson and
Khosrow Bakhtar.

REFERENCES

Bandis, S., A. Lumsden and N. Barton, 1981, "Experimental studies of
scale effects on the shear behaviour of rock joints," Int. J. of
Rock Mech. Min. Sci. and Geomech. Abstr. 18, pp. 1-21.

Barton, N. and Bakhtar, K. (1983) Instrumentation and Analysis of a
Deep Shaft in Quartzite, Proc. of 24th U.S. Symp. on Rock Mechanics,
Texas A&M University.

Bandis, S., A.C. Lumsden and N. Barton (1983) Fundamentals of Rock
Joint Dedormation. Int. J. Rock Mech. Min. Sci. and Geomech. Abstr.
Vol 20, No. 6, pp. 249-268.

Barton, N.,1976, "The shear strength of rock and rock joints," Int.
Jour.Rock Mech. Min. Sci. and Geomech. Abstr., Vol. 13, No. 9, pp.
255-279. Also NGI-Publ. 119, 1978.

Barton, N. and Bakhtar, K. (1987) Description and Modelling of Rock
Joints for the Hydrothermalmechanical Design of Nuclear Waste
Vaults. Atomic Energy of Canada Limited, TR 418, 430 p.

Barton, N., Bandis, S. and Bakhtar, K. (1985) Strength, Deformation
and Conductivity Coupling of Rock Joints, Int. J. Rock Mech. & Min.
Sci. & Geomech. Abstr. Vol. 22, Nr. 3, pp. 121-140.

Barton, N. (1985) Deformation Phenomena in Jointed Rock. 8th
Laurits Bjerrum Memorial Lecture, Oslo. Publ. in Geotechnique, Vol.
36, No. 2, pp. 147-167, (1986).

Snow, D.T., (1968) Rock fracture spacings, openings and porosities.
J. Soil Mech. Fdns Div. Am. Soc. Civ. Engrs. 94, SMI 73-91.

Makurat, A. and N. Barton (1987) Fluid flow in fractured rock. Final
Report for NTNF, BA 4.44.16784, Oslo, Norway.

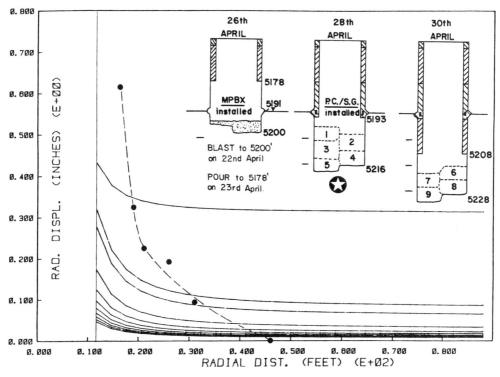

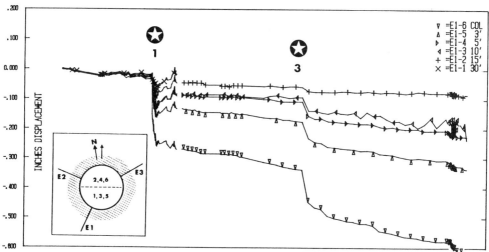

Figure 1. Example of non-elastic disturbed zone surrounding a deep
shaft excavation. Top: Calculated deformation curves for
constant E moduli varying from 5 to 60 GPa: Bottom:
Measured deformation for MPBXI installed perpendicular to
the steeply dipping bedding. (Barton and Bakhtar, 1983).

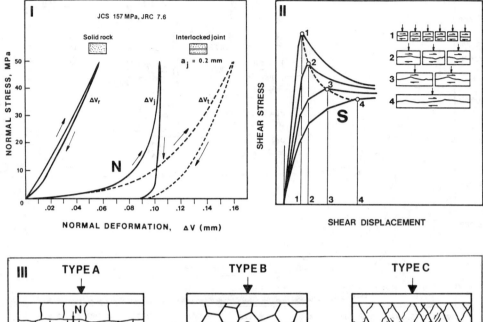

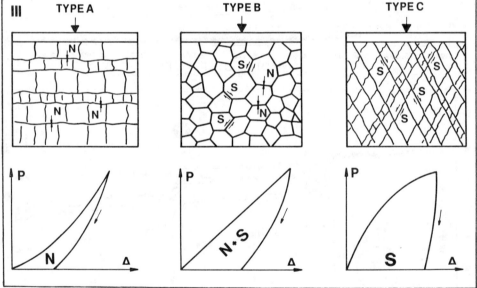

Figure 2. Normal and shear deformation components N and S for single
joints (I and II) cause concave, convex or linear stress-
deformation response for differently jointed rock masses
(III). (Bandis et al. 1981, 1983 and Barton 1986).

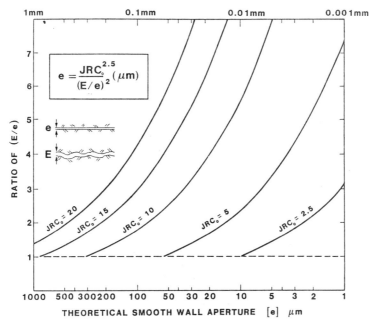

$$e = \frac{JRC_o^{2.5}}{(E/e)^2} (\mu m)$$

Figure 3.   Physical (E) and conducting apertures (e), and their approximate
relationship with joint roughness JRC. Joints with JRC = 15 are very
rough; JRC = 5 moderately rough. (Barton et.al. 1985).

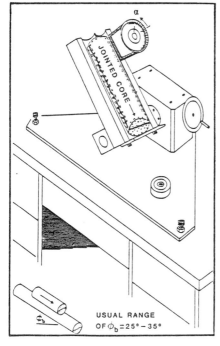

Figure 4.   Inexpensive tilt tests for
obtaining the statistical
variation in joint roughness
for the various sets of
joints involved.

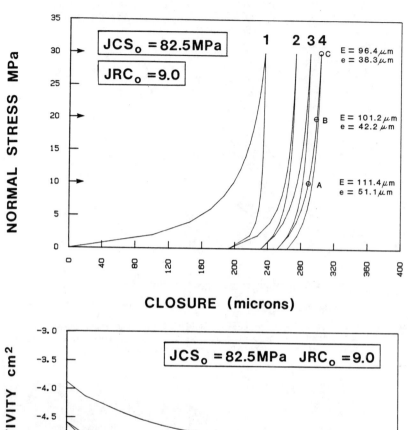

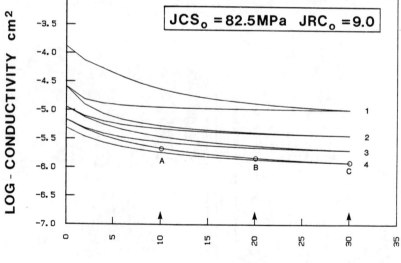

Figure 5. Stress-closure-conductivity modelling for joints in welded tuff. Note physical and conducting apertures (E, e). Characterization of the joints was made in tunnel exposures and on drill core.

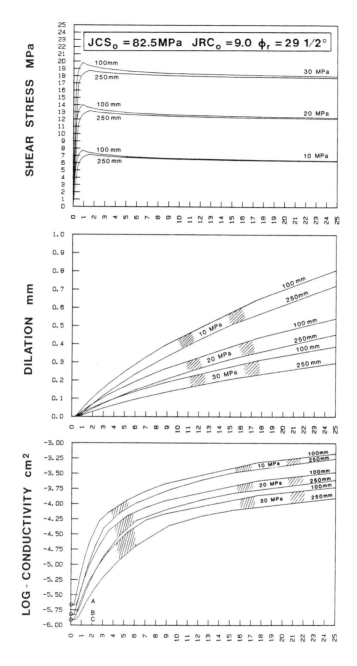

Figure 6.   Shear-dilation-conductivity modelling for joints in welded tuff.

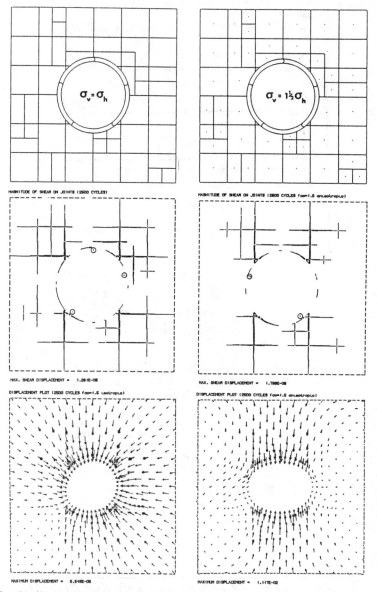

Figure 7. Studies of a hypothetical TBM (bored) sub-sea tunnel for a North Sea
oil field (Christianson and Hårvik, 1985). Line thickness (centre
diagram) is proportional to shear displacement magnitudes (max.
1.79 mm).

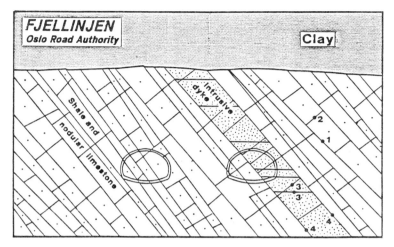

Figure 8. Two-dimensional geometry with specification of joint properties. Study of disturbed zones surrounding pre-injected
rock tunnels.

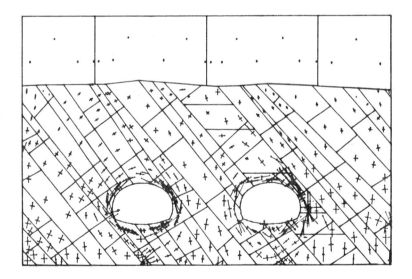

Figure 9. Stress redistribution caused by excavation (max. tangential
stress = 6,5 MPa)

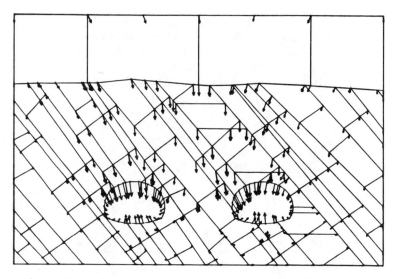

Figure 10. Deformation vectors showing a maximum of 5,5 mm.

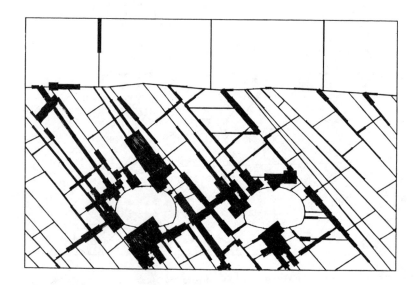

Figure 11. Location and magnitude of joint shearing, showing a maximum of 2,2 mm close to the left tunnel.

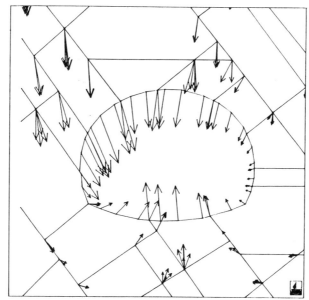

Figure 12. Details of behaviour for right hand tunnel. Deformations
of nearly 4 mm occur in the shale/limestone, but markedly
reduced values can be seen in the stiffer dyke.

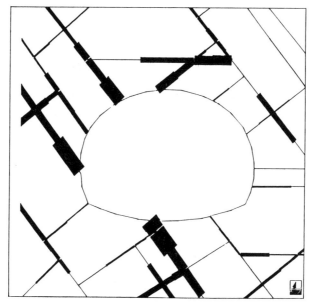

Figure 13. Details of behaviour for right hand tunnel. Joint shearing
occurs both above and below the tunnel (max. 1,6 mm).

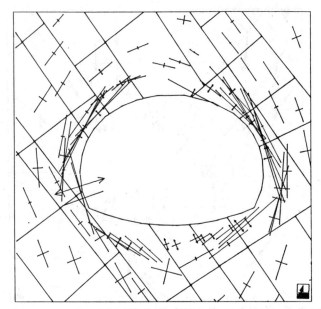

Figure 14. Principal stresses around left hand tunnel (max. 4,0 MPa).

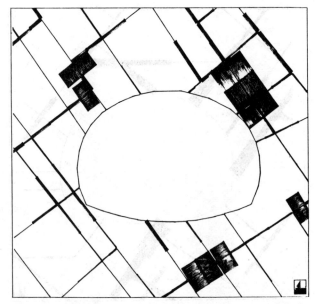

Figure 15. Distribution of conducting apertures around left hand tunnel. One line = 10 µm. Maximum aperture = 1,6 mm.

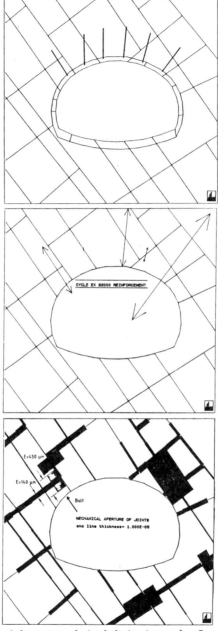

Figure 16. Bolt reinforcement of the left hand tunnel. Top: Bolt locations.
Middle: Bolt tension forces at joint crossings (max 13,7 tons).
Bottom: Detail of a joint that showed an aperture of 450 µm without
bolting and 140 µm when bolting was installed close to the face i.e.
early in the deformation process.

APPLICATION OF THE RESULTS OF EXCAVATION RESPONSE EXPERIMENTS AT CLIMAX AND THE COLORADO SCHOOL OF MINES TO THE DEVELOPMENT OF AN EXPERIMENT FOR THE UNDERGROUND RESEARCH LABORATORY*

W.F. Ubbes
Battelle Project Management Division
Willowbrook, Illinois, USA

J.L. Yow, Jr.
Lawrence Livermore National Laboratory
Livermore, California, USA

A. Hustrulid
Colorado School of Mines
Golden, Colorado, USA

ABSTRACT

Large-scale underground experiment programs to examine excavation response have been perfrmed at the Climax facility in Nevada and at the Colorado School of Mines. These two programs provided fundamental information on the behaviour of rock and the effects of excavation; on instrument performance and configuration; and on the relationship between test geometry and test behaviour. This information is being considered in the development of a major excavation response experiment to be carried out in the Canadian Underground Research Laboratory.

APPLICATION DES RESULTATS DES EXPERIENCES SUR LES EFFETS DES TRAVAUX D'EXCAVATION MENES A CLIMAX ET A L'ECOLE DES MINES DU COLORADO A LA MISE AU POINT D'UNE EXPERIENCE INTERESSANT LE LABORATOIRE SOUTERRAIN DE RECHERCHE

RESUME

Des programmes d'expériences souterraines à grande échelle destinées à étudier les effets des travaux d'excavation ont été réalisés à l'installation de Climax au Nevada et à l'Ecole des Mines du Colorado. Ces deux programmes apportent des informations fondamentales sur le comportement des roches et les effets des travaux d'excavation, sur les performances et la disposition des instruments ainsi que sur les relations entre la géométrie et le comportement lors des essais. Ces informations sont prises en compte dans l'élaboration d'une expérience de grande envergure sur les effets des travaux d'excavation qui doit être réalisée dans le laboratoire souterrain de recherche du Canada.

---

* Work performed under the auspices of the US Department of Energy by Battelle Memorial Institute under Contract DE-AC02-83CH-10139 and by Lawrence Livermore National Laboratory under Contract W-7405-ENG-48.

1. INTRODUCTION

1.1 Significance of the Problem

The underground openings of a repository may affect the conditions of the host rock in two ways: through the response to the excavation of the repository, and through the response to the thermal load resulting from the emplacement of heat-generating waste. The region affected by the combination of these perturbations is called the **disturbed zone**; the region affected predominately by excavation is called the **excavation response zone**. This paper deals only with excavation response. The response to excavation is a combination of the damage to the rock caused by blasting or other means of excavation and of the redistribution of stresses around the newly created opening. The zone of excavation response is characterized by the creation of new fractures, by the opening or closing of existing fractures, or by the creation of a reduced-modulus or plastic zone around the opening.

The excavation response zone is important for two reasons. First, it may form a preferred pathway for the migration of radionuclides away from a repository. Kelsall et al. [1] suggest that the excavation response zone strongly affects ground-water travel times and flow through and around shaft and tunnel seals. The parameters that control the flow are the hydraulic conductivity of the zone, and to a lesser degree, its extent. Second, the US Nuclear Regulatory Commission [2] recommends that the disturbed zone should be the inner boundary for ground-water travel time calculations. Therefore, it is important to know the **properties** and the **extent** of the excavation damage zone. In order to know the properties and extent of the excavation response zone, we need to develop tools and techniques for characterizing its properties, and we need to understand the fundamental mechanisms that govern the creation of the response zone in order to minimize its effects and predict its performance and extent.

1.2 Purpose of this Paper

This paper describes the development of an Excavation Response Experiment for the Canadian Underground Research Laboratory (URL) which will contribute to our understanding of the characteristics and mechanisms of excavation response. We will summarize the information gained from earlier excavation response studies at the Spent Fuel Test - Climax and the Colorado School of Mines (CSM), identify some of the information needs that still exist or were raised as a result of these studies, and show how the information gained from the Climax and CSM studies is being used to guide the development of the URL experiment to satisfy these information needs.

2. RESULTS OF PREVIOUS STUDIES

2.1 Spent Fuel Test - Climax [3]

The Spent Fuel Test - Climax was conducted to investigate the feasibility of storage of spent reactor fuel assemblies at a plausible repository depth in a typical granite rock. The Climax facility was excavated at a depth of 420 m in a partially saturated quartz monzonite intrusive known as the Climax Stock. Wilder and Yow [4] report that the rock contains four prominent and four less prominent joint sets, with a total frequency of 0.9 to 2.2 joints/m, and three sets of shear zones. A primary technical objective of the experiment was to simulate the effects of thousands of emplaced spent fuel elements using only a small number of spent fuel elements and electrical heaters. One of the secondary objectives, and the only one with which we are concerned in this paper, was to compare the magnitude of displacement and stress effects from mining alone with that of thermally induced displacement and stress changes that occur as the result of heating.

In order to determine the effects of mining, a "Mine-By" experiment was conducted during the excavation of the test facility. The Mine-By Experiment was carried out by mining two outer drifts, installing instrumentation to monitor stress and displacement in the pillar between the drifts, and then mining a central drift through the pillar. Plan and section views of the Climax facility, showing the configuration of the Mine-By Experiment, are shown in Figures 1 and 2. Because the Spent Fuel Test was the primary objective of the Climax program, the design of the Spent Fuel Test strongly affected the configuration, instrumentation layout, and schedule of the Mine-by Experiment. After the Mine-By Experiment, the main part of the Spent Fuel Test was carried out by installing spent fuel elements and electrical heaters in the three drifts.

The Climax Experiment subjected the rock to three loading phases: The mechanical unloading during the Mine-By test, the subsequent thermomechanical loading due to the spent fuel and electrical heaters, and the unloading during the cooldown following the removal of the spent fuel and heaters. Analyses by Wilder and Yow [5] show that the displacements caused by the mining/unloading are similar in magnitude to the displacements caused by the thermal/loading-unloading, but they are of a fundamentally different nature. The behavior during the thermomechanical loading phase appeared to be elastic, whereas during the Mine-By the deformations appeared to be controlled by the joints and shear zones in the rock. Also, the Mine-By response occurred with time as a function of mining and of yield in the rock mass; the thermal response occurred as a function of the heat transfer properties of the rock. Although rock mass behavior during cooldown was more difficult to assess because of curtailment of monitoring, the known differences may make the response to excavation in general harder to predict and to analyze than the response to heating.

The finite-element elastic continuum code ADINA was used to perform a pre-test estimate of the behavior of the Mine-By Experiment [6]. The ADINA results showed the pillars between the drifts expanding into the central drift, but the field measurements showed a narrowing of the pillars. The experiment was then modeled using the finite element model JPLAXD, which allows the inclusion of discrete joints, to see if the rock structure could account for the sign difference of the displacements [7]. However, the JPLAXD results for the pillars were also different in sign from the field measurements. Butkovitch [8] then showed that explosive expansion from the drill-and-blast mining of the central drift could account for some of the narrowing of the pillars. Although incomplete in some ways, this analysis indicates that models which simulate excavation merely through the removal of material do not account for all of the processes which control the deformation of the rock.

Predictions of the response to heating using ADINA showed fairly good agreement with the measured displacements from the thermal phase of the experiment. This tends to confirm that the rock fractures controlled the excavation response but not the heating response.

The following observations can be made about the Climax Mine-By Experiment:

- The effects of excavation are at least as significant as heating effects in rock response; excavation effects are probably harder to analyze due to the influence of rock structure

- The uncertainty in the modeling was due to a number of factors, the most important of which were the complexity of the rock structure, the initial failure to account for all important mechanisms in the excavation process, and the inadequate amount, and the failure, of important instruments. In particular, the absolute in situ stresses were uncertain at the time of the Mine-by, and stress monitoring instrumentation may have failed during the experiment.

2.2 Colorado School of Mines [9]

The Colorado School of Mines has carried out an investigation of excavation response in their Experimental Mine near Idaho Springs, Colorado. The objectives of this program were to develop and evaluate blast rounds intended to minimize blast damage to the rock, and to develop techniques for characterizing the nature and extent of excavation response in the rock surrounding an opening. The portion of the CSM Experimental Mine in which the program was carried out is situated in heavily foliated granitic gneiss with two major vertical fracture sets. A room 20 m long, 3 m high, and 5 m wide was excavated using blast rounds designed following the Swedish Langefors approach and Livingstone Cratering Theory. Vertical extensometers were installed in the roof next to the new face immediately following each round to attempt to capture the response to the excavation of the subsequent rounds. Following the excavation, six sets of seven 5 m boreholes were drilled radially outward from the room for the characterization of blast damage and excavation response. The configuration of the room and boreholes is shown in Figure 3.

Due to blast damage alone, the rock surrounding an opening would exhibit lower Rock Quality Designation (RQD), lower modulus, lower P and S wave velocities, and higher permeability. All of these characteristics except RQD are dependent on stress. During the removal of rock during excavation, the stresses that were previously carried by the opening are redistributed to the surrounding rock. This increase of stresses in the surrounding rock would lead to higher modulus, higher P and S wave velocities, and lower permeability. The favorable superposition of material and stress change characteristics could lead to a reduction in the deleterious effects or the extent of the excavation response zone. A characterization program was carried out in order to evaluate the success of the controlled blasting and to determine the characteristics of the excavation response zone. The characterization consisted of:

- RQD indexing through examination of core and borehole walls
- Modulus determinations using the CSM cell (dilatometer)
- Crosshole ultrasonic velocity measurements
- Permeability measurements by nitrogen injection.

The characterization showed that the controlled blasting was successful in limiting blast damage to within 0.5 to 1 m from the wall of the excavation. What is more interesting, however, is the comparison of the results of the different characterization techniques. The CSM cell measurements, the ultrasonic velocity measurements, and the permeability measurements in general showed good agreement with one another on the extent and degree of disturbance of the excavation response zone. The RQD indexing showed relatively poorer agreement with the other three techniques. Zones in which the RQD indicated a higher degree of induced fracturing did not necessarily exhibit the lowest modulus, the lowest wave velocities, or the highest permeability. This indicates that, in determining the properties of the response zone, the number of fractures is less important than whether or not the fractures are open. Therefore it is logical that direct measurements of modulus, wave velocities, and permeability are more reliable than RQD determination for evaluating excavation response. The vertical extensometers installed in the roof provided little quantitative information on response to excavation because most of the response to each blast round had already occurred by the time each extensometer was installed.

2.3 Observations on the Climax and CSM Experiments

If we are to use the Climax and CSM Experiments as guides in the development of future excavation response experiments, we may make the following observations:

1. In order to understand the phenomena, important parameters must be measured directly rather than inferred from some other parameter or effect. If permeability is an important parameter, we should measure permeability rather than determine the RQD. Stress change and displacements should both be determined explicitly rather than inferring one from the other.

2. It is essential that there be adequate numbers of instruments, and that the instruments be in the right locations and of sufficient resolution for measuring the critical parameters. There are three ways of installing instruments for evaluating excavation response: from a remote location prior to and in the direction of the eventual excavation; from within an opening as it is being excavated; and from within an opening after excavation has been completed. Each of these methods is probably necessary for capturing the total response to excavation. For example, both drift convergence (from within an opening) and pillar displacement (from a remote location) need to be measured to determine the complete response of the pillar.

3. The computer models that we use to evaluate the results of the experiment must adequately represent the processes that determine the behavior of the physical system. If our goal is to determine

the change in hydraulic conductivity due to excavation response, itmay not be adequate to calculate stress change or even change in fracture aperture, because the relationships between stress, aperture change, and conductivity might not be adequately understood for a deforming fracture.

4. The field environment that will contain the experiment must be fully characterized so that all characteristics and structural features that could affect the behavior of the experiment are known. The modelers must know the location and significance of these features so that they may correctly account for them in their analyses.

To put these observations in other words, we must measure the right parameters, we must measure them correctly, our computer models must deal with the parameters correctly in analyzing the problem, and we must understand the field situation that determines the behavior of the experiment. We will describe in the rest of this paper how the development of the Excavation Response Experiment for the URL is dealing with these issues.

3. PLANNING THE URL EXCAVATION RESPONSE EXPERIMENT

3.1 Summary and Objectives

The URL Excavation Response Experiment (ERE) is being developed as a part of a cooperative experimental program between Atomic Energy of Canada Limited and the United States Department of Energy. The actual planning of the experiment is being carried out by a group consisting of representatives from AECL, DOE, Lawrence Livermore National Laboratory (LLNL), Lawrence Berkeley Laboratory (LBL), and the University of Alberta. The exact configuration of the ERE has not yet been determined. Very generally, the experiment will probably consist of a drift several meters in diameter excavated parallel to, or intersecting at some angle, a hydraulically conductive fracture or system of fractures. The test will include several galleries overhead and alongside the central drift from which instrumentation will be installed to monitor the response of the rock an fractures to the excavation of the central drift.

The test development group has given particular attention to rigorously defining the objectives of the ERE so that it meets the programmatic needs of both AECL and DOE. Rigorous and concise definition of the experiment objectives is also an immense help to designing the experiment itself, as distractions from the primary needs arising from secondary or tertiary needs can be eliminated. The objectives for the URL ERE are:

- To evaluate our ability to predict the mechanical and hydrological response of the rock mass to excavation in a rock mass containing a range of fracturing, and to assess the limitations of our ability to predict, and to improve our ability to predict, these responses

- To determine the geomechanical and hydrological properties of the rock mass and their coupling

- To study the response of the rock mass in terms of fundamental mechanisms governing fluid flow and rock mass deformations. Specifically, to study and extend our understanding of:

  - excavation-induced fracturing
  - stress dependence of fracture permeability.

We will now discuss how the ERE is being developed to meet each of these objectives.

3.2 Planning the ERE

Objective: Evaluate and improve predictive ability

The first step in evaluating our modeling capability was to assess the state of the art. LLNL surveyed computer models available in the US and Canada to see if any of them had the capability of modeling excavation response in fractured rock. They found that, while several codes could handle discrete fractures or large deformations, none of them could explicitly treat the all-important hydrologic response and highly non-linear fracture deformation in a coupled fashion. LLNL then proceeded with the development of a three-dimensional hybrid boundary element-finite element code named GENASYS (Geotechnical Engineering Analysis System) which would allow the calculation of coupled deformation and fluid flow response of a fractured rock mass subject to excavation. AECL and the University of Alberta are improving existing finite element and boundary element codes for application to the ERE. This intimate involvement of the development of models in the process of designing the experiment ensures that the output of the experiment will be meaningful to the modelers.

In order to allow the planning group to check the progress of the model development, AECL made available the results of a small excavation response experiment they performed during the development of the 240 m level of the URL. The results of this experiment, called the Room 209 Experiment, are reported elsewhere in this workshop [10]. Each modeling group attempted, using the best available characterization information, to predict the results of the Room 209 Experiment prior to the release of the data from that experiment. This exercise provided extremely valuable practice for the development of the models and of the ERE. It should help ensure that the modelers request as input only those parameters that are capable of being measured in situ, and that those responsible for installing and operating the test instrumentation provide data in a form that is useful to the modelers.

Model development and improvement will continue until about 1990 to ensure that adequate codes are available prior to implementation of the experiment. Once the test location has been fully characterized, the modelers will use the information from the characterization to perform a blind prediction of the results of the ERE. Comparison of the predictions with the actual field results will constitute the most rigorous possible test of our ability to model the process of excavation response.

Objective: Determine geomechanical and hydrological properties

The ERE will affect a volume of rock on the order of 10^6 cubic meters. It therefore provides one of our best chances of determining large-scale rock mass mechanical and hydro-mechanical properties. This process starts with the intensive characterization of the test environment. All material properties, in situ conditions, and structural features that might affect the behavior of the experiment need to be determined and understood. In particular, any anisotropy in material properties needs to be understood to allow three-dimensional analyses. Comparison of test response to pre-test conditions will then help in confirming such large-scale parameters as fracture stiffness, rock mass deformation modulus, and stress and permeability tensors.

In order to capture the response of the experiment, the test instrumentation must measure the correct parameters, be in the correct locations, and be able to survive the excavation process. LLNL is now performing thorough pre-test scoping calculations to determine regions of maximum response and maximum gradients to guide the selection of instrument locations. Different drift and instrument configurations will be evaluated to determine which combination of drift, fracture, and stress directions provides the maximum measurable local hydrologic response.

Even though the experiment will seek to maximize hydrologic response, it is important to remember that the installation of instrumentation required to monitor a test of this magnitude is likely to cause a significant hydrologic response of its own. AECL is currently developing instruments which will combine displacement and hydrologic measurements in the same borehole, thus reducing significantly the number of boreholes required to instrument the ERE. The zone of greatest interest in determining excavation response is the zone closest to the opening. This zone is of greatest interest because it experiences the highest gradients and the greatest impacts due to blasting, which also makes this zone the harshest environment for instruments. AECL and DOE are evaluating displacement instrumentation in order to develop a system with a gage length short enough to capture the gradients and robust enough to survive the impacts that occur close to the excavated opening. These developments are described elsewhere in this workshop [11].

Objective: Study fundamental mechanisms governing rock mass response

As we have seen with the Climax Mine-By Experiment, the complexity of the test environment can contribute to the uncertainty in the interpretation of the test results. In order to avoid masking the fundamental response of the rock mass, the test environment for the ERE will be kept as simple as possible. The ground conditions at the URL range from massive intact rock to rock with several fractures per meter. The rock volume that will contain the ERE will be carefully selected in order to allow the observation of the effects of excavation on a single fracture and, if our modeling capabilities are sufficiently advanced at that time, on a simple fracture network. This will minimize the number, and the accompanying uncertainty, of simplifying assumptions required for the analysis.

The Climax and CSM Experiments showed that blasting and stress redistribution may produce distinct responses in the rock mass. If permitted by cost constraints and by test geometry, the ERE may use machine excavation in addition to drilling and blasting for part of the excavation. Comparison of the response to blasting to the response to machine excavation will allow the separation of the effects of blasting from the response to stress redistribution.

The mechanical and hydraulic response of the rock mass, and especially of any fractures, will be monitored during the excavation of the test drift. Following excavation, a characterization program will be performed from inside of the test drift to determine blast damage and near-field changes in material properties. Comparison of the pre-test conditions, the response during excavation, the post-test conditions, and the predicted response form the modeling groups will help us to determine the mechanisms which govern the creation and the characteristics of the excavation response zone.

REFERENCES

1. Kelsall, P. C., J. B. Case and C. R. Chabannes: A Preliminary Evaluation of the Rock Mass Disturbance Resulting From Shaft, Tunnel, or Borehole Excavation, Battelle Memorial Institute ONWI-411, November 1982.

2. Gordon, M., N. Tanious, J. Bradbury, L. Kovach and R. Codell: "Draft Generic Technical Position: Interpretation and Identification of the Extent of the Disturbed Zone in the High-Level Waste Rule (10 CFR 60)", US Nuclear Regulatory Commission, June 20, 1986.

3. Patrick, W. C.: Spent Fuel Test - Climax: An Evaluation of the Technical Feasibility of Geologic Storage of Spent Nuclear Fuel in Granite: Final Report, Lawrence Livermore National Laboratory UCRL-53702, March 30, 1986.

4. Wilder, D. G. and J. L. Yow, Jr.: Structural Geology Report: Spent Fuel Test - Climax, Nevada Test Site, Lawrence Livermore National Laboratory UCRL-53381, October 1984.

5. Wilder, D. G. and J. L. Yow, Jr.: Geomechanics of the Spent Fuel Test - Climax, Lawrence Livermore National Laboratory UCRL-53767, July 1987.

6. Butkovitch, T. R.: Mechanical and Thermomechanical Calculations Related to the Storage of Spent Nuclear-Fuel Assemblies in Granite, Lawrence Livermore National Laboratory UCRL-52985 Rev. 1, August 1981.

7. Heuze, F. E., T.R. Butkovitch and J. C. Peterson: An Analysis of the "Mine-by" Experiment, Climax Granite, Nevada Test Site, Lawrence Livermore National Laboratory UCRL-53133, June 1981.

8. Butkovitch, T. R.: "Spent Fuel Test - Climax: Mine-by Revisited", Lawrence Livermore National Laboratory UCID-20673, December, 1985.

9. Hustrulid, W. A. and W. F. Ubbes: "Results and Conclusions From Rock Mechanics/Hydrology Investigations: CSM/ONWI Test Site", Proc. Seminar Geological Disposal of Radioactive Waste: In Situ Experiments in Granite, OECD, Paris, 1983, 57-75.

10. Lang, P. A.: "Room 209 Excavation Response Test", Proc. Workshop on Excavation Response in Deep Radioactive Waste Repositories, OECD, Paris, In preparation.

11. Thompson, P. M., E. T. Kozak and C. D. Martin: "Rock Displacement Instrumentation and Coupled Pressure/Rock Displacement Instrumentation for Use in Stiff Crystalline Rock", Proc. Workshop on Excavation Response in Deep Radioactive Waste Repositories, OECD, Paris, In preparation.

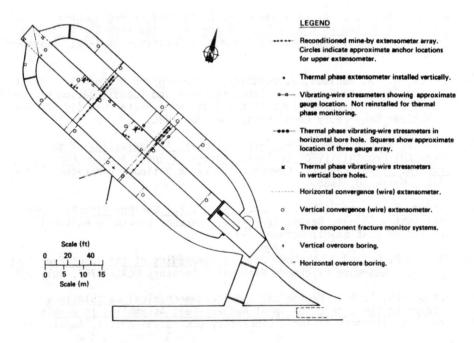

LEGEND

┄•┄•┄ Reconditioned mine-by extensometer array. Circles indicate approximate anchor locations for upper extensometer.

• Thermal phase extensometer installed vertically.

o—o— Vibrating-wire stressmeters showing approximate gauge location. Not reinstalled for thermal phase monitoring.

—•••— Thermal phase vibrating-wire stressmeters in horizontal bore hole. Squares show approximate location of three gauge array.

x Thermal phase vibrating-wire stressmeters in vertical bore holes.

┄┄┄┄ Horizontal convergence (wire) extensometer.

o Vertical convergence (wire) extensometer.

△ Three component fracture monitor systems.

+ Vertical overcore boring.

- Horizontal overcore boring.

Scale (ft)
0   20   40

0   5   10   15
Scale (m)

Figure 1 Plan view of the Climax facility, showing the two outer instrument drifts, the central mine-by drifts, and the location of the mine-by and thermal phase instrumentation.

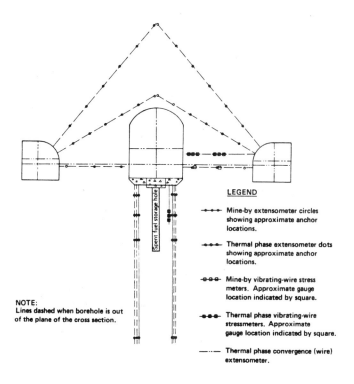

Figure 2 Section view of the Climax facility, showing the
mine-by instrumentation in the pillars between the
instrument drifts and the central drift.

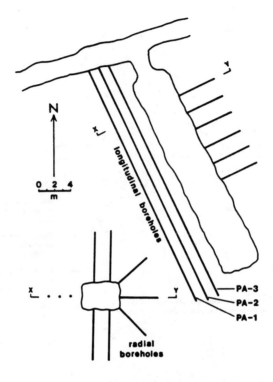

Figure 3   Plan and section views of the excavation response
room in the CSM Experimental Mine, showing the
location of the radial borehole arrays used to
characterize the excavation response zone.

ROCK DISPLACEMENT INSTRUMENTATION AND COUPLED HYDRAULIC PRESSURE/ROCK DISPLACEMENT INSTRUMENTATION FOR USE IN STIFF CRYSTALLINE ROCK

P.M. Thompson, E.T. Kozak and C.D. Martin
Atomic Energy of Canada Limited
Pinawa, Manitoba, Canada ROE 1L0

ABSTRACT

Precise rock displacement measurements in stiff crystalline rock are necessary to validate and calibrate numerical models designed to predict rock mass response to excavation. Aperture changes across fractures can theoretically increase the hydraulic conductivity of the fracture by the cube of the effective fracture aperture increase. By measuring the coupled mechanical/hydraulic response of discontinuities within a rock mass, we can increase our understanding of the fundamental mechanisms involved, and how they affect one another. This paper describes the Bof-ex/Pac-ex family of instruments, which have been applied successfully in stiff crystalline rock, and discusses some applications where the instruments have been used.

INSTRUMENTATION DE DETECTION DU DEPLACEMENT DES ROCHES ET INSTRUMENTATION COUPLEE DE MESURE DE LA PRESSION HYDRAULIQUE ET DU DEPLACEMENT DES ROCHES DESTINEES A ETRE UTILISEES DANS LES ROCHES CRISTALLINES DURES

RESUME

Il est nécessaire de mesurer précisément le déplacement des roches dans les formations cristallines dures pour valider et étalonner les modèles numériques conçus pour prévoir les effets des travaux d'excavation sur la masse rocheuse. Les modifications dans la section des fissures peuvent théoriquement augmenter leur conductivité hydraulique en raison du cube de l'augmentation effective de cette section. En mesurant simultanément la réponse mécanique et hydraulique des discontinuités à l'intérieur d'une masse rocheuse, on peut mieux comprendre les mécanismes fondamentaux en jeu et leurs interrelations. Les auteurs décrivent la série d'instruments Bof-ex/Pac-ex qui ont été utilisés avec succès dans une formation cristalline dure et examinent quelques exemples d'utilisation de ces instruments.

1. INTRODUCTION

Numerical models are being developed that are designed to predict rock mass response to excavation. Such response occurs as changes in stress, deformation or failure of intact rock, and dilation or closure of existing fractures in the rock mass. Since the most significant potential mechanism for radionuclide transport from a nuclear fuel waste disposal vault in plutonic rock is by migration through natural fractures surrounding the vault, we need an adequate understanding of fracture behaviour in response to excavation. Aperture changes across fractures can have a significant effect on the hydraulic conductivity, which theoretically increases as the cube of the effective fracture aperture. It is therefore important to combine mechanical measurements of fracture aperture changes with measurements of hydraulic pressure and conductivity changes. By measuring the coupled mechanical/hydraulic response of discontinuities within a rock mass, we can increase our understanding of the fundamental mechanisms involved.

At the Underground Research Laboratory (URL) in Canada, researchers are performing field measurements of excavation response in an effort to obtain real data to calibrate and validate the numerical models used in the disposal concept assessment [1,2]. Since 1983, when excavation of the shaft began, numerous field measurements of excavation response have been made during the URL Construction Phase Experimental Program [3]. An important aspect of this work has been the identification and development of the field instrumentation necessary to successfully achieve the experimental objectives. This paper describes instruments that have been developed and applied at the URL to measure displacements in the rock mass and across natural fractures, and also hydraulic pressure changes in these natural fractures.

2. INSTRUMENTATION REQUIREMENTS

There are three specific types of excavation response rock displacements that can occur. These need to be accurately measured in order to obtain a better understanding of the response of stiff, massive or sparsely jointed, crystalline rock to excavation and to assess the numerical models. The three displacement types are:

1. instantaneous deformation,

2. time-dependent deformation, and

3. aperture changes across natural fractures.

Instantaneous deformation is difficult to measure in stiff rock because the magnitudes are relatively small. Typical magnitudes at the URL are less than 0.5 mm, over 15 m [4]. This deformation is measured in individual segments of a borehole and each segment accounts for only a fraction of this total deformation. Accordingly, an overall measurement system accuracy of ± 0.01 mm, or better, is desirable [4]. This must be achieved in a harsh environment that is wet, dirty, and extremely close to the severe vibrations and concussion associated with the excavation blasts.

Time-dependent deformation of the rock is even smaller in magnitude than the instantaneous response and occurs over a much longer time span. In stiff granite, such deformation is in the order of μm and continues at a diminishing rate indefinitely. In the past such measurements could only be made in laboratories where conditions permit the use of sensitive transducers and logging equipment. Field measurement of such response requires extremely sensitive yet

- 258 -

stable instrumentation that can withstand the rigours of the harsh underground environment, including blasting adjacent to the installation. Since the measurements must take place over a number of years, the instruments must have the required longevity and/or the ability to be easily and quickly removed for repair, calibration, or replacement.

Because we are particularly interested in the mechanical and hydraulic response of natural water-bearing fracture to excavation, we need a high-resolution instrument that can hydraulically isolate a fracture intersecting the measurement borehole while recording any movements across the fracture. In stiff crystalline rock, fractures with high normal stiffnesses require an instrument sensitivity of 1 µm or better. The mechanical components need to be isolated from, or unaffected by, high hydrostatic pressures of several MPa within the fracture.

In addition to the above technical requirements, several specific features desirable for displacement instrumentation have been identified in previous URL measurements [4,5]:

- ease of installation;

- retrievability;

- ability to be continuously data-logged;

- many anchor points, especially close to the excavation wall;

- need for temperature monitoring at frequent intervals along the instrument;

- minimal thermal coefficient of expansion; and

- minimum borehole diameter to reduce drilling costs.

Some rock displacement instruments used and assessed at the URL are described in an earlier publication [5]. The instrument with the most potential for future use was identified as the Bof-ex manufactured by Roctest Limited of Canada. Bof-ex stands for BOrehole Fracture monitor-EXtensometer. This paper describes the Bof-ex system, how it is used at the URL, and modifications we have made to enable us to use the Bof-ex to obtain coupled mechanical/hydraulic measurements across water-bearing fractures.

3. THE BOF-EX

3.1 Operating Principle

The Bof-ex is a form of multiple-position borehole extensometer (MPBX). Unlike most MPBX designs, the Bof-ex measures relative displacements between adjacent anchors rather than transmitting all anchor displacements to a collar reference head. In this regard it is similar to the sliding micrometer [6]. The instrument system is shown schematically in Figure 1.

The anchors are installed in the measurement borehole one at a time, starting with the deepest anchor. Anchor details are illustrated in Figure 2. The mechanical locking system consists of a threaded anchor setting rod with a conical tip positioned axially through the anchor body. The threaded anchor setting rod is rotated during anchor setting using concentric installation rods. Two lugs on the anchor body are used to maintain anchor orientation. The conical tip of the threaded anchor setting rod acts against a tapered strut, which is part of a moveable anchor shoe forced radially against the borehole wall. Two additional non-moveable shoes are spaced symetrically around the anchor body and react against the force applied by the moveable shoe, helping to

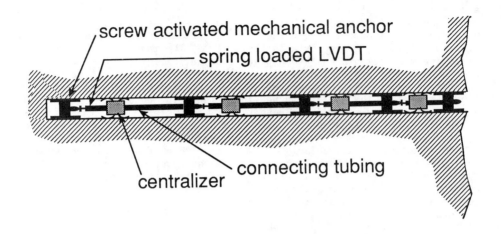

Figure 1:   Schematic Diagram of Bof-ex

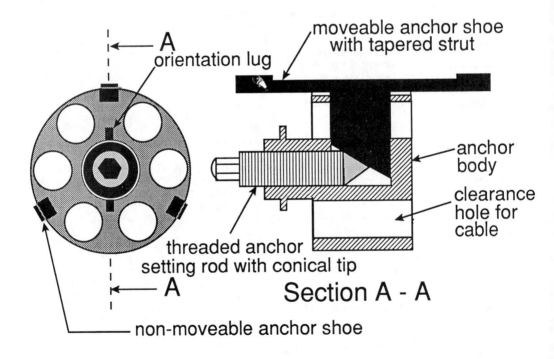

Figure 2:   Bof-ex Anchor Details

distribute the normal force around the circumference of the borehole. This anchor setting system is capable of exerting a 3.5 kN force against the borehole walls.

Each anchor is installed with its pre-assembled connecting tubing and LVDT displacement transducer. Periodically, a potted thermistor is attached to an anchor using electrical tape. Typically, 6 thermistors are installed along a 15-m-deep hole. The LVDT transducer is located at the far end of each assembly and is positioned so that the spring-loaded plunger contacts the previous anchor at approximately the mid-range of the transducer. The anchor is then tightened to lock it in place. Cables from displacement transducers and thermistors are fed through holes in the anchor bodies. The number of anchors is limited by the quantity of signal cables that can be accommodated by the anchor clearance holes. We have installed up to 11 anchors in an NQ (75-mm-diameter) borehole. The maximum length of installation is limited by the installation rods. The deepest anchor set to date at the URL is 33 m. Since all connecting tubing is pre-assembled to each anchor, only the deepest anchor must be set accurately. Anchor spacing can be as close as 300 mm, and within 100 mm of the borehole collar. Centralizers are installed with each anchor assembly to ensure the connecting tubing and transducer are positioned centrally in the borehole. A typical installation, including down-hole thermistors, can be completed in about four hours; however, several hours of assembly, preparation, and checking are necessary beforehand.

3.2 Advantages

Conceptually, the Bof-ex is a very simple instrument. Indeed, this simplicity is perhaps the main reason the instrument has proven superior to other rock displacement instruments during construction phase experiments

at the URL. Principal advantages over conventional MPBX designs are as follows:

1. The transducers are located in the measurement borehole. Provided hermetically sealed LVDT displacement transducers are used to prevent moisture problems, this location has the advantage of a relatively stable thermal environment and also provides protection from blast effects.

2. The 32-mm-O.D. stainless steel or Invar connecting tubing used to transmit displacements from one anchor to the next has a much greater section modulus than the typical 6-mm-diameter rods used in conventional MPBX designs.

3. Movements do not have to be transmitted from each anchor all the way to the collar reference head through friction points at each shallower anchor. This reduces the potential for stick-slip behaviour identified by Patrick et al. [7] in the Climax Spent Fuel Test.

4. Since the Bof-ex connecting tubes do not run from each anchor to a collar reference head, the magnitude of thermally induced strain is greatly reduced.

5. Anchor locations can be within 300 mm of one another and 100 mm of the excavation wall. Up to 11 anchors and 6 thermistors can be installed in an NQ borehole using standard wiring. Also, no special collar preparation is necessary.

6. The entire instrument is easily retrievable for re-use, repair, or calibration. It is felt that this feature will make the Bof-ex an attractive instrument for pre-closure monitoring at an actual nuclear waste disposal vault. In addition, the fact that the instrument can be re-used helps to

justify the relatively high
purchase price.

An apparent drawback to the
principle of differential line wise
displacement measurement employed by
the Bof-ex is that should a single
transducer malfunction, the full rock
displacement profile into the rock
mass cannot be measured. There will
be a missing section. In addition, if
one wishes to calculate total
displacements relative to the deepest
anchor, the measurement errors
associated with each transducer are
accumulated.

These drawbacks are no more
significant than the loss of a
transducer in a conventional MPBX, or
the errors associated with stick-slip
and bending in long rods. Kovari et
al. [6] indicated the significance of
differential observation. They
concluded that, when measuring
displacements in stiff rock, it is
generally most important to be able to
observe the movement across
significant discontinuities. Total
displacement is of interest, but it
can be obtained by simply measuring
convergence.

In most installations the
transducers do not malfunction and
total displacements can be calculated
by summing the incremental
displacements. Experience at the URL
has shown that when highly sensitive
transducers are selected, the errors
caused by adding each interval
displacement are insignificant. They
appear to be considerably less than
the errors resulting from a mechanical
transmission of the displacement from
the deepest anchor to the borehole
collar using a 6-mm rod that can stick
and bend as it passes through numerous
anchors.

3.3 Features

The Bof-ex system in use at the
URL incorporates a number of features
specified by AECL to help achieve

better results and to allow the Bof-ex
to be modified to provide coupled
hydraulic/mechanical measurements:

1. Stainless steel or Invar
   construction, instead of the
   standard aluminum, to provide
   greater strength, more corrosion
   resistance, and to reduce the
   thermal coefficient of expansion.

2. Six thermistors installed along
   each installation to monitor down-
   hole temperatures and allow for
   temperature compensation.

3. Centralizers with Teflon contact
   points to minimize friction
   between the centralizers and the
   connecting tubes.

4. Anchor shoes that act like stiff
   'leaf springs', creating an active
   anchor to maintain constant
   contact force with the borehole
   wall despite borehole deformations
   caused by blast vibrations or
   excavation-induced stress changes.

5. Schaevitz model GPD-121-250 DC-
   LVDT (Direct Current-Linear
   Variable Differential Transformer)
   displacement transducers. These
   transducers have a range of ±6 mm
   and a specified repeatability of
   0.6 μm. A version of this
   transducer with the same
   specifications out-performed other
   transducers in the Climax Spent
   Fuel test [8].

6. Centrally located connecting
   tubing to allow incorporation of
   hydraulic packer systems for
   coupled measurements.

4. THE PAC-EX

The Pac-ex (PACker-EXtensometer)
is a logical progression from the Bof-
ex: it allows monitoring of coupled
hydraulic/mechanical response of
natural fractures intersecting an HQ
(96-mm-diameter) measurement borehole.

Two versions of the Pac-ex have been designed, fabricated, and field-tested. Each has its own particular advantages and disadvantages. Both versions are shown schematically in Figure 3.

Laboratory testing to verify the proper operation of both Pac-ex versions is still in progress. Initial results appear promising; however, it should be noted that, since this instrument is a recent development, the field measurement results presented in this paper are considered preliminary.

4.1 Single-Packer System

In the single-packer version of the Pac-ex, a pair of standard Bof-ex anchors are used to span the fracture with a single transducer. A stiff spring is used on the transducer to counteract the sometimes high hydrostatic pressures that will apply pressure on the LVDT plunger. A pneumatic packer mounted on a hollow stainless steel mandrel, which extends above the borehole collar, is passed over the LVDT signal cable from the Bof-ex and positioned between the fracture and the borehole collar. The

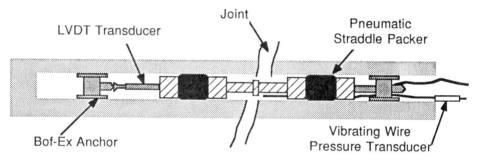

Double Packer Pac-ex

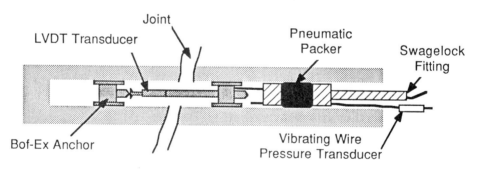

Single Packer Pac-ex

Figure 3:   Schematic Diagram of Pac-ex System

packer is then pressurized. The mandrel is sealed to the LVDT signal cable outside the borehole using a Swagelock fitting with a nylon ferrule. The hydraulic pressure within the fracture is transmitted to surface inside high-pressure stainless steel tubing. Pressure is monitored using a vibrating wire pressure transducer connected to this tubing outside the borehole. A heavy clamp is bolted to the rock at the borehole collar to hold the packer assembly in position.

This version of the Pac-ex has the advantage that the anchors can be mounted close to the fracture. A major drawback is that the entire mechanical portion of the instrument is subjected to fluctuating high water pressure, which may reduce the overall accuracy of the measurements or may cause reliability problems due to water leaking through seals or imperfections in the cable jacket. To date we have had no such failure with this system.

A second disadvantage is the relatively large volume of water stored in the packed off zone. This can have a deleterious effect on the hydraulic measurements.

4.2 Double-Packer System

The double-packer version of the Pac-ex uses a pneumatic straddle packer constructed on a hollow tube mandrel. This straddle packer assembly is installed across a fracture intersecting the measurement borehole. The length of the straddle packer can be adjusted to accommodate thicker 'fracture zones' if required. Standard Bof-ex anchors are installed on either side of the double packer, with the mechanical connecting tubing passing through the centre of the packer mandrel. The hydraulic pressure within the fracture is accessed through high-pressure stainless steel tubing, which passes up the borehole to the collar.

Pressure is monitored using a vibrating wire pressure transducer located outside the borehole.

With this system, it is possible to install a triaxial strain cell below the Pac-ex instrumentation in the same borehole. This reduces the drilling costs for instrumenting the rock mass to record displacements, hydraulic pressure changes, and in situ stress changes in the rock mass associated with excavation response. A second advantage of this version is that the mechanical components of the system are isolated from hydraulic pressure in the fracture. A third advantage is the smaller volume of water between the packers, which shortens the hydraulic response time. A disadvantage is the length of the straddle packer, which prevents spacing the Bof-ex anchors closer than about 1.3 m.

5. DATA LOGGING

Remote automatic data logging of rock displacement instrumentation is desirable at the URL for several reasons [4,5]:

- improved accuracy;

- more readings at relatively fast scans;

- errors due to manual data transfer/entry are eliminated;

- no operator measurement error;

- no time required at excavation face to take measurements, and therefore reduced costs associated with delaying contractor;

- digital computerized data is easily interpreted with greatly reduced labour cost component.

Compared to some technologies such as the sonic probe, the Bof-ex is a relatively easy instrument to data log. The output signal is d.c. voltage. In applications in horizontal excavations, such as the Room 209 Excavation Response Test, [9], the signal cables from the LVDT transducers can be run directly to a nearby data logger, which can be physically protected from damage during excavation using timber bulkheads. In a shaft, data logging is not as easy. It is not practical to locate a data logger within the shaft environment because of wet conditions and the inability to protect it from blast or other operational damage.

To overcome the problems of remote data logging of instruments located in the shaft, AECL designed a signal conditioner that can be located in a short percussion hole near the Bof-ex. The signal cables from the LVDTs and thermistors are routed through conduit from the collar of the Bof-ex borehole to a small junction box positioned within 3 metres. The cable to the signal conditioner is connected to the transducer signal and excitation cables inside this junction box. The waterproof signal conditioner is located in a 1-m-long, 96-mm-diameter percussion hole about 1-m from the junction box. From here, a three-conductor cable runs to a main junction box mounted higher up the shaft(4 to 20 m). One conductor is used to send an analog (4-20 mA) control signal from the data logger to the signal conditioner specifying the channel desired. A second conductor returns an analog ($\pm$20 mA) signal proportional to the transducer reading to the data logger. The third conductor is a common return for the control and output signals. A single multi-conductor cable then carries the signals up the shaft to an appropriate horizontal station or level where the data logger is located. The junction boxes, signal conditioners and borehole collars are protected from blast damage by 6-mm-thick steel plates. A schematic diagram of this set-up is shown in Figure 4. Fluke

model 2400 data loggers have been used successfully on the URL project.

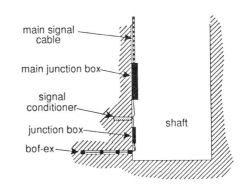

Figure 4: Data Logger Configuration

6. **FIELD MEASUREMENTS**

6.1 Bof-ex

The ability of the Bof-ex to accurately measure instantaneous response to excavation in stiff crystalline rock has been previously reported [5]. A plot of displacement versus excavation advance from the Room 209 excavation response test is shown in Figure 5. In this example, the total displacement measured after three blasts was 0.42 mm, and the incremental displacements were monitored with excellent resolution and stability. For example, the total displacement measured between 9.01 and 15.01 m in the borehole was 0.021 mm with $\pm$0.001 mm repeatability.

Measurements at the 324-m instrument array of the URL shaft extension, however, indicated a problem with the mechanical anchors used with the Bof-ex. Twelve of the 73 anchors installed slipped more than

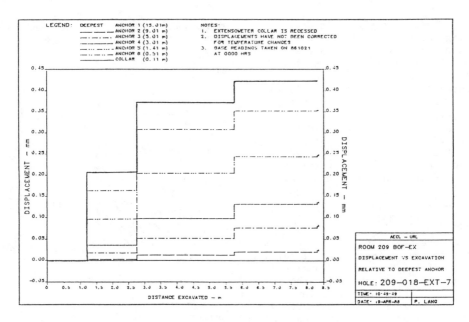

LEGEND: DEEPEST ANCHOR 1 (15.01m)
ANCHOR 2 (9.01 m)
ANCHOR 3 (5.01 m)
ANCHOR 4 (3.01 m)
ANCHOR 5 (1.41 m)
ANCHOR 6 (0.51 m)
COLLAR (0.11 m)

NOTES:
1. EXTENSOMETER COLLAR IS RECESSED
2. DISPLACEMENTS HAVE NOT BEEN CORRECTED
   FOR TEMPERATURE CHANGES
3. GAGE READINGS TAKEN ON 861021
   AT 0000 HRS

Figure 5: Displacement Measured With Bof-ex vs. Excavation Advance

a threshold value of 0.05 mm. This problem was quite unexpected since the instrument had operated so successfully in the previous trial in Room 209 at the 240 level [5,9]. A field test of some modified anchor designs has since been conducted with moderate success; however, some slippage, particularly of the collar anchor, still occurred on a reduced basis.

It has since been postulated that the anchors may work in a horizontal radial hole from a horizontal excavation, such as Room 209, because the moveable anchor shoe is positioned vertically and is actually held tighter in position by increasing circumferential stresses around the tunnel as the tunnel is advanced. These increasing circumferential stresses around the tunnel cause a reduction in the vertical diameter of the instrumented borehole. In a radial borehole from a vertical excavation, such as the shaft, the vertically positioned moveable anchor is located in a direction of increasing borehole diameter since it is perpendicular to the direction of the increasing circumferential stress. This is likely to loosen the anchor. In the next instrument array planned for 384-m depth in the URL shaft, this theory will be tested by installing the anchors with the moveable shoe positioned horizontally.

Other measures that have been taken to eliminate anchor slippage include:

1. Rounding the contact surfaces of the anchor shoes so they match the radius of the borehole wall. This should prevent crushing or spalling of the borehole wall due to high-stress concentrations.

2. Coating the contact surfaces of the anchor shoes with carbide spray welding to increase the coefficient of friction between the anchor shoes and the borehole wall.

3. Mounting a steel flange on the collar anchor to positively bolt the outer anchor to the rock surface.

4. Increasing the stiffness of the anchor shoes so that the 'active' component of the radial anchor force is optimized.

It is expected that the above measures will eliminate or reduce anchor slippage in the Bof-ex instrument. The effectiveness of these changes will be assessed after the 384-m array is complete.

Despite the anchor slippage problem at the 324-m instrument array in the URL shaft, we were still able to demonstrate the suitability of the Bof-ex to measure displacements in stiff crystalline rock. For example, by removing any instantaneous displacements associated with

excavation blast rounds, we were able to produce the plot of displacement versus time shown in Figure 6. The instantaneous displacements removed from the data were either immediate response to excavation, or anchor slippage, or a combination of both. The remaining displacement data is time-dependent rock response. Once again, the overall stability and repeatability of the measurements are quite remarkable. Deeper transducers in a stable temperature environment display an overall repeatability, which meets the transducer specification of 0.6 µm.

Even with anchor slippage occurring, we obtained measurements of instantaneous response to excavation that are not possible with other types of instrumentation. Should we achieve our goal of eliminating anchor slip entirely, the Bof-ex will be even more useful.

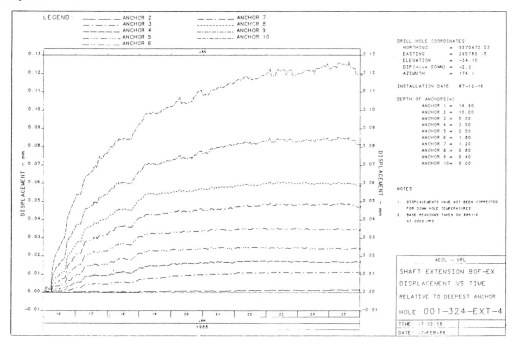

Figure 6: Time Dependent Deformation Measured with Bof-ex vs. Time

6.2 Pac-ex

An example of coupled displacement/hydraulic pressure measurements obtained with the Pac-ex instrument is illustrated in Figure 7. This plot shows measurements obtained in a borehole plunging at 72° from the URL shaft at the 283-m instrumentation array. This Pac-ex was the double-packer version and a CSIRO hollow inclusion triaxial strain cell was installed at the bottom of the hole to monitor excavation-related stress changes. In this example, the hydraulic pressure within a tight fracture was artificially cycled by pressurizing and depressurizing the packed off zone. The resulting decreases in fracture aperture varied from 2.6 to 1.7 μm. As one might expect, the normal fracture stiffness increases from one cycle to the next,

(806 to 1440 MPa/mm). The plot shows the high resolution achievable with the Pac-ex instrument.

Earlier instrumentation comprising straddle-packers installed across the Room 209 fracture [5,9] had shown a decrease in permeability in the fracture as the tunnel was excavated through it. This measurement contradicted numerical models that had predicted an increase in permeability. Obviously, the physics of the situation were not properly accounted for in the models. Unfortunately, the Pac-ex had not yet been developed prior to the Room 209 excavation. As a result, we were unable to collect displacement data that would further our understanding of the fundamental mechanisms involved. In future we will be able to deploy the Pac-ex in such a situation.

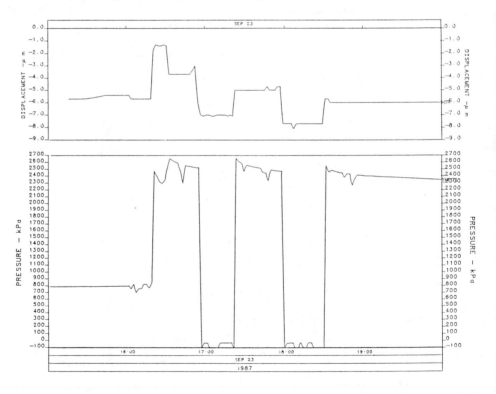

Figure 7: Coupled Displacement/Hydraulic Pressure Measurement Obtained With Pac-ex vs. Time

7. FUTURE DEVELOPMENTS

The anchor slip problem observed in the shaft extension has indicated there is a weakness in the mechanical anchor design. A new anchor is being designed that will have the following features:

- more 'active' spring range,

- stiffer spring mechanism, and

- more anchor surface in contact with the borehole wall.

Unlike the present anchor, this new design is intended to work satisfactorily regardless of its orientation with respect to the stress changes that take place during excavation.

Work is also under way on a version of the Pac-ex that will measure shear displacements in addition to normal displacements across fractures. This new design will take advantage of the simple Bof-ex mechanical anchor design. The concept is based on instruments that measure shear and normal displacements across fractures in the laboratory [10]. The deeper anchor will have a concentric conical reference surface onto which 3 or 4 axially mounted LVDT displacement transducers from the shallower anchor will react. Shear and normal components of displacement will be resolved from the combination of displacement measurements.

8. CONCLUSIONS

The Bof-ex rock displacement instrument and the Pac-ex coupled rock displacement/hydraulic pressure instrument are being used successfully in the URL experimental program. The Bof-ex has shown that it is capable of accurately measuring both instantaneous deformation and time-dependent deformation in response to excavation in stiff crystalline rock.

The Pac-ex can simultaneously measure both aperture changes and hydraulic pressure changes in a water-bearing fracture. Permeabilities can be measured before and after excavation and compared to measured displacements that took place across the fracture to cause the changes.

By measuring the coupled mechanical/hydraulic response of fractures within a rock mass, we are able to advance our understanding of the fundamental mechanisms involved, and how they interact. Model validation will only be successful when our instrumentation is capable of measuring the complex hydraulic/mechanical response of a rock mass.

9. REFERENCES

1. Simmons, G.R. and Soonawala, N.M. (editors). "Underground Research Laboratory Experimental Program", Atomic Energy of Canada Limited Technical Record, TR-153 (1982).

2. Simmons, G.R. (1986). "Atomic Energy of Canada Limited's Underground Research Laboratory for Nuclear Waste Management", IN: Proceedings of Geothermal Energy Development and Advanced Technology International Symposium, Tohoku University, Sendai, Japan, 1986 November 06-08. Also issued as Atomic Energy of Canada Limited Report, AECL-9283.

3. Thompson, P.M., Baumgartner, P. and Lang, P.A. (1984). Planned construction phase geomechanics experiments at the Underground Research Laboratory from the Proceeding of OECD/NEA and CEC Workshop on Design and Instrumentation on In Situ Experiments in Underground Laboratories for Radioactive Waste Disposal; Brussels, Belgium, May 1984.

4. Thompson, P.M. and Lang, P.A., (1986). Rock displacements measured during URL shaft sinking. Proc. 2nd International Conference on

Radioactive Waste Management,
Winnipeg, Canada, 1986 September.

5. Thompson, P.M. and Lang, P.A.
(1987). Geomechanical Instrumentation
Applications at the Canadian
Underground Research Laboratory. In
proceedings of the 2nd International
Symposium - field Measurements in
Geomechanics, Kobe, Japan, 1987 April.

6. Kovari, K., Amstad, C.H. and
Köppel, J., (1979). New developments
in the instrumentation of Underground
Openings, Proc. Rapid Excavation and
Tunnelling Conference, Atlanta.

7. Patrick, W.C., Rector, N.L. and
Scarafiotti, J.J. (1984).
Instrumentation Report No. 3:
Performance and Reliability of
Instrumentation Deployed for the Spent
Fuel Test - Climax. Lawrence
Livermore National Laboratory Report
No. UCRL-53637.

8. Patrick, W.C., Carlson, R.C. and
Rector, N.L. (1981). Instrumentation
Report No. 2: Identification,
Evaluation, and Remedial Actions
Related to Transducer Failures at the
Spent Fuel Test - Climax. Lawrence
Livermore National Laboratory Report
No. UCRL-53251.

9. Lang, P.A. (1988). Room 209
Excavation Response Test. For
Presentation at this Workshop.

10. Sun, Z., Gerrard, C. and
Stephansson, O., (1985). Rock Joint
Compliance Tests for Compression and
Shear Loads. International Journal
Rock Mechanics Min-Sci & Geomechanics
Abstract. Vol. 22, No. 4, pp.
197-213, 1985.

MAPPING OF THE EXCAVATION DAMAGE ZONE AROUND THE
CIRCULAR ACCESS SHAFT AT ATOMIC ENERGY OF CANADA LIMITED'S
UNDERGROUND RESEARCH LABORATORY

Everitt, R.A., P. Chernis, D. Good and A. Grogan
Atomic Energy of Canada Limited
Pinawa, Manitoba

ABSTRACT

The excavation damage about the circular shaft extension was mapped
during the excavation of a station at 300-metre depth. The reduction in
damage zone thickness compared to that seen in the previous rectangular shaft
(300 mm versus 1500 mm) is attributed to improvements in excavation methods,
to conversion to a circular cross section, and to increasing isotropy of
stresses with depth; however, this is based on preliminary interpretation of
new stress data. Damage distribution may be correlated with rock fabric and
with the principal stress orientations. This study will benefit subsequent
investigations directed at the assessment of excavation damage on vault design
and performance. It is recommended that this experiment be repeated under
higher stress conditions and with different shaft profiles.

RELEVE DE LA ZONE DE PERTURBATIONS LIEES À L'EXCAVATION AUTOUR DU
PUITS D'ACCÈS CIRCULAIRE DU LABORATOIRE SOUTERRAIN DE RECHERCHE
DE L'ENERGIE ATOMIQUE DU CANADA LIMITEE

RESUME

Les perturbations liées à l'excavation autour du prolongement du puits
circulaire ont fait l'objet d'un levé pendant les travaux d'excavation d'une
station située à 100 mètres de profondeur. La diminution de l'épaisseur de la
zone perturbée par rapport à celle constatée dans le puits rectangulaire
antérieur (300 mm contre 1 500 mm) est attribuée aux améliorations apportées
aux méthodes d'excavation, à l'adoption d'une section circulaire et à
l'augmentation de l'isotropie des contraintes en fonction de la profondeur.
Toutefois, ces conclusions se fondent sur une interprétation préliminaire de
nouvelles données sur les contraintes. La distribution des perturbations peut
être corrélée à la texture de la roche et aux principaux axes de contraintes.
Cette étude profitera aux recherches qui serait menées ultérieurement en vue
d'évaluer les conséquences des perturbations dues aux travaux d'excavation sur
la conception et les performances de la casemate. Il est recommandé que cette
expérience soit répétée en présence de contraintes plus et avec des sections
de puits variées.

1.0 INTRODUCTION

The Underground Research Laboratory (URL) of Atomic Energy of Canada Limited (AECL) is being constructed in the Lac Du Bonnet batholith, approximately 2.5 kilometres north of its southern contact and 14 kilometres east of the town of Lac du Bonnet, Manitoba. Prior to beginning the construction of the circular shaft extension in 1987 July, the URL consisted of a rectangular shaft excavated to a depth of 255 m, with access levels at depths of 130 m and 240 m. The 4.6-m-diameter shaft extension is being excavated to a depth of 445 m, with additional access levels at 300 m and 420 m. Experiments relating to the Canadian concept of nuclear fuel waste disposal in plutonic rock (Simmons [1]) are under way at this facility.

One of these experiments, referred to as the Excavation Damage Assessment (EDA) Program, is investigating the nature and causes of excavation damage in granite. The objectives of this program and an overview of its methods and results are presented by Lang [2].

Excavation damage refers in this case to the various scales of fracturing induced in a rock mass as a result of excavation. As a group, these fractures are generally distinguishable from natural fractures by their lack of natural mineral infilling and mineralogical wall rock alteration, and by their symmetry with respect to blasting centres or room geometry [3]. The causative processes are not fully understood but likely incorporate instantaneous fracturing associated with detonation, followed by subsequent time-dependent propagation of existing and new surfaces by stress relief.

This damage is of concern to the Canadian Nuclear Fuel Waste Management Program (CNFWMP) as the resultant changes in rock mass permeability and stability are prominent considerations in the assessment of vault design. A rare opportunity to view a cross section of the damage zone around a circular shaft was provided during shaft extension by the excavation of a station at 300-m depth. The purpose of this paper is to describe the dimensions and characteristics of the damage zone as seen at that location.

In this paper, all locations in the excavations are given by their depth, in metres, below the shaft collar which is at elevation 289.9 m above average mean sea level.

2.0 GEOLOGICAL SETTING

According to Brown et al. [4], the Lac du Bonnet batholith is late Archean and post-kinematic, with a crystallization depth of between 10 km and 16 km. The URL is situated near the centre of the batholith outcrop; however, the distribution of alteration and xenoliths suggests that excavation is proceeding through the original roof zone of the batholith.

The most common rock type is a grey granite. In fractured and marginal facies of the batholith, the colour has been changed to pink as a result of the combined effects of primary and secondary alteration. Xenolithic, leucocratic and heterogeneous zones are common and within the URL lease define a strong northeast-striking, southeast-dipping planar fabric (Figure 1). Concordant with this fabric are major low-dipping thrust faults and associated splays.

The tabular plates demarcated by these fault zones are crosscut by one or more sets of subvertical fractures, the pattern and frequency of which varies with proximity to the bounding faults and with local rock type.

The 300 level is situated in unfractured grey granite below a thrust fault and minor splay referred to as Fracture Zones 2 and 1.9, respectively (Figures 1 and 2). The grey granite is crosscut by several generations of consanguineous dykes, the most noteworthy of which is a granodiorite swarm. At the 300 level this granodiorite is the most common lithology. The host phase of the batholith is represented by inclusions within these dykes and by variously disrupted blocks between them.

3.0 DATA COLLECTION

Special efforts were made during the excavation of the 300 level to optimize the amount and quality of exposure available for characterizing the damage zone around the circular access shaft. The shaft was excavated 15 m (about 3 diameters beyond the floor of the station) to accommodate any time-dependent behaviour in the formation and propagation of the excavation-damage, and to avoid the effect of stress concentration. The station was then excavated using controlled pilot and slash techniques for the walls, floor, and crown. This was done to duplicate the quality of excavation characteristic of the shaft, and to avoid inadvertent excavation of damaged rock. Two surfaces were then made available for observation of the damage zone about the shaft perimeter:

1. a temporary 90° brow formed by the intersection of the shaft walls and the station crown (once photographed, this was trimmed to a more stable 45° slope), and

2. the station floor.

A third area, referred to as the station lip and lip pocket, provided a sample of the damage zone beneath the floor of the station as seen from the shaft walls.

The mapping methods were derived from those currently in use for the characterization of the shaft extension [5]. In brief, permanent survey control points are placed on the excavation surface which is then photographed in colour and stereo using a pair of 35-mm Rolleiflex cameras. Prints are compiled into photomosaics, which are then used underground as a mapping base.

Photogrammetric methods have not yet been applied to these photographs, although this may be attempted if required. The chief value of the photographs is in their highly detailed and unbiased depiction of the **fresh** excavation surface. This record may be referred to in future to quantify the effects of time on the remaining parts of this exposure.

Geological features are examined at the exposure, identified, and their characteristics recorded on the photomosaics and on coded field notes. For fractures, these characteristics include:

1. orientation – dip direction and dip,
2. length – in metres to the nearest decimetre,
3. aperture – in mm,
4. kinematic indicators,
5. roughness – estimated from standard profiles recommended by the ISRM [6],
6. planarity – estimated from standard profiles recommended by the ISRM [6],
7. seepage – general categories recommended by ISRM and modified by Brown and Everitt [7], and
8. mineralization and alteration.

Although it may be possible to distinguish in outcrop certain specific fractures as having originated either through stress relief or through blasting (chiefly on the basis of location and pattern), most fractures are likely to be the result of a combination of processes. Accordingly, no general attempt was made to classify the observed damage as to causative mechanism.

For reasons of access and safety, characterization of the excavation damage on the temporary brow and in the lip pocket was limited to mapping of the general pattern and extent of induced fractures from the photomosaics. Photographs of the lip and lip pocket were taken from the centre of the Galloway stage used for shaft sinking. Photographs of the brow were taken by putting a remotely operated camera on the end of an angle iron, which was then rotated by regular increments about the shaft centre. For scheduling reasons, it was impossible to photograph the lip prior to the excavation of the lip pocket.

4.0 EXCAVATION DAMAGE

The pattern of excavation-induced fracturing and micro-fracturing on the floor, brow, lip and lip pocket of the 300 level are shown in Figure 3. The extent and type of damage varies markedly within each of the three locations cited, and from one location to the next.

4.1 Excavation Damage in the Brow

Of the three exposures studied, the brow displayed the least amount of damage. Examination in situ and later using the photomosaics revealed only a few minor fractures, which either radiated from remnant half barrels or cut partially across the base of small protuberances on the shaft wall. Unfortunately, these fractures appear to be only the remnants of a more extensive damage zone seen on the floor and described in Section 4.2. Evidence for this statement is provided by the gap (up to 80 mm) between the present shaft surface and rock bolt plates. The remainder of the damage zone was presumably lost during slashing of the station crown.

4.2 Excavation Damage on the 300 Station Floor

An annulus of damage, up to 300 mm wide, was mapped on the floor of the station. Within this zone, micro-fracturing (lengths < 50 mm) is dominant in the north and south quadrants, while mesoscopic fractures predominate in the remainder. The pattern for both scales of fracturing is circumferential, but fractures parallel to the primary flow foliation in the granodiorite dyke are preferred over other directions. This is most apparent on the east side of the shaft. The most noteworthy characteristics of the mesoscopic fractures are outlined below: Additional information is presented in Table I.

1. the absence of natural infillings and blasting residues;
2. the absence of any visible offset or granulation indicative of shear displacement;
3. the absence of any detectable seepage;
4. the termination of the fractures either in rock or against other fractures of the same set;
5. the extremely tight apertures of the fractures (all were considerably less than 0.5 mm).

4.3   Excavation Damage in the Lip and Lip Pocket

The greatest concentration of induced fractures and the greatest thickness of fractured rock was observed on the walls of the lip pocket beneath the floor of the 300 level. Subhorizontal fractures were confined to a discontinuous zone up to 1000 mm thick. The estimated maximum length of these fractures is 1500 mm. Significantly, this set was rare in the remainder of the lip, where the circular profile of the shaft had been retained. Infillings were absent and there was no evidence of offset. All fractures ended within the rock or curved into another of the same set. Unfortunately, the intersections between these fractures and the subvertical induced fractures (observed on the station floor) were hidden from view. This would have established the relative age of these two sets, and whether one was intra-block fracturing and, therefore, generated by a pattern of stress relief independent of the macroscopic scale.

4.4   Summary of Observations

Excavation of the 300 level station has provided two cross sections through the damage zone about the circular shaft: one on the floor of the level and one on a temporary brow. The former provided the most complete exposure with up to 300 mm of damaged rock preserved. While the annulus of damage itself has a circular outline and is concentric about the shaft, the internal structure of the zone is markedly asymmetric. Micro-fractures (trace lengths < 50 mm) are restricted to the north and south quadrants of the shaft while mesoscopic fractures (trace lengths between 290 and 1750 mm) are confined to the east and west. The pattern for both scales of fracturing is largely concentric (i.e., parallel to the shaft walls) and is locally enhanced by rock fabric.

In addition to the damage zone about the shaft, information was also obtained on the damage associated with the excavation of the station. Subhorizontal fractures are most prominent on the walls of the lip pocket. Their development is markedly reduced where the circular outline of the shaft has been retained.

5.0   DISCUSSION

The geological and geomechanical factors thought to contribute to, or control, the pattern of damage about the shaft and station are reviewed here. Much of the information necessary for a complete evaluation, such as local stress conditions, and survey information for the preliminary and final excavation surfaces, was not available at the time of writing. Accordingly, the interpretations presented are tentative.

5.1   Origin of the Damage Zone

The causative processes behind the development of the damage zone are not fully understood but likely incorporate instantaneous fracturing associated with detonation, followed by subsequent time-dependent propagation of existing and new surfaces by stress relief. No unequivocal criteria are known, or were discovered during the course of this investigation, to distinguish fractures formed by one mechanism or the other.

The stress magnitudes and orientations in the shaft extension have been reviewed by Martin [8] (this workshop). According to the author

1. the maximum principal stress direction is likely oriented 040° (to 055° for the region of the 300 level, pers. comm.);

2. stress magnitudes are not known with certainty at this time, but the maximum principal compressive stress could be in excess of 100 MPa. This is sufficient to cause shaft-wall failures;

3. time – dependent behaviour recorded by extensometers in the shaft is likely the result of micro-crack creation and/or propagation under the high ambient stresses.

With this information, and assuming elasticity and Coulomb failure criteria, the zones of micro-cracking in the north and south quadrants would appear to be extensional failure, generated perpendicular to the minimum compressive stress in the horizontal plane. This view is supported by the absence of any visible evidence of shear. Blasting shock and natural micro-fabrics (if present) may have contributed to the formation of these fractures but are likely of secondary importance. (The existence of micro-fabrics will be investigated at a later date as part of a general petrofabric study of the rock mass.)

The origin of the larger fractures in the east and west quadrants is more problematic in that they appear to be orthogonal to the maximum principal compressive stress in the horizontal plane. The absence of any evidence of shearing suggests that these are either extension or release fractures. They may have propagated during blasting, from the perimeter holes, along the planes of the flow banding in the dyke. Alternatively, they may be stress-relief fractures generated by complex patterns of stress distribution related to the room geometry. It is not possible at this time to critically evaluate these hypotheses.

Whatever the actual mechanism, or combination of mechanisms, it appears that at least part of the observed fracture population was initiated inside the rock. This is consistent with more extensive observations by Maury [9] and supports the view cited by Martin [8] (this workshop) that, "...with an elastic modulus changing with confining radial stress, the maximum tangential stress occurs inside the shaft wall.".

5.2 Comparison of Damage Extent with other URL Excavations

Excavation-induced fractures have been observed throughout AECL's Underground Research Laboratory (URL). In the rectangular access shaft, the zone of induced fractures extends up to 1500 mm behind the wall of the excavation. In the access tunnels, 240 m below surface, the extent of the damage zone varies, but is in the order of 500 mm. In both of these areas, the majority of induced fractures are average between 50 mm and 1500 mm in length.

In comparison to these earlier excavations, the 300 level shows a marked reduction in both the size of the individual fractures, and the dimensions of the total damage zone. This is attributed to the adoption of a circular shaft cross section, and to careful blast design. This reduction in damage may also be due, in part, to increasing isotropy of stresses with depth as indicated by recent and preliminary in-situ stress determinations (D. Martin, pers. comm.). Verification of this hypothesis will follow more detailed stress analysis. The higher frequency of fracturing in the 300 level lip pocket, compared to the rest of the lip, is attributed to the angular design of the pocket, to the lower level of excavation control exercised , and to the increase in the total area of excavation.

6.0 CONCLUSIONS AND RECOMENDATIONS

Coordination of the mapping with the excavation of the 300 level station has provided a rare opportunity to view, in section, the annulus of damage present around a circular shaft. The results indicate a marked reduction in the thickness of the damage zone compared to that seen previously in the rectangular shaft (300 mm versus 1500 mm), in spite of higher overall in situ stresses. These differences are attributed to improvements in excavation methods, conversion to a more stable circular cross-section, and to the increasing isotropy of stresses with increasing depth.

As a group, the excavation-induced fractures were distinguished from natural fractures by their lack of natural mineral infilling and mineralogical alteration, and by their symmetry with respect to blasting centres or room geometry. Two scales of induced fracturing were recognized: micro-fractures having trace lengths less than 50 mm, and mesoscopic-scale fractures having trace lengths between 290 mm and 1750 mm. The former appear to be parallel to the maximum principal compressive stress in the horizontal plane and are interpreted as extensional in origin. The mesoscopic fractures are perpendicular to this direction and are of uncertain origin.

The information obtained from this study will benefit subsequent geomechanical, hydrological and geophysical investigations directed at the assessment of the role of excavation damage on vault design and performance. It is recommended that this experiment be repeated under different combinations of stress conditions and shaft profiles, so that the effects of these parameters on the damage zone can be observed.

REFERENCES

[1] Simmons, G. : "Atomic Energy of Canada Limited's Underground Research
     Laboratory For Nuclear Waste Management". Proc. of "Geothermal Energy
     Development and Advanced Technology". International Symposium. Tohoku
     University, Sandai, Japan. 1986 Nov. 06 to 08.

[2] Lang, P.A. : Room 209 Excavation Response Test in the Underground Research
     Laboratory.

[3] Everitt, R.A. and A.Brown. : "Subsurface Geology of the Underground
     Research Laboratory: An Overview of Recent Developments". Proceedings of
     the 20th Information Meeting of the CNFWMP, Winnipeg, 1985. Atomic Energy
     of Canada Limited Technical Record*, TR-375.

[4] Brown, A., C.C. Davison and R.A. Everitt. : "Geoscience Research at
     Canada's Underground Research Laboratory", Trip 10 (Guidebook).
     Geological Association of Canada, Mineralogical Association of Canada
     Annual Meeting, 1986, Ottawa, Ontario.

[5] Everitt, R.A., A.E. Chapman, A.V. Grogan, D. Laderoute, A. Brown. :
     "Geological Characterization for the URL Shaft Extension". Atomic Energy
     of Canada Limited Technical Record*, in prep.

[6] Brown E.T. (ed.). : "Rock Characterization, Testing and Monitoring.", ISRM
     Suggested Methods. Pergamon Press (1978).

[7] Brown A. and R.A. Everitt. : "Geologic Mapping for URL- Procedures in :
     The Underground Research Laboratory Underground Construction Phase
     Experimental Program". Atomic Energy of Canada Limited Technical Record*,
     TR-225 (in prep.).

[8] Martin, C.D. : "Shaft Excavation Response In A Highly Stressed Rock Mass."
     Proc. Excavation Responses in Deep Radioactive Waste Repositories -
     Implications For Engineering Design And Safety Performance, OECD,
     Winnipeg, 1988, (this workshop).

[9] Maury, V. : "Observations, Researches and Recent Results About Failure
     Mechanisms Around Single Galleries.", Proc. of the International Congress
     on Rock Mechanics, Montreal, Canada. vol.2, pp. 1119-1128.

* Unrestricted unpublished report available from SDDO, Atomic Energy of Canada
   LimitedResearch Company, Chalk River, Ontario K0J 1J0

	MINIMUM	MAXIMUM	MEAN	STANDARD DEVIATION
LENGTH	290 mm	1750 mm	560 mm	330
APERTURE	?	<0.5 mm	N/A	N/A
ROUGHNESS	4 (JRC = 6-8)	5 (JRC = 8-10)	4.5 (JRC = 8)	0.5
PLANARITY	5*	5*	N/A	N/A
	* undulating and rough			
INFILLINGS	none			
SEEPAGE	none			

TABLE I

Summary of the characteristics of the mesoscopic scale excavation-
induced fractures seen on the floor of the 300 level station.
The classification for roughness and planarity
is that recommended by Brown [6].

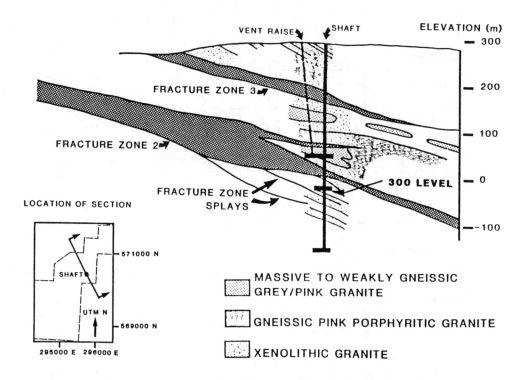

Figure 1 **GEOLOGICAL SETTING OF THE UNDERGROUND RESEARCH LABORATORY. THE LOCATION OF THE STUDY AREA (300 LEVEL) IS SHOWN**

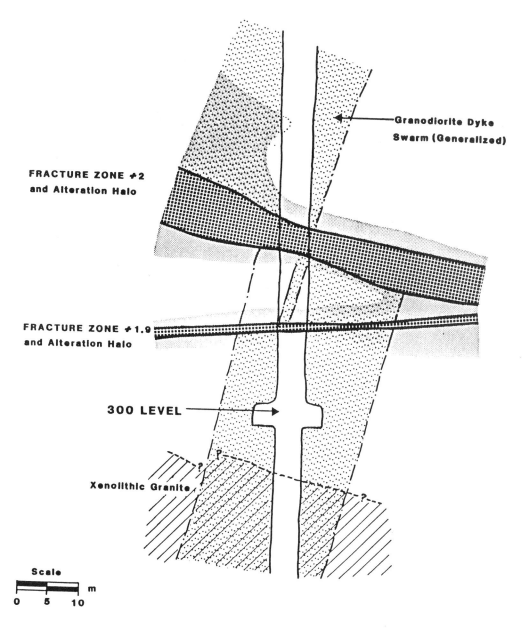

Granodiorite Dyke Swarm (Generalized)

FRACTURE ZONE #2 and Alteration Halo

FRACTURE ZONE #1.9 and Alteration Halo

300 LEVEL

Xenolithic Granite

Scale

m

0   5   10

Figure 2   NE-SW SECTION THROUGH SHAFT EXTENSION SHOWING THE GENERALIZED GEOLOGY IN THE AREA OF THE 300 LEVEL

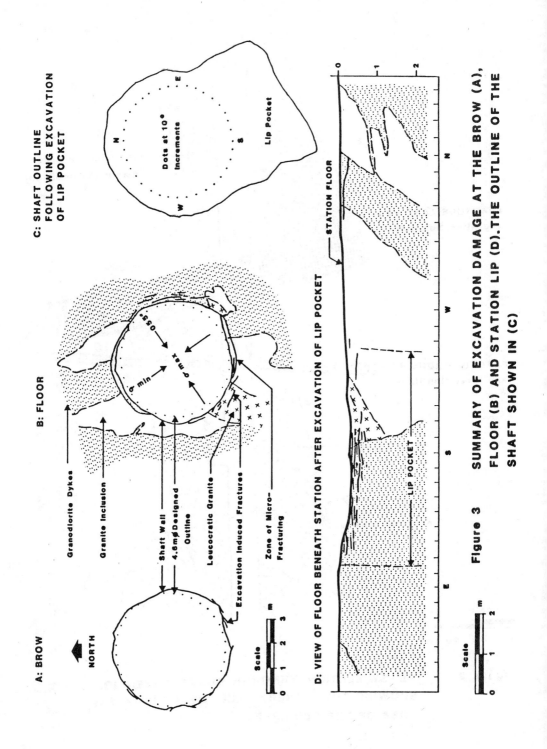

A: BROW

NORTH

B: FLOOR

C: SHAFT OUTLINE
FOLLOWING EXCAVATION
OF LIP POCKET

Dots at 10°
Increments

Lip Pocket

Granodiorite Dykes

Granite Inclusion

Shaft Wall
4.8m∅Designed
Outline

Leucocratic Granite

Excavation Induced Fractures

Zone of Micro-
Fracturing

σ min

σ max

σ₀₅₅

Scale

0 1 2 3 m

D: VIEW OF FLOOR BENEATH STATION AFTER EXCAVATION OF LIP POCKET

STATION FLOOR

LIP POCKET

Scale

0 1 2 m

Figure 3 SUMMARY OF EXCAVATION DAMAGE AT THE BROW (A),
FLOOR (B) AND STATION LIP (D).THE OUTLINE OF THE
SHAFT SHOWN IN (C)

THE ASSESSMENT OF EXCAVATION DISTURBANCE SURROUNDING
UNDERGROUND OPENINGS IN ROCK

R. Koopmans and R.W. Hughes
Ontario Hydro Research
Toronto, Ontario, Canada

ABSTRACT

A borehole dilatometer was used at several sites to assess the extent of excavation disturbance surrounding underground openings in rock. Results from field and laboratory programs indicate that the deformation moduli obtained were influenced by the in situ stress field. This suggests that the dilatometer is sensitive enough to profile deformation moduli as dictated by in situ rock mass characteristics.

EVALUATION DES PERTURBATIONS LIEES AUX TRAVAUX D'EXCAVATION AUTOUR
DES OUVERTURES SOUTERRAINES DANS UNE FORMATION ROCHEUSE

RESUME

Un dilatomètre a été utilisé dans les trous de forage de divers sites afin d'évaluer l'étendue des perturbations liées aux travaux d'excavation que l'on observe autour des ouvertures souterraines pratiquées dans les formations rocheuses. Il ressort des programmes sur le terrain et en laboratoire que les modules de déformation obtenus étaient influencés par le champ de contraintes in situ. Apparemment, le dilatomètre est suffisamment sensible pour tracer le profil des modules de déformation qui découlent des caractéristiques de la masse rocheuse in situ.

INTRODUCTION

The borehole dilatometer has been used to determine the extent of excavation disturbance surrounding underground openings in rock. In so doing, the effectiveness of various blasting techniques can be assessed. The zone of disturbance is determined by conducting borehole tests at relatively close intervals and profiling the results with depth from the excavation wall. Since in situ stresses directly affect the deformation modulus of the rock mass, increases and decreases in stress levels correspond to increases and decreases in moduli. In general, lower moduli are obtained in disturbed zones which have become stress relieved due to rock relaxation and/or the presence of rock discontinuities while higher values are obtained in relatively undisturbed zones which have not become stress relieved due to an absence of rock relaxation and/or rock discontinuities.

TESTING LOCATIONS

The assessment of excavation disturbance was conducted at two sites in Canada. Two programs were undertaken within the Underground Research Laboratory (URL) located in crystalline rock near Pinawa in south-eastern Manitoba, while another was conducted within the sandstone rock mass surrounding two parallel circular tunnels in the Donkin Mines near Sydney, Nova Scotia.

URL - Shaft Collar

At the URL site, four horizontal Ex (3.81 cm diameter) boreholes were tested in the shaft collar to depths of approximately 15 metres into the rock mass surrounding the shaft. The shaft was excavated to a size of 3 metres by 5 metres using drill and blast techniques, and testing was undertaken at a depth of 15 metres below the surface. The first five metres were tested at 0.25 metre intervals, while the next 10 metres were tested at one metre intervals.

Donkin Mines

Testing was conducted in the sandstone rock mass between two parallel 7.5 metre diameter circular tunnels. One tunnel was excavated using a tunnel boring machine while the second tunnel was mined using drill and blast techniques. An Ex (3.81 cm) borehole was drilled approximately 42 metres from one tunnel to the other. Tests were carried out from within each tunnel at 0.25 metre intervals to a depth of 5 metres and 0.5 metre intervals at depths in excess of 5 metres.

URL - Room 209

Testing was undertaken in room 209 at the 240 metre level in the Underground Research Laboratory in Manitoba. This room is oriented in a direction perpendicular to the maximum in situ stress direction. Four 10 metre long EWG holes (3.81 cm) were drilled at right angles to the tunnel axis at chainage 28 metres in room 209. These holes, which were oriented vertically up, vertically down and horizontally into each wall, permitted testing in the tangential stress direction. The EWG stress meter holes were oriented parallel to the axis of room 209. These holes were extended 1.5 metres to allow dilatometer testing in the radial stress direction. In the 10 metre long holes, tests were conducted at 0.25 metre intervals from 0 to 5 metres and at 0.5 metre intervals from 5.0 metres to 10.0 metres. In the 1.5 metre section at the bottom of the stress meter holes, three tests were undertaken at 0.25 metre intervals.

BOREHOLE DILATOMETER SYSTEM

The borehole dilatometer is shown schematically in Figure 1 and discussed in detail by Koopmans and Hughes [1]. The system is highlighted by an intensifier which is operated by a manual or electrically driven hydraulic pump to allow controlled pressure loading and unloading of the dilatometer. The intensifier has a 6:1 piston differential ratio between oil and water chambers respectively. This allows lower oil pressures to be administered from the hydraulic pump through flexible hoses to the intensifier where the pressure is increased six times to allow water inflation of the dilatometer.

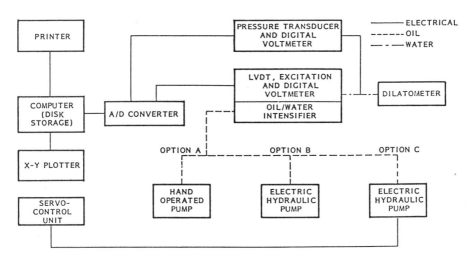

FIGURE 1
SCHEMATIC OF COMPUTERIZED DILATOMETER SYSTEM

A linear variance differential transformer (LVDT) has been fastened to the top of the intensifier and measures linear movement of the high pressure water piston during loading and unloading. Linear movement represents volume change and is determined by calibration. A pressure transducer has also been connected to the water side just outside of the intensifier to measure changes in pressure. Analog signals from both the

LVDT and pressure transducer are sent to an A/D converter for conversion to digital signals and then into a computer. The dilatometer itself is similar to the CSM cell [2] and consists of a high-strength, stainless steel heat-treated shaft and an adiprene membrane capable of applying pressures in excess of 70 MPa. Lead beads or other fillers may be placed in the membrane cavity to reduce water volume and thereby increase the system stiffness. High-pressure, thick-walled tubing is used to connect the dilatometer to the intensifier.

TESTING PROCEDURE

Prior to testing, the dilatometer is calibrated in both an aluminum and steel thick-walled cylinder. Each cylinder has an inner diameter of 3.81 cm, an outer diameter of 11.44 cm and an overall length of 30.48 cm. Five cycles are run for each calibration with the last four recorded. The first cycle is used to seat the dilatometer in the cylinder. After calibration, the dilatometer is placed in the borehole using placement rods. Once again five cycles are used with the last four recorded. Due to the very small volume changes which occur within the dilatometer, especially in high modulus rock, it has been determined that more accuracy is obtained by averaging the slopes of the four curves. Pressures in the order of 2 MPa are used to seat the dilatometer while pressures of approximately 32 MPa are used at the top of each cycle. Borehole testing progresses by moving the dilatometer from the collar to the bottom of the borehole. Upon completion of testing, the calibrations are taken once again to verify the original calibrations. With the aid of a computer, both pressure and volume readings are recorded every second and stored on disc.

Data Reduction

The tangent moduli were calculated for all test locations at all three projects. In addition to the tangent moduli, the secant moduli were also calculated for those results obtained in Room 209 in the URL. The tangent modulus was determined from the straight line portion of the curve, while the secant value was determined from the slope between the maximum applied pressure and the seating pressure (2 MPa). The slope values for all four runs were averaged for each test. A computer program was used to determine the deformation modulus. The complete derivations of equations for the calculation of deformation modulus of rock is provided by Hustrulid and Hustrulid [2]. The only change in these equations is that the calculations are now expressed as pressure/volume instead of pressure/turn. The intensifier moves in a linear direction and eliminates the need to read rotational turns inherent in the original CSM system.

DISCUSSION

Stress Effects on Deformation Modulus

Laboratory testing [3] has indicated that deformation moduli can be substantially affected by changes in levels of confining pressure for various rock types including sandstone, limestone and granite (Figure 2).

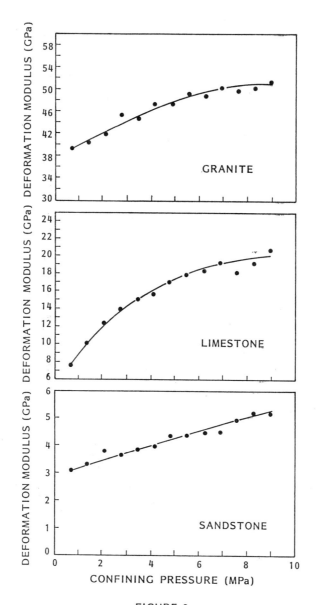

FIGURE 2

THE EFFECT OF VARYING STRESS LEVELS
ON DEFORMATION MODULI IN GRANITE, LIMESTONE
AND SANDSTONE ROCK BLOCKS

With increasing stress, there is a tendency for the deformation modulus to level out. This is most likely due to the closure of microfractures which are open when the rock block is in a relaxed state. Based on these results, one could expect lower in situ deformation moduli for rock masses which are influenced by low stress magnitudes and higher deformation moduli for higher stressed zones. The laboratory simulations on the rock blocks were also noted in the test results obtained from the three field programs.

Shaft Collar - URL

Figure 3 shows the borehole deformation moduli profiles for all four boreholes in plan view around the URL shaft collar [4]. In general, each borehole shows a tendency for the deformation modulus to increase with depth from the shaft wall and level out at a constant value between one and two shaft diameters. The cause for the undulating appearance of the profiles close to the shaft wall is probably due to variations resulting from blast-induced rock block movement. As indicated by Lang [5], the initial shaft sinking operation lacked proper control and inspection, which resulted in some questionable workmanship (ie, virtually 0% half barrels were achieved during shaft sinking). This may help to explain why a reasonably large zone of excavation disturbance was obtained during this testing program.

Average results for the dilatometer testing in all four boreholes was 54 to 63 GPa in the undisturbed zone of the rock mass including fractures and 63 to 67 GPa in the undisturbed rock mass excluding fractures. Cross-hole ultrasonics provided values in the order of 65 - 70 GPa for the rock mass surrounding the shaft [6].

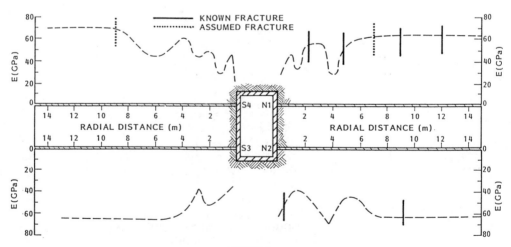

FIGURE 3
DEFORMATION MODULUS WITH DEPTH FROM SHAFT WALL

Average values for both biaxial cell testing [7] and uniaxial compression testing [9] were determined for samples obtained from borehole URL-1 which was located near the URL shaft. These results were 54 GPa and 73 GPa respectively. The comparison indicated that the dilatometer results were reasonably good especially when one considers that dynamic testing and compression testing of good quality core samples typically provide higher moduli than static in situ testing.

Donkin Mine - Nova Scotia

The tangent deformation moduli between tunnels 2 and 3 in the Donkin mines is shown in Figure 4 [4]. Results suggest that high moduli are encountered in the rock mass close to the tunnel boring machine (TBM) excavation, while lower moduli are obtained in the rock mass close to the drill and blast excavation. In the TBM excavation, reasonably constant values averaging 9.92 GPa are found at a depth greater than about 6 metres. In the drill and blast excavation, moduli gradually increase to a depth greater than about 6 metres after which they increase quickly before dropping off to 9.92 GPa. Since the moduli also leveled out at an average of 9.92 GPa, with depth from the TBM excavation, it is assumed that this value is constant between the two tunnels once approximately one tunnel diameter is exceeded.

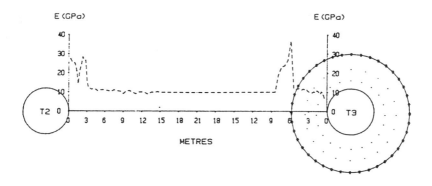

——— Extent of assumed excavation disturbance
T2 - Mined using tunnel boring machine
T3 - Mined using drill and blast
---- Deformation modulus

FIGURE 4

DEFORMATION MODULUS WITH DEPTH FROM
TUNNELS 2 AND 3 IN DONKIN MINES

Based on a limited number of USBM and CSIRO stress measurements taken in the drill and blast tunnel [9], it was noted that lower radial and tangential stresses are obtained to a depth of approximately one tunnel diameter (7 to 8 metres), after which the stresses increase rapidly and level out with increasing depth. Hammer seismic surveys conducted in both tunnels provide lower dynamic moduli (12 to 30 GPa) in the drill and blast tunnel and higher dynamic moduli (23 to 34 GPa) in the TBM tunnel.

Room 209 - URL

The profiles for the secant and tangent deformation moduli trends for each 10 m long borehole are shown in section view in Figures 5 and 6.

Borehole 209 - 029 - DIL 1 (horizontal south wall) shows that the extent of the excavation disturbance is about 0.5 metres. The average moduli calculated were 39.9 GPa for the secant and 63.0 GPa for the tangent. The most notable variance in the results occurs between 6.5 and 8.5 metres, where extremely high moduli were obtained. This zone appears to be so highly stressed that the dilatometer tests performed are not valid. This occurs when the rock mass has a higher stiffness than the testing equipment and results in equipment expansion rather than rock mass expansion.

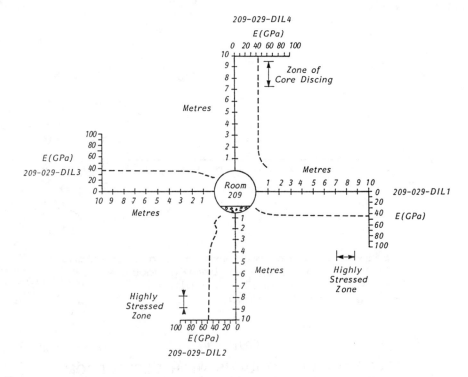

FIGURE 5

CROSS-SECTION OF ROOM 209 SHOWING SECANT
MODULUS TREND WITH DEPTH FROM TUNNEL WALL

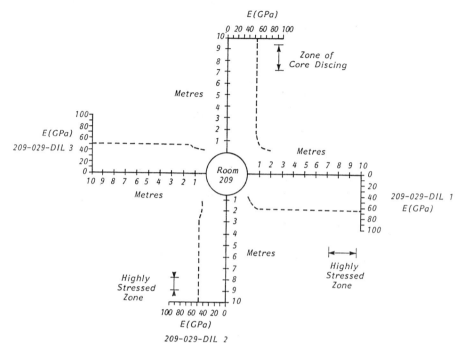

FIGURE 6

CROSS-SECTION OF ROOM 209 SHOWING TANGENT
MODULUS TREND WITH DEPTH FROM TUNNEL WALL

Borehole 209 - 029 - DIL 2 (vertical down) shows that the extent of the excavation disturbance is in the order of 3 metres in the rock mass. The average moduli calculated were 41.5 GPa for the secant and 53.7 GPa for the tangent. A highly stressed zone was noted at a depth from 7.5 to 9.0 metres. The greater depth of excavation disturbance experienced in this borehole is most likely due to the more heavily charged lifter holes which were used in the floor during excavation.

Borehole 209 - 029 - DIL 3 (horizontal north wall) shows that the extent of excavation disturbance is approximately one and a quarter metres. The average moduli calculated were 35.1 GPa for the secant and 49.4 GPa for the tangent. The results obtained in this borehole appear to be fairly consistent with depth with no unusual variances.

Borehole 209 - 029 - DIL 4 (vertical up) shows excavation disturbance to a depth of about one metre. Average moduli of 41.2 GPa for the secant and 55.5 GPa for the tangent were calculated. High moduli were found at the collar of the borehole. This would appear to result from the existence of high compressive stresses in the crown of the tunnel. These values drop off and level out at approximately one metre. At a depth between 6.5 metres and 9.5 metres, a zone of lower moduli was found. Core logs indicated the existence of core discing in this area [10]. The lower moduli obtained suggest the possibility that fractures exist in the rock mass and only simulate core discing after the core has been removed from the borehole.

The deformation moduli obtained in the stress meter holes, paralleling the tunnel axis are shown in Table 1. For comparative purposes, moduli at similar depths from the perpendicular dilatometer boreholes are shown in this table as well. It should be emphasized that accurate comparisons between the two directions is not possible since the exact testing locations are not known. Keeping this in mind, we can generalize and say that the tangent moduli in both directions are reasonably similar, with only minor variations noted. On the contrary, the secant moduli in the parallel direction are significantly lower than those obtained in the perpendicular direction. The reason for this is probably due to radial stress relaxation associated with the opening of fractures oriented in a direction parallel to the tunnel axis. Subsequent to the closing of these fractures, typical tangent moduli are obtained. The lowest secant and tangent moduli in the stress meter holes were obtained in the floor, thus further suggesting that the more heavily charged lifter holes were responsible for this disturbance.

CONCLUSION

The borehole dilatometer has been used to assess excavation disturbance in the rock mass surrounding underground openings. Results from the field and laboratory programs indicate that the deformation moduli obtained are influenced by the in situ stress field. This suggests that the borehole dilatometer is an effective tool in measuring rock mass characteristics and features. It also provides a good indication of the extent of excavation disturbance and a possible means for assessing various excavation techniques and procedures.

ACKNOWLEDGMENTS

The authors wish to thank Mr. Peter Lang, Mr. Paul Thompson, Mr. G.R. Snider and the Rock Mechanics staff at AECL for the excellent co-operation they gave us during our testing program in Room 209 at the URL.

TABLE **1**

TABLE **1**

COMPARISON OF MODULI PARALLEL AND
PERPENDICULAR TO THE TUNNEL AXIS

Location of Borehole	Direction to Tunnel Axis	Depth (m) of Test	Secant Modulus (GPa)	Tangent Modulus (GPa)
South Wall	Parallel	1.0	29.5	67.1
South Wall	Perpendicular	1.0	39.3	63.3
South Wall	Parallel	2.0	31.9	69.7
South Wall	Perpendicular	2.0	40.6	73.2
Floor	Parallel	1.0	21.7	51.7
Floor	Perpendicular	1.0	33.3	42.0
Floor	Parallel	2.0	22.3	45.8
Floor	Perpendicular	2.0	38.6	64.0
North Wall	Parallel	1.5	28.2	59.7
North Wall	Perpendicular	1.5	35.2	55.2
North Wall	Parallel	2.5	28.0	65.6
North Wall	Perpendicular	2.5	34.1	54.6
Roof	Parallel	1.0	24.7	62.5
Roof	Perpendicular	1.0	42.1	54.1
Roof	Parallel	2.0	26.9	53.7
Roof	Perpendicular	2.0	46.1	55.9

REFERENCES

1. Koopmans, R. and Hughes, R.W.   An Automated and Computerized Dilatometer System to Measure Deformation Modulus of Rock, Western States Mining Expo-84, Reno, Nevada, 1984.

2. Hustrulid, W. and Hustrulid, A.   The CSM Cell-A Borehole Device for Determining the Modulus of Ridigity of Rock, 15th Symposium on Rock Mechanics, South Dakota, 1975.

3. Koopmans, R. and Hughes, R.W.   The Effect of Stress on the Determination of Deformation Modulus, 27th US Symposium on Rock Mechanics, Tuscaloosa, Alabama, June, 1986.

4. Koopmans, R. and Hughes, R.W.   Determination of Near Field Excavation Disturbance in Crystalline Rock, 9th International Symposium on the Scientific Basis for Nuclear Waste Management, Stockholm, 1985.

5. Lang, P.A.   Overview of the URL Excavation Response and Excavation Damage Assessment Programs, Workshop on Excavation Response in Deep Radioactive Waste Repositories, Winnipeg, 1988.

6. Wong, J., Hurley, P.A. and West, G.F.   Small-Scale Cross-Borehole Seismology for Mapping of Dynamic Elastic Moduli and Detection of Cracks within Rock Masses.   Prepared for Atomic Energy of Canada, April 1984.

7. Koopmans, R.   Near Surface In-Situ Stress Measurements - Lac du Bonnet Batholith, Manitoba.   Ontario Hydro Research Division Report No. 82-73-H, 1982.

8. McKay, D.A.   Lac du Bonnet Geomechanical Test Results, Ontario Hydro Research Division Report No. 82-441-K, 1982.

9. Canmet Contract Report No. 265 W 23440-2-9159, Measurement and Analysis of Rock Deformation and Support System Response in the Drill and Blast Project (UP-G198).   Prepared by Golder Associates, Vol. 2, 1985.

10. AECL-Drill Core Logs for Boreholes 209-029-DIL 1, 209-029-DIL 2, 209-029-DIL 3 and 209-029-DIL 4.   Logged by D.M. Routliffe (unpublished data), 1987.

ROOM 209 EXCAVATION RESPONSE TEST
IN THE UNDERGROUND RESEARCH LABORATORY

P.A. Lang
Atomic Energy of Canada Limited
Pinawa, Manitoba, Canada   ROE 1L0

ABSTRACT

An in situ excavation response test was conducted at the Canadian Underground Research Laboratory (URL) in conjunction with excavation of a tunnel (Room 209) through a near-vertical water-bearing fracture oriented almost perpendicular to the tunnel axis. Encountering a fracture with such desirable characteristics provided a unique opportunity during construction of the URL to try out instrumentation and analytical methods for use in the Excavation Response Experiment (ERE) planned as one of the major URL experiments.

The test has produced a valuable data set for validating numerical models, both current and future.

Four modelling groups predicted the response that would be monitored by the instruments. The predictions of the mechanical response were generally good. However, the predictions of the permeability and hydraulic pressure changes in the fracture, and the water flows into the tunnel, were poor. It is concluded that we may not understand the mechanisms that occur in the fracture in response to excavation. Laboratory testing, and development of a contracting joint code, has been initiated to further investigate this phenomenon.

Preliminary results indicate that the excavation damaged zone in the walls and crown is less than 0.5 m thick and has relatively low permeability. The damaged zone in the floor is at least 1 m thick and has relatively high permeability. The damage in the floor could be reduced in future excavations by using controlled blasting methods similar to those used for the walls and crown.

The test provided a valuable instrumentation evaluation. Good and poor instruments were identified, and instrument types were selected for future experiments. Several modifications and improvements were made.

Back calculation of the rock mass properties and fracture characteristics is currently under way. Some preliminary results are presented.

The test has been a valuable learning experience for all involved. Future experiments in the URL will be more successful and more cost effective than they would have been without this test.

ETUDE EXPERIMENTALE DES EFFETS DES TRAVAUX D'EXCAVATION
DANS LA CHAMBRE 209 DU LABORATOIRE SOUTERRAIN DE RECHERCHE

RESUME

Une étude expérimentale des effets des travaux d'excavation in situ a
été réalisée au Laboratoire souterrain de recherche du Canada (URL) en liaison
avec l'excavation d'une galerie (chambre 209) traversant une fissure aquifère
subverticale quasiment perpendiculaire à l'axe de la galerie. La présence
d'une fissure possédant des caractéristiques aussi intéressantes a constitué
une occasion unique de tester, lors de la construction de l'URL,
l'instrumentation et les méthodes d'analyse qui sont utilisées dans
l'expérience sur les effets des travaux d'excavation (ERE) qui est l'une des
expériences les plus importantes prévues dans l'URL.

Cet essai a permis d'obtenir une série de données précieuses pour la
validation des modèles numériques, tant actuels que futurs.

Quatre groupes de modélisation ont formulé des prévisions quant aux
effets des travaux d'excavation qui seraient enregistrés par les instruments.
Les prévisions relatives à la réponse mécanique ont été généralement bonnes.
Toutefois, celles concernant les changements de perméabilité et de pression
hydraulique dans la fissure et la circulation de l'eau dans la galerie ont été
mauvaises. Les chercheurs ont conclu qu'on ne comprenait peut-être pas les
mécanismes qui interviennent dans la fissure sous l'effet des travaux
d'excavation. Pour approfondir l'étude de ce phénomène, des essais en
laboratoire ont été entrepris ainsi que l'élaboration d'un programme de calcul
relatif aux diaclases de contraction.

D'après les résultats préliminaires, la zone perturbée par les travaux
d'excavation dans les parois et la couronne n'atteint pas 0,5 mètre
d'épaisseur et est relativement peu perméable. Dans le mur, la zone perturbée
a au moins un mètre d'épaisseur et une perméabilité relativement élevée. Les
perturbations dans le mur pourraient être réduites dans les prochaines
excavations par l'emploi de méthodes contrôlées de travail aux explosifs
analogues à celles utilisées pour les parois et la couronne.

Cet essai permis une évaluation très utile de l'instrumentation. On a
pu faire la différence entre les instruments adaptés et ceux qui ne l'étaient
pas et certains ont été retenus pour les expériences futures. Plusieurs
modifications et améliorations ont été apportées à ces instruments.

On procède actuellement à une analyse rétrospective des propriétés de
la masse rocheuse et des caractéristiques de la fissure. Quelques résultats
préliminaires sont présentés.

Cet essai a été riche d'enseignements pour tous ceux qui y ont
participé. La réalisation de cet essai contribuera à rendre plus fructueuses
et plus efficaces par rapport à leur coût les expériences qui seront exécutées
à l'avenir dans l'URL.

1. INTRODUCTION

The Underground Research Laboratory (URL) [1] is being excavated in the Lac du Bonnet batholith. Unusually large volumes of unfractured rock, not commonly seen in hard rock mines or civil underground structures, are being revealed by the URL excavations and exploratory boreholes. If an unfractured rock mass is selected for a nuclear waste disposal vault, any leakage from the vault would have to pass either through the unfractured rock until it reached water-bearing fractures, or through the tunnel seals and excavation-damaged zones around the tunnels. In such rock masses, the effect of even apparently minor fractures that intersect the vault excavations, if connected to a fracture network, could be significant in controlling the leakage of radio-nuclides from a vault. These potential leakage paths are shown schematically in Figure 1. The behaviour of previously unfractured rock in the excavation-damaged zone and the behaviour of pre-existing fractures need to be understood. Numerical models that can simulate the behaviour in these zones are needed.

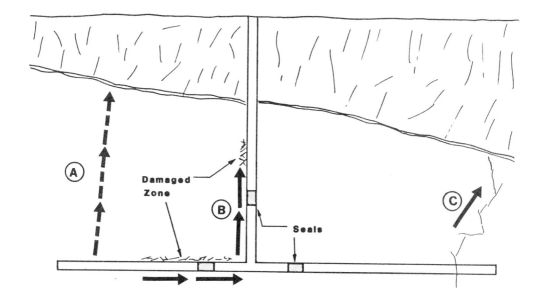

Figure 1. Potential leakage paths from a disposal vault constructed in unfractured or sparsely fractured rock. a) Leakage through intact rock, b) through and around tunnel and shaft seals, c) through interconnected fractures.

Several large-scale experiments will be conducted in the URL to study various aspects of isolating nuclear waste by burying it deep in stable rock masses in the Canadian Shield. One of these large-scale experiments is the Excavation Response Experiment (ERE), commonly called a mine-by test in other programs. The ERE will investigate the possible formation of a zone of increased permeability around the excavation (called the excavation-damaged zone) and the effects of excavation on the permeability of pre-existing fractures intersecting the excavations. Procedures will be developed for excavation response tests at a disposal vault. Excavation response measurements will be required at a disposal vault to obtain design parameters on the rock mass, to verify design assumptions, and to validate and calibrate numerical models specifically for the particular site.

Exploratory drilling from tunnels at the 240-m level in the URL (see Figure 2) located a near-vertical water-bearing fracture (the Room 209 Fracture) in what was otherwise a virtually unfractured rock mass. This combination of conditions offered a unique opportunity to determine the properties of the rock mass and fracture in situ, to evaluate the response of these to excavation of a nearby tunnel, and to evaluate excavation methodology, instrumentation and numerical modelling codes. In effect we had the opportunity to conduct a trial run for the ERE, and at the same time gain useful information in time for our Concept Assessment program. This test, called the Room 209 Excavation Response Test, began in 1985 August; Figure 2 shows the dates for the various phases of excavation, and Figure 3 the

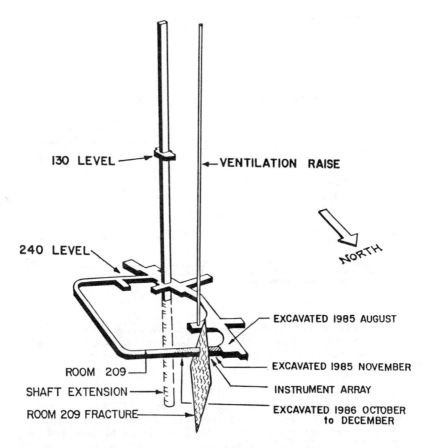

Figure 2. Isometric view of the URL showing the Room 209 Fracture, instrument array, and the excavation dates.

sequence of main activities. Analysis of results and post-excavation characterization are still under way.

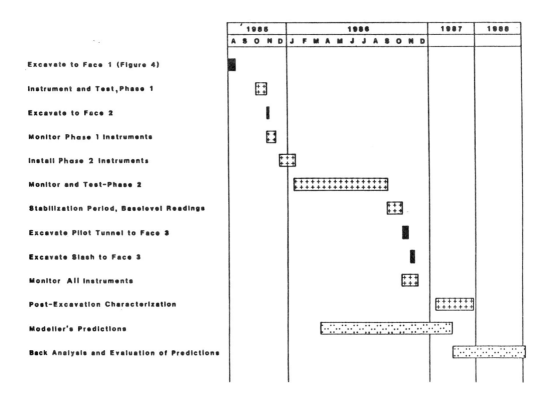

Figure 3. Schedule of the main excavation, instrumenting, field testing and monitoring activities.

2. TEST OBJECTIVES

The objectives of the test can be summarized as follows:

1. to conduct a trial run for the ERE; this would test instrumentation, data acquisition and management, excavation quality control, numerical modelling and general organizational requirements,

2. to test instruments for use in upcoming tests, e.g., the shaft extension (Figure 2) instrument arrays,

3. to determine the engineering properties of the rock mass on the scale of the rock mass; for use in designing and analyzing other experiments in the URL and as part of the ongoing program to develop a data base of the engineering properties expected to be required for designing and licensing a nuclear waste disposal vault in Canadian Shield rocks,

4. to characterize the fracture for future large-scale tests that will be done in it, and

5. to evaluate the excavation-induced damaged zone around the tunnel.

3. **TEST PLAN**

    The test plan was as follows:

1. Characterize the geological, hydrogeological, and in situ rock properties, stress and temperature conditions in the vicinity of the test.

2. Instrument the fracture and an adjacent volume of unfractured rock to monitor the response of the rock mass and fracture to excavation of the tunnel. The parameters selected for monitoring were stress changes, displacements and temperatures in the unfractured rock; water pressures and permeability changes in the fracture; and water flows into the tunnel from the fracture.

3. Attempt to predict using numerical models the responses that would be measured by the instruments. We planned to have four separate modelling groups, each using a different approach, make blind predictions of the responses. The information collected during characterizing and instrumenting would be provided to the modellers to calibrate their models.

4. Excavate the tunnel through the instrument array while monitoring the response. We planned to excavate a pilot tunnel first then enlarge (slash) it to full size in order to a) minimize the wall damage, and b) obtain two sets of response data.

5. Evaluate the numerical modelling codes by comparing the predicted and measured responses.

6. Evaluate our understanding of the effect of stress changes on fracture permeability, and of fundamental mechanisms such as fluid flow in fractures and fracture formation and propagation in the excavation-damaged zone.

7. Back-analyze the measured responses to determine the engineering properties of the rock mass.

8. Evaluate the performance of the various types of instruments used and the data acquisition system.

    During the test we recognized that additional field-testing was required to better characterize the properties of the fracture, to check that the stress field used for modelling applied throughout the volume of rock being modelled, and to determine the properties and extent of the damaged zone.

4. ARRAY LAYOUT AND SEQUENCE OF ACTIVITIES

4.1 Sequence of Activities

The sequence of the main excavation, drilling, instrumenting, field-testing and monitoring activities was as follows (Figure 3 and 4):

1. Excavate to Face 1 (Figure 4 and 5).

2. Drill boreholes N1, N2, S1, and S2 (Figure 5) beside the north and south walls and install CSIRO triaxial strain cells ahead of the fracture and straddle packers across the fracture. Drill the pilot borehole down the center of the tunnel alignment and install straddle packers. Determine the in situ stresses using the USBM gauge in three orthogonal boreholes.

3. Conduct single-hole permeability tests of the fracture in each packed off section.

4. Excavate from Face 1 to Face 2 and flatten the tunnel face. This 7.5-m distance was advanced by three 2.3-m-long pilot rounds, two slash rounds and a face flattening round (Figure 5). Measure the permeability of the fracture in boreholes N1, N2, S1, and S2 after each excavation step, and monitor the piezometric pressures in the fracture and strains at the strain cells.

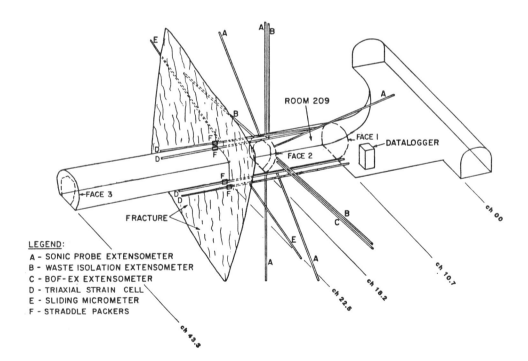

Figure 4. Isometric view of the Room 209 Instrument Array.

LEGEND:
A - SONIC PROBE EXTENSOMETER
B - WASTE ISOLATION EXTENSOMETER
C - BOF-EX EXTENSOMETER
D - TRIAXIAL STRAIN CELL
E - SLIDING MICROMETER
F - STRADDLE PACKERS

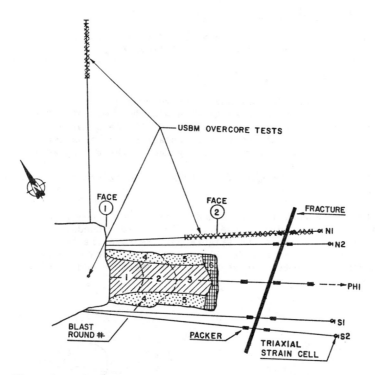

Figure 5. Plan view of the Room 209 Instrument Array showing the phase 1 boreholes and instrumentation, and the excavation sequence from Face 1 to Face 2.

5. Drill boreholes R1, R2, F1, and F2 (Figure 6b) in the roof and floor and install triaxial strain cells and straddle packers. Drill extensometer holes (EXT-1 to -9) and install extensometers and convergence pins. Drill borehole SLM-1A from Room 208 (Figure 6a) and install Sliding Micrometer casing. Drill a "far-field" monitoring borehole (OC1) and conduct overcoring tests with the Modified CSIR method [2] to determine the complete stress tensor in one borehole, and install straddle packers and monitoring system. Conduct cross-hole pressure interference tests between the nine straddle packer zones to characterize the permeability distribution in the fracture.

6. Excavate the pilot tunnel from Face 2 to Face 3, a distance of 25 m, then follow with the slash. Monitor all instruments throughout, and measure the permeability after each excavation step. Measure water inflows during drilling of the pilot and slash rounds through the fracture, and following excavation of the pilot and slash rounds through the fracture. Obtain a detailed survey profile of the face after each round. Conduct detailed geological mapping of each face, and stereophotography and mapping of the walls, of the pilot tunnel.

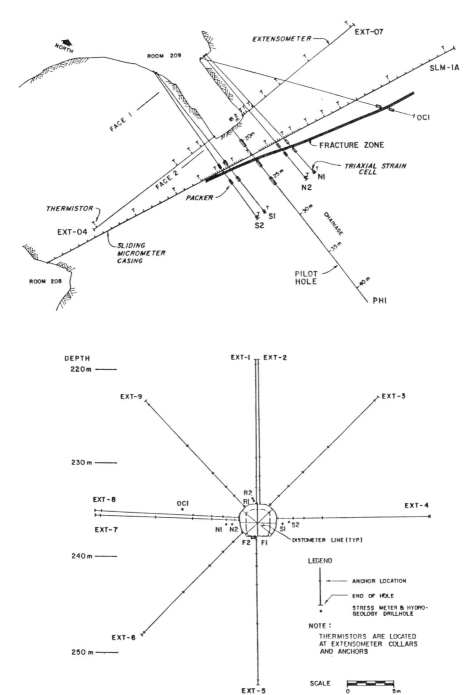

Figure 6. Plan and section through the instrument array showing the layout of the instruments.

7. Survey the final tunnel profile and conduct stereophotography and geological mapping of the final tunnel surfaces. Overcore the triaxial strain cells, conduct biaxial testing on the cores and calculate the stress changes from the strains recorded during excavation.

9. Conduct the following post-excavation characterization. Replace some of the straddle packers with Pac-ex instruments [3]; determine the normal stiffness of the fracture by varying the hydraulic pressure in the fracture while monitoring the normal displacements across it. Install weirs in the floor each side of the fracture to facilitate measurement of the rate of water inflow from the fracture. Conduct overcore testing each side of the fracture to determine the stress perturbation around the fracture. Conduct overcore testing at other locations to verify that the stresses measured before the test, and used by the modellers for their predictions, apply throughout the volume of rock modelled in the test. Conduct dilatometer tests to determine the modulus variation with increasing radial distance from the walls of the tunnel.

10. Present the modeller's prediction reports, and the experimental data report (1987 June) [8],[10].

11. Compare the measured and predicted responses and evaluate the models. Back-analyze the measured responses to determine the large-scale properties and material behaviour of the rock mass.

12. Draw conclusions from the test and make recommendations for the ERE, for other experiments in the URL, and for the Canadian Nuclear Fuel Waste Disposal Concept Assessment program. Report these in 1989 January.

4.2 Instrumentation

The following instruments were installed in the array (Figures 4, 5 and 6):

– Eight CSIRO Hollow Inclusion triaxial strain cells to monitor the strain tensor as the tunnel was excavated past. These were installed 9 m, just more than two tunnel diameters, ahead of the tunnel face and approximately 2 m and 1 m outside the planned final excavation lines. After excavation was complete, the cells were overcored. The strains recorded during overcoring were converted to stress changes using the Young's modulus and Poisson's ratio determined from biaxial tests on the overcored rock cylinder and cell.

– Eight convergence points to provide four diametral lines of convergence measurements. The convergence pins were installed 0.4 m behind the tunnel face; i.e., the same as the extensometers. Convergence was measured with an ISETH Distometer.

– Nine multiple position borehole extensometers. Three different types were installed for comparison: the Irad Gage Sonic Probe rod-type extensometer, the Terrametrics Waste Isolation tensioned-rod-type extensometer with hydraulic anchors, and the Roctest Bof-ex extensometer. All were installed in 15-m-long boreholes collared 0.4 m behind the tunnel face and oriented radially to the tunnel center line. The Sonic Probe had five anchors, the Terrametrics had four and the Bof-ex had seven.

- One ISETH Sliding Micrometer [4] in a borehole drilled from a parallel tunnel. The borehole intersected the Room 209 center-line 3.5 m ahead of the tunnel face. The rock displacements were measured as the tunnel face approached the casing and readings were continued after the tunnel had intersected it.

- Thermistors at each extensometer anchor, each triaxial strain cell, and at the walls, crown and floor of the tunnel at the extensometer array. The thermistors at the instruments were to enable corrections to be made for thermal strains in the instruments, and to monitor temperature changes in the rock mass. The thermistors in the crown, floor and walls were to monitor air temperatures so we could determine the thermal diffusivity and thermal coefficient of expansion of the rock mass.

- AECL-designed pneumatic straddle packers straddling the fracture in ten boreholes: the eight triaxial strain cell holes, the probe hole down the center of the tunnel, and OC1. Geocon vibrating wire piezometers connected to a Geocon data acquisition system monitored piezometric pressures in the fracture zone. Tubes to the packed-off zone enabled us to conduct permeability tests within the fracture and collect water samples [5].

5. INFORMATION FOR MODELLERS

The modellers were provided with information on the geometry of the existing excavations, the planned excavation sequence, the planned geometry of each blast round, and on the geology, rock properties, in situ stresses, rock temperatures, instrumentation layout, and fracture characteristics [5]. As well they were provided with reports by Gale [6] on results of laboratory testing to determine the normal stress permeability relationships of URL fractures, and by Barton and Bakhtar [7] on the characterization and modelling of rock joints with the Barton-Bandis Model. Selected information is provided below.

5.1 Geology

The URL is being excavated in granite. Room 209 is located in a wedge of rock between two sheared zones (called Fracture Zone 2 and Fracture Zone 2.5, see Figure 7). The near vertical Room 209 Fracture Zone, striking almost perpendicular to the tunnel direction, intersects the tunnel line at chainage 22.5 m from the start of the room. In the vicinity of the tunnel the Room 209 Fracture Zone reaches a maximum thickness of 0.4 m and consists of from one to six, discontinuous, en echelon fractures. It dies out to the south of Room 209 and just below the floor (Figure 8). To the north, it splays into more fractures and becomes a thicker and more permeable zone. It is hydraulically connected to the overlying sheared zone but not to the underlying sheared zone, or at least not in the vicinity of Room 209. Apart from this isolated fracture zone the rock mass in the Room 209 area is, essentially, unfractured.

Several phases of the granite are recognized (medium-grained host rock, intruded by granodiorite dykes, pegmatite dykes and leucocratic segregations). However, these do not appear to have significantly different mechanical properties, so no attempt was made to incorporate the different rock types into the models.

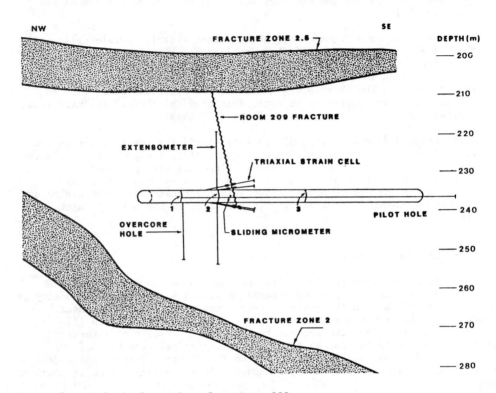

Figure 7. Geological section along Room 209.

5.2   Rock Properties, In Situ Stresses and Temperatures

The intact rock properties, from laboratory testing on 45-mm diameter drill-cores, are listed in Table I.

TABLE I

INTACT ROCK PROPERTIES

- Unconfined compressive strength   182 ±10 MPa
- Tensile strength (Brazilian)   9.1 ±0.4 MPa
- Young's modulus   69.1 ±1.7 GPa
- Poisson's ratio   0.24 ±0.02
- Coef. linear thermal expansion at 25°C: (2.5 ±0.7) x $10^{-6}(°C)^{-1}$

These results are presented as the mean ± the standard deviation.

The Young's modulus is confining pressure dependent and the rock is anisotropic. The modelling groups were provided with plots of Young's modulus versus stress level, with information on the anisotropy, and with values of rock mass modulus determined from back calculation of rock mass response measured at instrument arrays in the shaft. Each group selected a different value of Young's modulus for use in their model. All groups assumed isotropic conditions since the available time and funding did not allow the greater complexity of anisotropic conditions to be modelled.

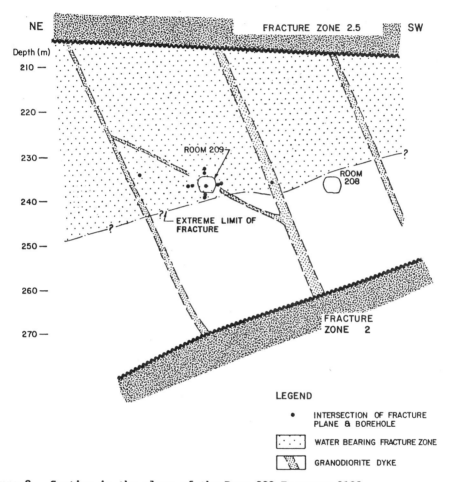

Figure 8. Section in the plane of the Room 209 Fracture [10].

The in situ stresses were measured in two independent programs: one using the USBM overcoring method in three orthogonal boreholes (22 tests) and one using the AECL modified CSIR overcoring method [2] in one borehole (five tests). The two methods produced comparable results. The average stresses from each program are listed in Table II, and are shown on Figure 9 together with the orientation of the Room 209 Fracture Zone, Fracture Zone 2 and tunnel

direction. Note the correlation between the orientation of the principal
stresses and the major structural features.

TABLE II

IN SITU STRESSES

n	Sigma 1			Sigma 2			Sigma 3		
	MPa	Plunge	Trend	MPa	Plunge	Trend	MPa	Plunge	Trend
CSIR	5	26.5	16	231	16.4	40	127	9.1	46	338
USBM	22	30.0	15	213	17.1	31	113	12.9	54	325

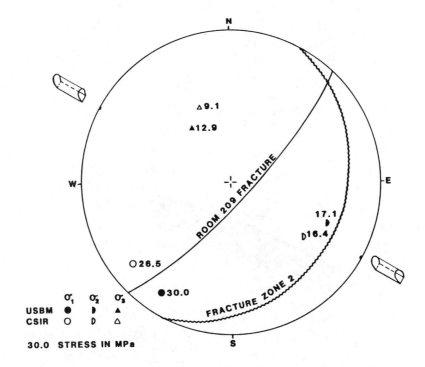

Figure 9. Lower hemisphere stereo net showing the orientation and magnitudes
of the principal stresses in the Room 209 area, and the orientation of the
Room 209 Fracture and Fracture Zone.

The ambient rock temperature at the 240 level is 8.2°C. The average air
temperature in the tunnels is approximately 14°C, so a temperature gradient
has developed in the rock around the tunnels. Therefore, the temperatures in
the rock near the tunnel depend on the time since it was excavated. By the
time excavation began from Face 2 towards the instrument array, the gradient

of rising temperatures had penetrated to beyond the end of the extensometers, i.e., 15 m from the tunnel walls. However, no measurable temperature increase was recorded at the triaxial strain cells 9 m ahead of the tunnel face. The temperature contours around the tunnel at the extensometer array were provided to the modellers. The temperature changes at the fracture were not measured but could be interpolated if required.

5.3 Fracture Characteristics

The modellers were provided with the following information on the Room 209 Fracture for calibrating their models:

- the geological extent and boundary conditions as described above,

- the piezometric pressure response and permeability changes in the fracture measured in the north wall boreholes as the tunnel face was advanced 7.5 m from Face 1 to Face 2 (Figure 10),

- the results of cross-hole hydraulic interference tests (Figure 11),

- the Barton-Bandis rock joint index properties determined on 45-mm, 63-mm, and 200-mm diameter drill cores collected from various locations around the URL, and

- the reports by Barton and Bakhtar [7] and Gale [6].

5.4 Excavation Method and Sequence

Excavation was by drill and blast method using a pilot tunnel and slash sequence. The sequence for the section of tunnel from Face 1 to Face 2 is shown on Figure 10. For the section from Face 2 to Face 3 (Figure 4) the pilot tunnel was excavated the full 25 m and then the slashing followed. All pilot and slash rounds were 2.8 m in length except the first two pilot rounds. These were shorter, 1.5 m, to provide additional data points on the displacements measured by the extensometers and convergence instruments. These extra data points are valuable for defining the rapid changes that occur very close to the tunnel face.

The tunnel face was conical after each blast. The geometry and dimensions of the rounds were specified to the excavation crews and provided to the modellers. In general, the excavation crews were able to achieve an excavation profile very close to the designed profile. The deviation of the actual and planned excavation profile was generally less than 300 mm. However, a few instances of poor drill alignment, survey errors, or incorrect blast implementation occurred. The designed blast round geometry, drill hole pattern and firing sequence are shown on Figure 12.

The pilot and slash method was used primarily to reduce the damage to the walls and crown due to blasting; it had the added advantage of providing two sets of response data. No attempt was made to minimize damage in the floor. The blast holes in the floor (called lifters) were fully charged with tamped Forcite 75 explosive (all explosives were supplied by Canadian Industries Limited (CIL)). The final perimeter of the walls and crown were charged with Primaflex and the outer row of pilot tunnel holes were loaded with Xactex. The rounds were delayed with up to 18 delays of long period

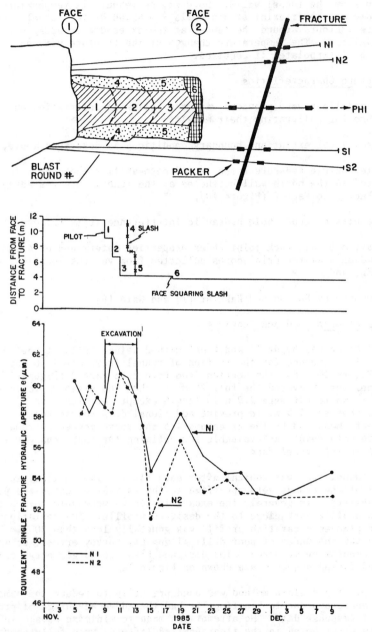

Figure 10. Change in effective hydraulic aperture of the fracture, measured in boreholes N1 and N2, as the tunnel face was advanced from Face 1 to Face 2; i.e., from 12 m before to 4 m before the fracture (Figure 5). Note the decrease in permeability of the fracture zone as the tunnel approached it [5].

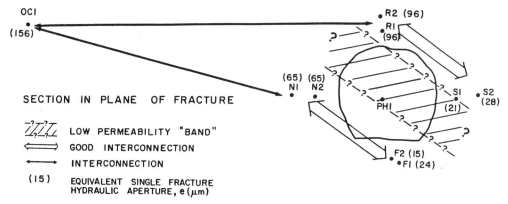

SECTION IN PLANE OF FRACTURE

LOW PERMEABILITY "BAND"

GOOD INTERCONNECTION

INTERCONNECTION

(15)   EQUIVALENT SINGLE FRACTURE
       HYDRAULIC APERTURE, e (μm)

Figure 11.   Interborehole pressure interference test observations and permeability distribution in the plane of the fracture [5].

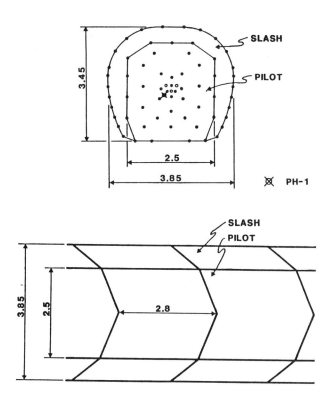

Figure 12.   Designed blast round geometry, blast hole pattern and firing sequence.

detonators. All blast holes were 32-mm diameter; three 64-mm diameter relief ("burn cut") holes were uncharged.

The excavation rate was controlled to allow time for the hydraulic response in the fracture zone to stabilize after each blast. The first four pilot tunnel rounds and first three slash rounds, i.e., those up to and penetrating the fracture zone, were taken at a rate of one round every two days. All the other rounds were taken at the rate of one round per day.

6. MODELLING

The modellers were initially given 12 months to conduct their predictions. They were later given a three-month extension. Characterization information was being provided to them during the first 9 months and clarifications continued for another month.

Table III summarizes the pertinent information about the modelling groups and the types of codes being used.

The prediction reports were submitted by the AECL, University of Alberta, and Lawrence Berkley Laboratories in 1987 June. These reports are included in [8]. The Lawrence Livermore National Laboratories Report is not yet complete. Selected predictions are compared with measured response data in Section 7.2.

7. EXPERIMENTAL RESULTS

7.1 Instrument Performance

Details of the instrument performance are provided in [9],[10] and [3]. The main points are summarized below.

CSIRO Triaxial Strain Cells

All eight strain cells survived and all recorded responses that intuitively look correct. The strains were converted to stress changes using the Young's modulus and Poisson's ratio determined from the biaxial tests on the core. The strain versus time plots show that the glue was drifting (possibly due to moisture absorption [11]). These plots, and the Bof-ex extensometer displacement plots, indicate that the rock was creeping as well (see Figure 16; notice creep on Bof-ex plot immediately after blasts on October 21, 24, and 27). Since we could not separate the rock creep from the glue drift we removed both effects when converting the strains to stress changes. We did this by using only the strains immediately before and after a blast for calculating the stress changes; the strains between blasts were ignored [10]. Figure 13 [12] shows the measured and predicted stress changes for S1, the south wall hole nearest the tunnel. Note the good comparison of measured and predicted orientations on the stereonets, and the comparable trends for the magnitudes. All cells measured a larger than predicted reduction in stress parallel to the tunnel. The discrepancies could be due to glue drift, to the "saw tooth" longitudinal tunnel profile that is caused by drill and blast excavation methods, to assumptions used in the models, or to the constitutive material models used. We consider these discrepancies to be

TABLE III
MODELLING GROUPS AND THEIR APPROACH

	Atomic Energy of Can. Ltd. Applied Geoscience Branch Computation & Analysis Section	University of Alberta Department of Civil Engineering	Lawrence Berkley Laboratories, Earth Sciences Division	Lawrence Livermore National Laboratories
Modelling Group				
Principal Investigator	T. Chan	P. Kaiser	R. E. Goodman	A. Wijesinghe
Name of Numerical Code	ABAQUS (stress & displacements) MOTIF (fracture flow & pressure)	SAFE (stress & diplac.) ADINAT (fracture flow & pressure)		GENESIS
Type of Code and Mesh	3-D, Finite Element, Full Mesh continuum (i.e., no joint element). Actual in situ stress & fracture geometry incorporated. Stress changes projected on fracture plane & permeability changes calculated using empirical data.	3-D, Finite Element Half Mesh, with joint element. In situ stress & fracture geometry approximated	Discontinuous deformation analysis. 2-D with face advance coefficients to adjust for 3-D effects of tunnel face	3-D, Coupled finite element and boundary element
Fracture Hydrogeology Calculations	Uncoupled fluid flow analysis using planar elements in the plane of the fracture. Fracture orientation correctly modelled.	Planar element using uncoupled fluid flow (flow net in plane of fracture). Fracture approximated as normal to tunnel	Fluid flow calculations based on flow net in fracture plane	Fully coupled joint element

rather minor and the overall agreement between the predicted and measured responses to be very good considering the limitations of an exercise of this type. Figure 14 [10] summarizes the stress changes recorded by the eight cells, transformed into the plane of the section.

The CSIRO triaxial strain cell is the only instrument currently available that can monitor the complete stress change tensor in one instrument and one borehole. However, a better method of correcting for glue drift is required. We are investigating two methods of correction but have applied neither to the data presented here.

Because these instruments utilize an arrangement of quarter bridge strain gauges sharing common ground wires and a common third wire for temperature compensation in the lead wires, they are extremely sensitive to moisture leakage in the cables, voltage variations in the data loggers, or any sort of ground loops. Thus good wiring and data logging are essential to obtain stable readings.

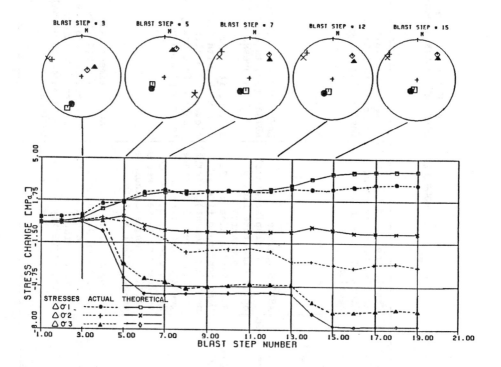

Figure 13. Comparison of measured and predicted principal components of stress change versus blast step number at the triaxial strain cell in borehole S1. The stereo nets show the orientation of the principal components of stress change and the plot shows the magnitudes of change. Blasts 1 to 10 were pilot rounds and 11 to 19 were slash rounds [12].

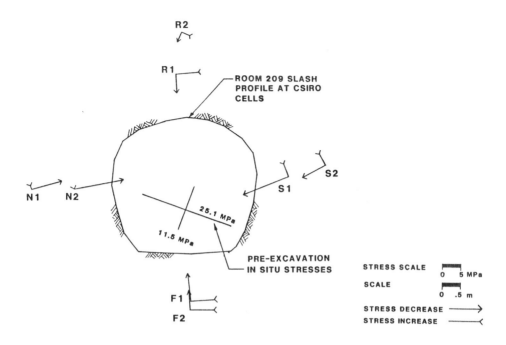

Figure 14. Principal components of stress change in the plane of the section [10].

Convergence Measurements

The ISETH Distometer was reliable and produced far superior repeatability than we had achieved previously with tape extensometers. The Distometer uses Invar wire and a better anchoring system than the tape extensometers. Figure 15 [10] shows the Room 209 convergence measurements. The repeatability of these readings is ±0.04 mm. Well trained and diligent personnel are required to obtain such repeatable readings.

Extensometers

The Bof-ex extensometer worked exceptionally well, giving a precision of ±0.001 mm between adjacent anchors. The Sonic Probe extensometers suffered from anchor slip in some anchors and in general was approximately 50 times less precise than the Bof-ex. The difference in the quality of the Sonic Probe and Bof-ex data can be seen on Figure 16. The Terrametrics Waste Isolation extensometers were difficult to install, were limited to only four anchors in a 75-mm-diameter borehole, had to have the collar anchor recessed 200 mm, and could not accommodate any anchors between this collar anchor and 1.7-m depth; thus, displacements in the zone of most interest could not be investigated in detail. For these reasons, and due to a poor choice of transducer, the Terrametrics extensometers produced virtually no usable results in this test.

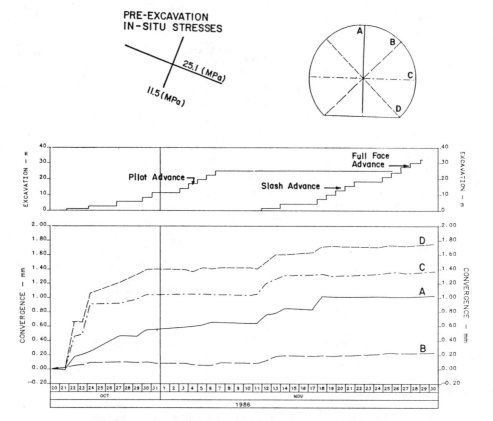

PRE-EXCAVATION
IN-SITU STRESSES

25.1 (MPa)

11.5 (MPa)

Figure 15. Convergence measurements during tunnel advance from Face 2 to Face 3. The pre-excavation in situ stresses and the orientation of the convergence lines are shown at the top of the figure, and the excavation advance is shown in the center [10].

The Sliding Micrometer gave a repeatability of ±0.03 mm. It suffered from having material collecting on the spherical measuring seats in the casing. Readings were very time consuming; a 50-m-long borehole can take several hours when calibration of the probe and flushing the borehole to clean the seats are included. Data entry and assimilation is also time consuming. This instrument has the advantage of enabling a complete record of the strains on both sides of the tunnel to be obtained as the tunnel advances towards and then passes through the instrument. This enabled us to capture the total radial displacements and compare them with the partial displacements recorded by the extensometers. Unfortunately some anchors were damaged when the tunnel was blasted through the instrument, so the record is not complete.

Based on the results of this test we decided to use Bof-ex extensometers exclusively in the shaft extension.

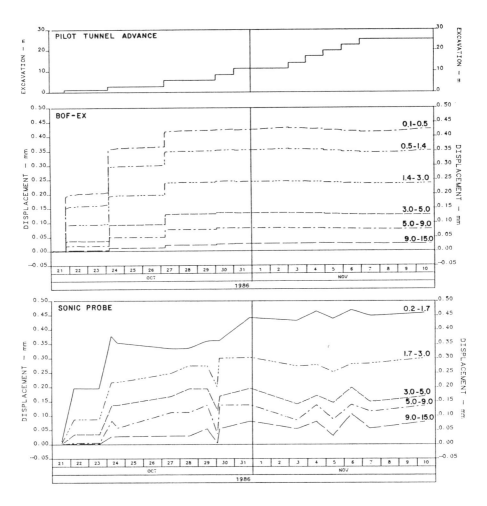

Figure 16. Comparison of results from the Bof Ex and Sonic Probe extensometers [10].

Thermistors

About 50% of the thermistors failed. This was due to a potting problem that has since been rectified. The thermistors meet the manufacturers specification of giving readings within 0.2°C of each other. They appear to have remained stable for over two years and give good resolution. They are inexpensive and we have found them satisfactory and cost effective for ambient temperature measurements.

Hydrogeological Monitoring System

This system consisted of the straddle packers, vibrating wire pressure transducers, and data logger. The system evolved during the test. The original packer systems were not stiff enough to enable satisfactory measurements of the hydrogeologic characteristics of the fracture zone. This system was replaced with a stiffer packer system. The system in place during excavation from Face 2 to Face 3 worked well, although one packer developed a leak and the data from this hole was lost. Blast damage in the floor extended down to the boreholes beneath the floor and allowed the pressure to drain from these holes, preventing further permeability testing [5],[10].

Water flows from the fracture were measured intermittently by collecting drips from the crown in a graduated cylinder and by estimating the seepages on the walls. No estimate could be made in the floor because it was covered with muck. A better system of measuring the water flows into the tunnel from the fracture was needed.

7.2 Comparison of Measured and Predicted Response

This section provides some preliminary comments on comparison of the measured and predicted responses. The comparisons are currently under way and are due to be completed by 1989 January.

Stress Changes

An example of measured and predicted stress changes at one of the triaxial strain cells is shown in Figure 13 and discussed in Section 7.1.

Displacements

The predicted (using the code ABAQUS [13]) and measured displacements along the Bof-ex extensometer are plotted against blast step number in Figure 17 [12]. This plot is of the displacements between adjacent anchors. The measured displacement between the first two anchors (at 0.1-m and 0.5-m depth) is four times larger (at 0.08 mm) than the predicted displacement (0.02 mm); this is probably due to blast damage. Apart from this case, the predicted and measured responses are all within 0.03 mm, and within a factor of two, of each other. The Young's modulus of the rock mass, back calculated from these displacements, is plotted against radial distance from the tunnel wall in Figure 18 [12]. The decreasing modulus towards the tunnel is considered to be mainly due to the confining pressure dependence of the modulus rather than to excavation damage, except within the first 0.5 m from the tunnel wall. The far-field modulus of 65 GPa is close to the modulus obtained on laboratory-sized samples of intact rock at similar confining stress.

Permeability Changes

Figure 19 [10] shows the single fracture hydraulic aperture measured in the nine monitoring holes as first the pilot tunnel and then the slash were advanced from Face 2 to Face 3. Note the decrease and recovery in hydraulic aperture in the roof and north wall holes as the face of the pilot tunnel advanced past the fracture. A similar response was measured as the slash went past except that the aperture in the roof did not recover. The "far-field" monitoring hole (OC1) at 12.8 m from the excavation showed no change. The two

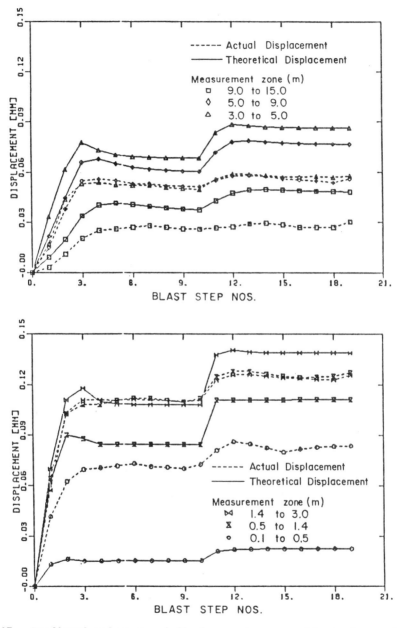

Figure 17. Predicted and measured displacements versus blast step number in Ext 7 (the Bof ex). Blasts 1 to 10 were pilot rounds and 11 to 19 were slash rounds. The plot is of the displacements between pairs of adjacent anchors. The anchor depths in the borehole are shown in the legend [12].

boreholes below the floor lost pressure when the excavation passed over the
packed-off zone, due to blast-induced fractures intersecting the borehole. A
packer in one of the south wall holes developed a leak. The other south wall
hole shows trends similar to the north wall holes. None of the permeability
changes predicted by the modelling groups were close to the measured changes.
Joint elements are to be added to the AECL model, a hydraulic component is to
be added to Goodman's model, and Kaiser is developing a contracting joint
element, along the lines of Sun et al's work [14], in an attempt to better
simulate the measured response.

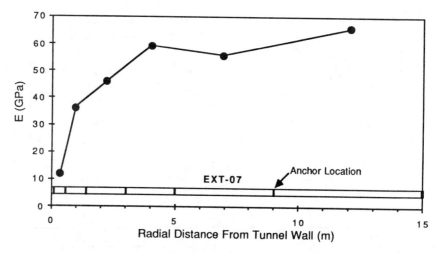

Figure 18. Rock mass Young's modulus versus radial distance from the tunnel
wall. The modulii were back calculated from displacements measured betweeen
adjacent anchor pairs in extensometer #7. The anchor locations are shown
[12].

Piezometric Pressure Response

Figure 20 shows the measured and predicted (from MOTIF [15]) pressure
decreases in the fracture. The pressure decrease is measured relative to the
pressure that was in the fracture before excavation through it. The pressure
change caused by excavating the pilot tunnel and by the slash are shown.
Figure 20 indicates that the model grossly over-estimated the pressure
decrease except in the south- and the floor- boreholes. The reason for the
large drop in pressure in the floor boreholes was explained above; i.e.,
blast-induced fractures intersected the boreholes. The closer agreement
between the predictions and measured response in the south wall holes is
fortuitous. Geological and hydrogeological evidence indicates that the
fracture terminates a short distance to the south of the tunnel. This
boundary condition was not modelled but it has caused a greater actual
drawdown than would otherwise have been observed. Thus the predicted response
is closer to the measured response at this location.

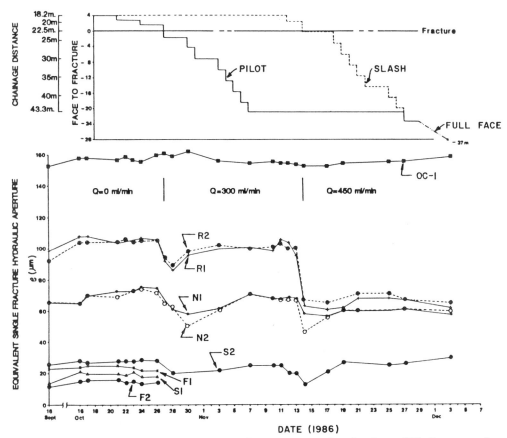

Figure 19. Measured changes in hydraulic aperture in the Room 209 Fracture in boreholes N1, N2, R1, R2, S1, S2, F1, F2, and OC1 as the tunnel face was advanced from 4 m before the fracture to 21 m past the fracture. The top part of the figure shows the excavation advance on the same time scale. The "Q" values indicate the rate of flow from the fracture into the tunnel [10].

Rate of Water Inflow

Water inflows (seepages) were measured from the blast holes that intersected the fracture (i.e., during drilling of the round that would advance the pilot tunnel through the fracture). They were measured from the roof and estimated from the walls after first the pilot tunnel and then the slash had penetrated the fracture; no estimate of inflows in the floor was possible. The measured and estimated flow rates were 1000 mL/min, 350 mL/min, and 450 mL/min from the blast holes, pilot tunnel and slash respectively. The predicted (from MOTIF) flow rates from the pilot tunnel and slash were, respectively, 2077 mL/min and 3565 mL/min [12]. No prediction was made of flows from the blast holes. The decrease in inflow rate from the fracture after the tunnel had advanced through it, compared to the flow rate from the blast holes, supports the other evidence that a decrease in hydraulic aperture

- 321 -

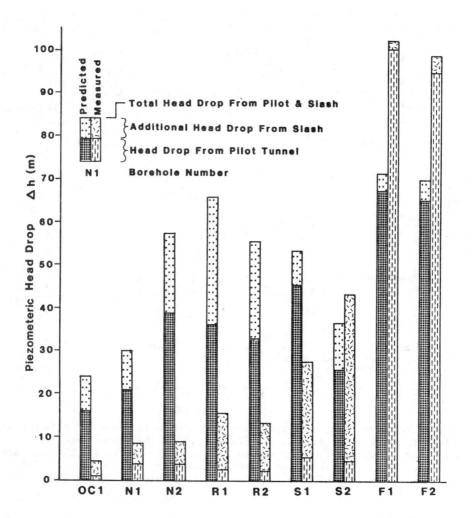

Figure 20. Comparison of predicted (using the MOTIF code) and measured piezometric pressure decreases in the fracture for two stages; after the pilot tunnel, and after the slash [10] and [12].

occurred in the near field of the excavation as the excavation advanced through the fracture.

Summary

The models predicted the stress changes and displacements in the intact rock fairly well but failed to predict, with any consistency, the permeability changes in the fracture. The MOTIF code generally over-predicted the piezometric pressure response by about a factor of five, and the water inflows by a factor of seven (part of the discrepancy in inflow rates may be accounted for by unknown inflows in the floor). The modelling groups plan to enhance their models during the year.

8.   POST-EXCAVATION CHARACTERIZATION

A number of activities were performed after excavation was completed to gather information that will improve the interpretation of the test. These are outlined below.

Water Inflows

As pointed out earlier, the method of monitoring the rate of water inflow from the fracture was inadequate. To improve it, we recently installed weirs in the floor of Room 209 each side of the fracture. One weir has an outlet to facilitate water flow rate measurements. Unfortunately, water is leaking out under the weirs through the damaged zone in the floor; this will be rectified in the future.

Fracture Stiffness

Pac-ex (combined packer and extensometer) instruments [3] were installed in some of the triaxial strain cell holes in place of the straddle packer systems. Water pressures in the fracture zone were raised and lowered while the displacements across the zone were monitored (Figure 21). From these measurements the normal stiffness of the zone was calculated to be about 500 MPa/mm.

Laboratory Testing of Coupled Mechanical, Thermal and Hydraulic Properties of the Fracture

Cored samples of the fracture, 200 mm in diameter, are being collected for laboratory testing. Two laboratories, Memorial University of Newfoundland and the Norwegian Geotechnical Institute, are conducting tests to determine the relationships between fracture permeability, mechanical and hydraulic aperture, normal and shear stress and normal and shear displacements, at temperatures between 8°C and 85°C.

Stresses

Additional overcore stress measurements confirmed that the stresses measured during the first two stress measurement programs are applicable throughout the volume of rock being modelled.

We also determined the stress concentration around the tunnel. A borehole was drilled normal to the tunnel direction near the plane of the triaxial strain cells. Overcoring tests began about 0.5 m from the tunnel wall and continued at 0.4 m intervals to 4.5 m depth; testing was discontinued until 7 m depth and thereafter continued at 0.5 m spacing. Measuring the stress profile near the triaxial strain cells will allow us to compare the stress profile around the excavation with the stress changes measured during excavation.

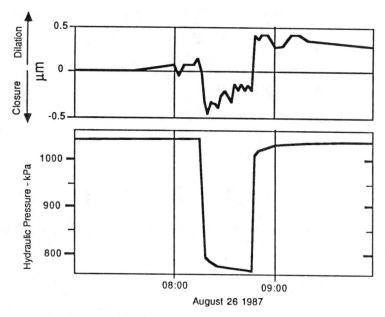

Figure 21. Mechanical response of Room 209 fracture to varying the hydraulic pressure. The response indicates that the normal stiffness of the fracture is about 500 MPa/mm at this location.

We also drilled a borehole through the fracture and conducted stress measurements each side of the fracture to determine if there is a significant perturbation in the stress field around the fracture. We found that the stresses were only perturbed within about 1 m of the fracture at the measurement location.

Excavation Damage Assessment

The extent of the excavation damage zone is being assessed in several ways: cross-hole seismic tomography between parallel boreholes drilled radially from the tunnel, radar surveys from the tunnel, dilatometer testing in boreholes drilled radial and parallel to the tunnel [16] and by comparing the measured displacements and stress changes at the extensometer and strain cell array with the displacements and stress changes predicted by elastic theory. We are developing methods for measuring the permeability in the damaged zone but have not conducted any field testing yet.

The leakage of the zone isolated by straddle packers under the floor indicates the damaged zone extends at least 1 m below the tunnel floor at this location. Leakage of water under the weirs at only about 0.3 m head indicates that the permeability of this damaged zone, or at least the zone near the floor, is relatively high. Combining the extensometer and convergence data, and comparing the predicted and measured displacements along the Bof-ex extensometer (Figure 17) indicates that the damaged zone in the walls and crown is less than 0.5 m thick. Visual inspection of the boreholes in the walls and crown shows no apparent fractures beyond about 0.2 m, which suggests that the permeability of the damaged zone in the walls is relatively low. The damage in the floor could be reduced by using similar blasting control on the floor as was used on the walls and crown.

9. WHAT HAVE WE LEARNT FROM THE ROOM 209 EXCAVATION RESPONSE TEST?

1. The experiment has been valuable as a trial run for planning and conducting future experiments in the URL. Future experiments will be better planned, more cost effective, and better able to achieve their objectives as a result of this test.

2. The finite-element elastic models used in this experiment predicted the displacements and stress changes quite well. However, the models used for predicting the hydrogeologic response did poorly. In fact, in the case of permeability change, they did not even predict the trend correctly. We have concluded that the models and/or the codes used do not correctly simulate the physical processes that occur in a fracture subjected to excavation-induced displacements.

3. It is important to measure displacements and permeability changes in a fracture at the same location so we can get a mechanical measurement to check the effective hydraulic apertures we infer from hydraulic testing. It is desirable to measure both normal and shear displacements of the fracture for validating the constitutive relationships used in the models.

4. A better method of measuring the flow rates from the fracture is required.

5. Better methods are required for ensuring that excavation procedures and blast designs are followed more closely.

6. Modelling took far longer than expected, and the planned program had to be reduced significantly to complete the prediction reports in an acceptable time.

7. It is important to define evaluation criteria and success criteria before the start of a model evaluation exercise.

8. The resources required to plan the test, install and maintain the instruments and data acquisition system, conduct the field testing, compile, analyze and report the results were far greater than envisioned when the test began.

9. Confining pressure dependent Young's modulus needs to be introduced into the models. The effect of not including anisotropic properties in the models needs to be assessed.

10. It is difficult to define error bars for much of the experimental data, especially the stress changes. A comprehensive assessment of the measurement uncertainty associated with all the field measurements is essential. It must be given a much higher priority in the URL program if the planned experimental program is to achieve its objectives with the necessary credibility.

11. Instrument performance:

-   The data acquisition system generally worked well. On a few occasions power cables were disconnected or other mishaps occurred.

-   The Distometer worked well for measuring convergence.

-   50% of the thermistors failed due to a potting problem. This has since been rectified. The thermistors gave good resolution and stability, 0.2°C interchangeability, and were cost effective.

-   The vibrating wire pressure transducers worked well.

-   The straddle-packer hydrogeological monitoring, testing and sampling system worked well. The system installed initially was not stiff enough to allow the permeability of the fracture zone to be measured, but this was replaced before excavation through the zone began.

-   All the CSIRO triaxial strain cells survived and worked as expected. They provided the complete strain tensor, which has since been converted to the complete tensor of stress change. However, they suffer from a glue drift problem. Tests are under way in the URL to evaluate how we can overcome or correct for this problem.

-   The Bof-ex worked exceptionally well and surpassed all expectations. It provided readings with a precision of ±1 μm. Based on the results in Room 209 it was selected for use in the shaft extension.

-   The Irad Gage Sonic Probe extensometer was not precise enough to measure the very small displacements that occurred in the stiff granite. As well, some anchors slipped during blasting.

-   The Terrametrics Waste Isolation Extensometers gave virtually no useful data, due partly to poor choice of transducer. The Terrametrics extensometers were difficult to install, and the number of anchors and locations at which anchors can be installed are quite restricted.

-   The Sliding Micrometer gave a repeatability of approximately ±30 μm in most measurement zones. It suffered from having material collect on the spherical reading seats. The instrument is very labour intensive, both for reading and data reduction. The Sliding Micrometer has an advantage over the other extensometers in that we could excavate through the casing and continue to obtain data from the remaining casing afterwards.

13. CONCLUSIONS

In conclusion, the five objectives for the Room 209 Excavation Response Test, listed in Section 2, these have been achieved as follows:

1. We have conducted a very successful trial run for the Excavation Response Experiment (ERE) and for other planned experiments. All groups involved have gained a great deal from the experience, and future experiments will be more successful and more cost effective than they would have been without this test.

2. A successful instrument evaluation was completed and from this we selected instruments and layouts to use in the shaft extension program. The instruments selected were the Bof-ex extensometer, the CSIRO triaxial strain cell, Iseth Distometer, the AECL-designed straddle-packer system and Geocon vibrating wire pressure transducer, the Fluke 2400 data logger system for logging all but the vibrating wire transducers and a Geocon data logger for the vibrating wire transducers.

3. Back calculation of the measured responses to determine properties of the rock mass and fracture is currently under way. Preliminary calculations of rock mass modulus variation around the tunnel are included (Figure 18).

4. Characterization of the fracture is continuing. The fracture has been well characterized for future large-scale tests in it and a good data base has been obtained. This data base will be valuable for validating models, both current and future. It has already indicated that the models used in this test did not properly simulate the hydrogeologic behaviour of a fracture zone as it was subjected to excavation induced changes.

5. Evaluation of the excavation damage around the room is continuing. Initial indications are that the damaged zone is less than 0.5 m thick in the walls and crown and the permeability of this zone is not high. The damaged zone in the floor is at least 1 m deep. The permeability of this zone, or at least the upper part of it, is high. The damaged zone in the floor could be reduced in future by using controlled blasting methods similar to those used for the walls and crown.

14. ACKNOWLEDGEMENTS

Most of the experimental information and figures presented in this paper are drawn from reports by: Davison and Kozak; Everitt; Kuzyk, Babulic, Bilinsky and Karklin; and Spinney [5],[10]. The modelling information and most of the figures comparing modelling and experimental data are drawn from reports by: T. Chan, P. Griffiths and B. Nakka; G. Shi, R. Goodman and P. Perie; and P. Kaiser, D. Chan, D. Tannant, F. Pelli and C. Neville, [8],[12].

15. REFERENCES

1. Simmons, G.R., (1986): "Atomic Energy of Canada Limited's Underground Research Laboratory for Nuclear Waste Management", Proceedings of Geothermal Energy Development and Advanced Technology, International Symposium, Tohoku University, Sendai, Japan.

2. Thompson, P.M., Lang, P.A. and Snider, G.R., (1986): "Recent Improvements to In Situ Stress Measurements Using the Overcoring Method", Proceedings of the 39th Canadian Geotechnical Conference, Ottawa.

3. Thompson, P.M., Kozak, E.T. and Martin, C.D., (1988): "Rock Displacement Instrumentation and Coupled Hydraulic Pressure/Rock Displacement Instrumentation for Use in Stiff Crystalline Rock", Proceedings of the OECD/NEA Workshop on Excavation Response in Deep Radioactive Waste

Repositories - Implications for Engineering Design and Safety Performance, Winnipeg, Canada, 26 - 28 April 1988.

4. Kovari, K., Amstad, C.H. and Koppel, J., (1979): "New Developments in the Instrumentation of Underground Openings", In Proc. Rapid Excavation and Tunnelling Conference, Atlanta.

5. Lang, P.A., Everitt, R.A., Kozak, E.T. and Davison, C.C., (1988a). "Underground Research Laboratory Room 209 Instrument Array; Pre-Excavation Information For Modellers", Atomic Energy of Canada Limited Report AECL-9566-1.

6. Gale, J.E., (1985), "Stress Dependent Hydraulic Properties of Fractures in Lac du Bonnet Granite - Phase 1", Preliminary report to AECL, unpublished.

7. Barton, N., and Bakhtar, K., (1987), "Description and Modelling of Rock Joints for the Hydrothermo-mechanical Design of Nuclear Waste Vaults", Atomic Energy of Canada Limited Technical Record, TR-418*.

8. Lang, P.A. (editor), (1988b), "Underground Research Laboratory Room 209 Instrument Array; Modellers' Predictions of the Rock Mass Response to Excavation", Atomic Energy of Canada Limited Report AECL-9566-2.

9. Thompson, P.M. and Lang, P.A., (1987), "Geomechanical Instrumentation Applications at the Canadian Underground Research Laboratory", Proceedings of the 2nd International Symposium - Field Measurements in Geomechanics, Kobe, Japan.

10. Lang, P.A., Kuzyk, G.W., Babulic, P.J., Bilinsky, D.M., Everitt, R.A., Spinney, M.H., Kozak, E.T. and Davison, C.C., (1988c), "Underground Research Laboratory Room 209 Instrument Array; Measured Response to Excavation", Atomic Energy of Canada Limited Report AECL-9566-3.

11. Walton, R.J. and Worotnicki, G. (1986), "A Comparison of Three Borehole Instruments for Monitoring the Change of Rock Stress with Time", Proceedings of the International Symposium on Rock Stress and Rock Stress Measurements, Stockholm, Sweden 1-3 September, 1986.

12. Chan, T., Griffiths, P.M. and Nakka, B, (1988), Interim Progress Report, "Finite Element Modelling of Geomechanical and Hydrogeological Responses to the Room 209 Heading Extension Excavation Response Experiment: II. Post-Excavation Analysis of Experimental Results", unpublished.

13. Hibbitt, Karlson and Sorrensen, Inc., (1984), "ABAQUS: Theory, Users and Example Problems Manuals", Providence, Rhode Island, U.S.A.

14. Sun, Z., Gerrard, C. and Steffansson, O., (1985), "Rock Joint Compliance Tests for Compressions and Shear Loads", International Journal of Rock Mechanics and Mining Sciences & Geomechanical Abstracts, 22 (4), pp. 197-213.

15. Chan, T., Guvanasen, V. and Reid, J.A.K, (1987), "Numerical Modelling of Coupled Fluid, Heat and Solute Transport in Deformable Fractured Rock", Coupled processes associated with nuclear waste repositories, Proc. Int.

Symposium Berkely, California, Ed. C.F. Tsang, pp. 605-625, Academic Press, Orlando.

16. Koopmans, R. and Hughes, R.W., (1988), "The Assessment of Excavation Disturbance Surrounding Underground Openings in Rock", <u>Proceedings of the OECD/NEA Workshop on Excavation Response in Deep Radioactive Waste Repositories - Implications for Engineering Design and Safety Performance.</u> Winnipeg, Canada, 26 - 28 April 1988.

---

* Unrestricted, unpublished report, available from SDDO, Atomic Energy of Canada Limited Research Company, Chalk River, Ontario KOJ 1J0

SHAFT EXCAVATION RESPONSE IN A HIGHLY STRESSED ROCK MASS

C.D. Martin
Atomic Energy of Canada Limited
Pinawa, Manitoba, Canada, ROE 1L0

ABSTRACT

The extension of the Underground Research Laboratory access shaft from 255- to 445-m depth is being excavated in massive grey granite. The granite is essentially unfractured, except for a major low-dipping shear (Fracture Zone 2) about 1.5 m thick, and associated minor splays (Fracture Zone 1.9). Observations, hydraulic fracturing and overcore results indicate that unusually high in situ stresses may be associated with large volumes of unfractured granite at depth. In areas where the in situ stress magnitudes are sufficiently high, time-dependent behaviour is observed. It is proposed that the time-dependent phenomena are probably related to the formation of new microcracks and propagation along existing microcracks. In areas where stress magnitudes are very high and anisotropic, the formation of microcracks is amplified and discontinuities are formed, resulting in shaft-wall failure. The depth of this failure zone extends to at least one metre beyond the shaft wall between Fracture Zones 1.9 and 2. The impact of these high stress magnitudes on excavation and vault designs must be evaluated to ensure the satisfactory performance of a nuclear waste disposal vault.

EFFETS DES TRAVAUX D'EXCAVATION DANS UN PUITS CREUSE DANS UNE MASSE ROCHEUSE SOUMISE A DE FORTES CONTRAINTES

RESUME

De la cote 255 à 445 mètres, le prolongement du puits d'accès du Laboratoire souterrain de recherche est aménagé dans un granite gris massif. Ce granite est en général non fissuré, à l'exception d'une importante zone de cisaillement faiblement inclinée (Fracture Zone 2) d'environ 15 mètres d'épaisseur à laquelle sont associés des évasements de faible ampleur (Fracture Zone 1.9). Les résultats des observations, de la fissuration hydraulique et des sur-carottages indiquent, qu'en profondeur, les grands volumes de granite non fissuré sont en général soumis à de fortes contraintes in situ. Dans les zones où l'ampleur des contraintes in situ est suffisamment élevée, on observe le comportement en fonction du temps. L'auteur émet l'hypothèse que les phénomènes dépendant du temps sont probablement liés à la formation de nouvelles microfissures et à une propagation le long de microfissures existantes. Dans les zones où les contraintes sont très élevées et anisotropiques, la formation de microfissures est amplifiée et des discontinuités se forment entraînant un éboulement de la paroi du puits. Entre les Fractures Zones 1.9 et 2, l'éboulement affecte une épaisseur d'au moins un mètre à partir de la paroi du puits. Les répercussions de ces contraintes de forte intensité sur la conception de l'excavation et de la casemate doivent être évaluées si l'on veut s'assurer que la casemate servant au stockage définitif des déchets nucléaires se comporte de façon satisfaisante.

1. Introduction

Atomic Energy Canada Limited (AECL) is assessing the concept of disposal of nuclear fuel waste deep in plutonic rock. One of the justifications for considering the plutonic rock option is that there are large volumes of competent plutonic rock with extremely low porosity and permeabilitiy (Dormuth et al., 1987). This concept appears at first to be very attractive since it limits access of groundwater to the waste and makes effective use of natural barriers. Experience gained at the Underground Research laboratory (URL) near Pinawa, Manitoba, however indicates that unusually high in situ stresses may be associated with large volumes of extremely low porosity and permeabilitiy (essentially unfractured) granite at depth. The impact of these high stresses on excavation and vault designs must be evaluated to ensure the satisfactory performance of a nuclear waste disposal vault.

The extension (presently under construction) of the URL access shaft from 255- to 445-m depth is being excavated in massive grey granite. The granite is essentially unfractured except for a major low-dipping shear (Fracture Zone 2), about 1.5 m thick, and several associated minor shears (splays), Figure 1.

The shaft intersected Fracture Zone 2 between 273- and 274.5-m depth and a splay (Fracture Zone 1.9) at about 288-m depth. The rock sandwiched between Fracture Zones 1.9 and 2 was stressed high enough to induce failure in the northwest area of the shaft wall from 276- to 288-m depth. The shaft is being excavated by carefully controlled full face blasting techniques. During shaft sinking slabs up to 1.8 m x 1.5 m x 0.1 m thick were observed to "pop" off the shaft wall, leading to the development of a V-shaped failure zone in the wall of the shaft. TRIVEC (Kovari et al., 1979) instrumentation and a Bof-ex extensometer (Thompson et al, 1987) were used to monitor the nonelastic displacements observed in the failure zone. The mode of failure was observed to be a combination of shear and tensile failure. The nonelastic movements extend to a depth of at least 1 m.

Below Fracture Zone 1.9, excavation in the massive grey granite continued to show signs of stress instability. Spalling of the shaft walls was frequently observed and side-wall failures were observed between 340- and 350-m depths. Extensometers installed at the 324- and 352-m depths also recorded time-dependent behaviour not observed in the excavations above

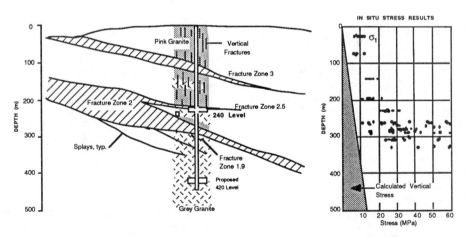

Figure 1: Generalized geological section and σ_1 magnitudes obtained from overcoring. Scattering in σ_1 magnitudes is discussed in Section 3.

the 240 level. Again the depth of the excavation disturbed zone displaying this phenomena was generally less than 1 m.

Although very carefully controlled full face blasting techniques are being used to excavate the URL access shaft, a time dependent excavation disturbed zone is being created, which extends to at least 0.5- to 1.0-m depth. Such a disturbed zone may require special attention for shaft sealing.

2. Geology And Geotechnical Properties

The URL is located within the Lac du Bonnet granite batholith. Two major regional thrust faults (referred to as Fracture Zone 2 and Fracture Zone 3) are present between the surface and the 300-m depth in the vicinity of the access shaft, Figure 1. These dip about 25° southeast. In addition to the major fracture zones, several prominent splays are also present, Figure 1. Between the surface and Fracture Zone 2.5, the rock is pink granite and contains a prominent near-vertical fracture set striking about 030-040°. Between Fracture Zone 2.5 and the proposed 420 Level, the rock is essentially unfractured grey granite, except in the immediate vicinity of the fracture zones where pink discolouration occurs. The pink colour is associated with groundwater flow along the fracture zones.

Grey granite is used to describe the general

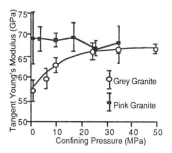

Figure 2: Tangent Young's Modulus at 50% peak strength measured in triaxial compression tests on Lac du Bonnet Granite, (Katsube and Hume, 1987)

rock type below Fracture Zone 2.5; however, detailed geological mapping carried out by site geologists have revealed the following major rock types: xenolithic granite, granodiorite dikes, pegmatite dikes, leucrocratic granite and fine-grained grey granite (Everitt et al., 1985). The granodiorite dikes, pegmatite dikes, and the leucrocratic granite contain discontinuous (trace length < 1.5 m) fractures striking about 040°.

Mechanical testing programs on the Lac du Bonnet granite have been carried out by CANMET over the past ten years. Because the rock is essentially unfractured, mechanical testing programs at ambient temperatures have not been difficult to perform. It has been recognized for some time, however, that all samples of the grey granite are subject to sample disturbance in the form of "microcracking". This phenomenon is illustrated on Figure 2, which shows the tangent modulus as a function of confining stress (Katsube and Hume,1987). Standard geotechnical properties of the Lac du Bonnet granite are listed in Table I.

3. In Situ Stress Setting

3.1 Overcoring

In situ stress measurements at the URL were carried out as part of the general characterization of the rock mass. The results of in situ stress measurements in the vicinity of the URL shaft above Fracture Zone 2.5 have been published by Lang et al. (1986). A summary

TABLE I: Geotechnical Properties of Lac du Bonnet Granite at the Underground Research Laboratory (Katsube and Hume, 1987)

	Pink Granite	Grey Granite
Uniaxial Compressive Strength (MPa)		
Range	134 - 248	147 - 198
Mean	200	167
Young's Modulus (GPa)		
Range	53 - 86	46 - 64
Mean	69	55
Hoek & Brown Failure Parameters		
m	31.17	30.54
s	1	1
r	0.955	0.924

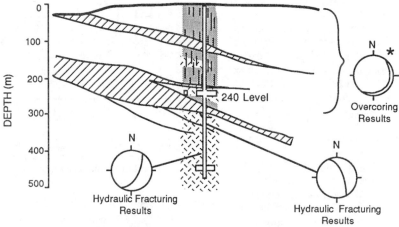

* Lower Hemisphere Stereo Net Showing Orientation Of
  Sigma 1- Sigma 2 plane

Figure 3: Summary of fracture orientations produced by the hydraulic fracturing program and the orientation of
Sigma1-Sigma 2 plane obtained from the overcoring program.

of the σ_1 magnitudes are presented in Figure 1.
For comparison, the σ_1 magnitudes obtained
above and below Fracture Zone 2, from the
240 Level, are also shown. As illustrated on
Figure 1, the in situ stresses around a major
fracture zone are quite variable. This should be
expected because of the highly variable geometry and mechanical properties of the fracture
zone on a local scale and the locked-in stresses
associated with past shearing. It should be noted that, above Fracture Zone 2, σ_1 is subhorizontal and oriented parallel to the near-vertical
joint set which trends about 040° (Figure 3) and
σ_3 is subvertical. On the 240 Level σ_1 magnitude is about 30 MPa, σ_2 about 15 MPa and
σ_3 about 12 MPa.

3.2 Hydraulic Fracturing and Observations

Attempts were made to measure the in situ
stresses below Fracture Zone 2, but with overcoring methods, the 86-mm-diameter core
disced and hydraulic fracturing methods produced only subhorizontal fractures. Between
Fracture Zones 2 and 1.9, back calculation of
σ_{max} orientation based on observed shaft wall
failures indicated a trend of 040° (see Section
5.2). This observed stress orientation is consistent with the σ_1 orientations measured on the

240 Level using standard overcoring techniques.

One subhorizontal fracture produced by the
hydraulic fracturing between Fracture Zone 1.9
and 375-m depth indicates that σ_{max} is oriented
about 330°, Figure 3. Back calculation of σ_{max}
orientation based on the observed shaft-wall
failure at about the 350-m depth also indicates a
trend of 330°. Below about the 375-m depth,
however, hydraulic fracturing results indicate
that σ_{max} orientation is about 040°, which is
consistent with the σ_1 orientations measured on
the 240 Level using standard overcoring meth-
ods. The ratio of the horizontal σ_{max} to σ_{min}
below Fracture Zone 2, based on hydraulic
fracturing results, is about 2 and the magnitude
of σ_{max} is estimated at about 100 MPa, Table
2. The high stress magnitudes indicated by hydraulic fracturing have not been verified.

3.3 Summary

In summary, above Fracture Zone 2, σ_1 is
subhorizontal and oriented parallel to the near
vertical joint set trending about 040°, (Figure 3)
and σ_3 is subvertical. On the 240 Level, σ_1
magnitude is about 30 MPa, σ_2 about 15 MPa,
and σ_3 about 12 MPa. Below Fracture Zone 2,
in the plane perpendicular to the shaft axis,

TABLE II: In Situ Stress Results from Hydraulic Fracture Measurements (Doe, 1987)

Tests Conducted Above Fracture Zone 2 - Subvertical Fractures

Depth (m)	σ_{max} (MPa)	σ_{min} (MPa)	Az	Dip
241.8	31	17	010°	70E
245.7	28	17	-	-
248.8	37	20	022°	67E
260.4	39	21	022°	77E
269.3	45	24	024°	84E
278.2	40	21	028°	61W

Tests Conducted Below Fracture Zone 2 - Subhorizontal Fractures

Depth (m)	σ_{max} (MPa)		σ_{min} (MPa)	
	Max.	Min.	Max.	Min.
336.6	116	92	63	40
390.8	103	78	61	36
391.8	114	84	73	42
425.1	116	92	67	40
427.6	110	81	72	42

σ_{max} orientation varies from 040° to 330°. Magnitudes below Fracture Zone 2 have not been measured except by hydraulic fracturing. Intrepretation of these results indicates that σ_{max} could be about 100 MPa. These results have not been verified; none-the-less, stress magnitudes are sufficiently high to cause localized shaft-wall instability.

4. Blast Round Design

Full face blasting is used to excavate the 4.6-m-diameter circular access shaft at the URL. The blasting pattern consists of a near centrally located burn cut, 3 concentric rings of blast holes and a perimeter ring. Blast round lengths vary from 1 to 3.4 m without changing the blast pattern. There are a total of 77 holes drilled in this pattern: 3 89-mm-diameter relief holes and 74 38-mm-diameter blast holes consisting of 3 cut holes, 11 helpers, 28 production holes and 32 perimeter holes. The maximum spacing between the perimeter holes is 450 mm. All cut and helper holes, and the inner most production ring are loaded with sticks of 32 mm x 400 mm Forcite 75%. The outermost production ring is loaded with 25 mm x

200 mm Forcite 75% to provide decoupling and the perimeter is loaded with 19-mm-diameter Xactex and a 25-mm-diameter Forcite toe charge of 3 sticks. Magnadet detonators with long period delays from #1 to #18 are used to initiate the blast. The blast pattern is designed to minimize energy impacted to the surrounding rock that may result in overbreak or fracturing.

5. Observations During Shaft Sinking
5.1 Overbreak

The quality of the shaft wall obtained with the blast design described in the previous section is very good. As shown on Figure 4, the amount of overbreak beyond the 4.6-m design line is very small (average from 262- to 352-m

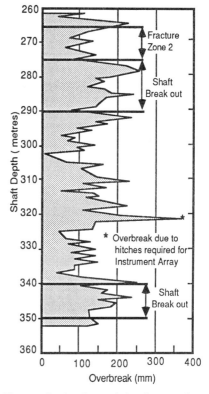

Figure 4: Overbreak recorded at about 1 m intervals in the circular shaft between 262- and 352-m depth.

depth was 131 mm). Also, the percentage of drill hole traces remaining on the shaft wall after excavation varies from about 10 to 80%. The amount of overbreak did not increase significantly during excavation through Fracture Zone 2 or the zones containing the shaft-wall failures. It would appear that the blast design is probably as close to optimum as practical and it is effective in minimizing the energy impacting the wall rock.

5.2 Shaft-Wall Failures

Excavation from 255-m depth into the highly stressed unfractured rock mass caused frequent "spalling" of the shaft walls and on several occasions larger "pop ups" of the shaft bottom. In all cases, the volume of rock was less than 0.03 m³. When excavation began below the bottom of Fracture Zone 2, however, failure of the shaft wall was noticed immediately. This failure process, which involved onion-peel-like slabs measuring up to 1.8 m x 1.5 m x 0.1 m thick (0.3 m³) popping off the shaft wall, continued until the excavation reached Fracture Zone 1.9, Figure 1. This process resulted in a V-shaped notch, which developed on the northwest side of the shaft to a depth of 0.4 m beyond the excavation design line and was confined between Fracture Zone 1.9 and the base of Fracture Zone 2, Figure 5.

Once below Fracture zone 1.9, only the

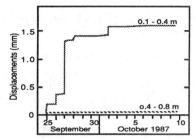

Figure 6: Displacements recorded by the extensometer installed in the side wall failure zone. The anchor depths in the borehole are shown.

normal "popping" was observed. A Bof-ex extensometer was installed in the failure zone after the failure process had slowed and conditions were safe, Figure 5. The 45-mm core obtained from drilling the extensometer borehole disced to a depth of about 0.5 m. The movement recorded by the extensometer is shown on Figure 6. The extensometer indicated that the failure was occurring between 0.1- and 0.4-m depth.

The mode of failure, based on visual observations, is believed to be a combination of shear and tensile failure as described by Maury (1987). Santarelli and Brown (1987) demonstrated that the maximum concentration of the tangential stress can be located inside the shaft wall when the modulus E is a function of the confining stress, Figure 2. The formation of the onion-skin-like slabs can indicate, according to Maury, the kind of stress distributions described by Santarelli and Brown.

The development of sidewall V-shaped notches was also observed between 340- and 350-m depth. In this case the notches developed on both sides of the shaft and were oriented northeast-southwest suggesting that σ_{max} was oriented in a northwest-southeast direction. Preliminary mapping indicates that the development of these zones is confined to particular geological domains. The depth and extent of these zones is presently under investigation.

5.3 Geological Mapping At 300 Station

The shaft was excavated for 15 m (about 3 diameters) beyond the invert of the planned 300

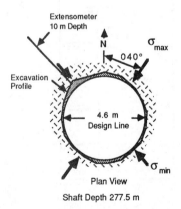

Figure 5: Orientation of the V-shaped failure zone which occurred in the shaft between Fracture Zones 2 and 1.9

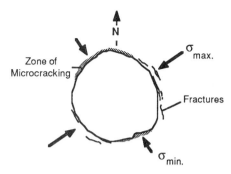

Figure 7: Zones of microcracking and fracture traces around the shaft mapped at the 300 Station

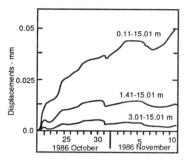

Figure 8: Time-dependent displacements recorded by EXT 7 in Room 209 on the 240 m Level. The anchor depths in the borehole are shown. Displacements which occurred during a 10 minute period after each blast have been removed.

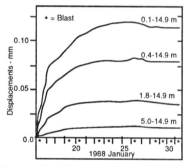

Figure 9: Time-dependent displacements recorded by extensometer installed at 324 m depth. The anchor depths in the borehole are shown. Displacements which occurred during a 2 minute period after each blast have been removed.

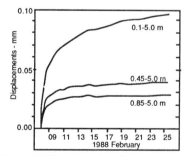

Figure 10: Time-dependent displacements recorded by extensometer installed at 352 m depth. The anchor depths in the borehole are shown. No blasting occurred over the period shown.

Station excavation. Then the station was carefully excavated using pilot and slash techniques. This provided a unique opportunity to observe the zone around the shaft disturbed by the shaft sinking operation. Detailed mapping of the shaft perimeter from the 300 Station excavation was carried out. The results are summarized on Figure 7 (Everitt et al., 1988). Two zones of microcracking (fractures < 50 mm long) were mapped. These occurred on the northwest -southeast walls. As was the case with the failure zone shown on Figure 5, this would indicate that σ_{max} is oriented about 040°. Of particular interest is the fact that the zone of microcracking only extended to a depth of about 200 mm.

Fractures, generally parallel to the excavation, were also recorded, Figure 7. These tended to be controlled, to some degree, by the lithology banding (Everitt et al.,1988). The fractures did not occur in the zones of microcracking and they extended up to 300 mm from the shaft perimeter. If conventional hypothesis, ie. elasticity and Coulomb failure criteria, were applied than these cracks should not form since the maximum tangential stress would occur at the shaft wall. The formation of these fractures tend to support the view that with an elastic modulus changing with confining radial stress, the maximum tangential stress occurs inside the shaft wall.

6. Time-Dependent Behaviour

Time-dependent behaviour has been recorded by extensometers installed on the 240 Level,

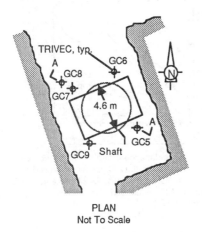

PLAN
Not To Scale

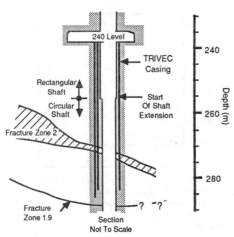

Figure11: Plan and section showing the location of TRIVEC instruments used to monitor the shaft excavation between Fracture Zones 2 and 1.9

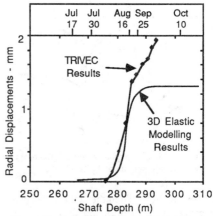

Figure 12: Radial displacement between Fracture Zones 2 and 1.9 recorded by GC7 TRIVEC compared to the 3D elastic finite element solution.

TRIVEC instruments installed from the 240 level, around the shaft through Fracture Zone 2, and extensometers installed from the shaft at the 324-m and 352-m depths. The phenomena was first noticed on the 240 level, with extensometers that provided repeatability to within ± one micrometre, Figure 8. As shaft excavation deepened and stress levels increased the time-dependent phenomena became more apparent in the instrumentation plots. Figures 9 and 10 show the time dependent displacements recorded by the Bof-ex extensometers installed at the

324- and 352-m depths, respectively. Figure 9 shows the time-dependent displacements after the elastic displacements due to excavation have been removed (displacements that occurred during a two minute period after each blast has been removed). Figure 10 shows the displacements that occurred over a 17 day period after a single 2-m round had been excavated. Both figures indicate that the majority of the time-dependent displacements occurred between 0.1- and 0.5-m depth in from the shaft wall.

TRIVEC instrumentation installed 1 m and 2 m from the shaft wall recorded the complete displacement as shaft excavation passed, Figure 11. Radial displacements recorded by GC7 between Fracture Zones 2 and 1.9 are shown on Figure 12. Also shown on Figure 12 is the response calculated using a 3-dimensional finite element elastic code. Time-dependent displacements were recorded by the instrumentation; the displacements recorded by the TRIVECS are much greater than those recorded by the extensometers. These results may reflect both time-dependent displacements and displacements associated with the formation of the V-shaped notch that occurred between Fracture Zones 2 and 1.9.

The physical phenomena responsible for the time-dependent displacements recorded by the shaft instrumentation is probably related to high in situ stresses which cause propagation of existing microcracks and the formation of new microcracks. It is unlikely that the behaviour is related to true plastic deformation with associated flow rules. In areas where the in situ stress magnitudes are sufficiently high (i.e., between Fracture Zones 1.9 and 2), the formation of microcracks is amplified and discontinuities are formed resulting in failure of the rock at the shaft wall. The depth of this failure zone extends to at least 1 m between Fracture Zones 1.9 and 2. Additional information on the extent and permeability of these zones will be obtained during the shaft characterization program scheduled to commence 1988 September.

7. Conclusions

Observations made during the construction of Underground Research Laboratory shaft extension from 255- to 352-m depth, lead to the following conclusions:

a) Shaft excavation using full face drill and blast methods can be carefully controlled to produce minimal damage to the shaft wall and reduce overbreak. Even under excellent quality control, however, a significant excavation-disturbed zone will be created in a rock mass with high in situ stresses.

b) In highly stressed rock, the excavation-disturbed zone can extend at least 1 m beyond the shaft wall and result in localized failure at the wall of the shaft. The mode of failure is likely a combination of shear and extension. The impact of a confining stress-dependent modulus on the observed failure process is not clear.

c) Time-dependent behaviour is clearly evident in the URL. At present, it is suggested that formation of new microcracks and the propagation of existing microcracks is responsible for the time-dependent behaviour. It is also suggested that this behaviour under higher stress and anisotropic conditions will lead to the formation of discontinuities and subsequent instability of the shaft wall .

The impact of high stress magnitudes on excavation and vault designs must be evaluated to ensure the satisfactory performance of nuclear waste disposal vaults in highly stressed rock.

References

Doe, T. 1987. Hydraulic fracturing stress measurements in the shaft probe hole at the Underground Research Laboratory, Manitoba.. Report prepared for Batelle Memorial Institute, Office of Waste Technology Development, Willowbrook, Illinois

Dormuth, K.W. and Nuttall, K. 1987. The Canadian Nuclear Fuel Waste Management Program. Radioactive Waste Management and the Nuclear Fuel Cycle. vol. 8(2-3), pp. 93-104.

Everitt, R.A. and Brown, A. 1985. Subsurface geology of the Underground Research Laboratory: An overview of recent developments. Proceedings of the Twentieth Information Meeting of the Canadian Nuclear Fuel Waste Management Program, Winnipeg. vol.1 pp 146-181.

Everitt, R.A., Chernis, P., Good, D. and Grogan, A., 1988. Mapping of the excavation damage zone around the circular access shaft at the Atomic Energy of Canada Limited's Underground Research Laboratory. Paper Presented at Workshop on Excavation Responses in Deep Radioactive Waste Repositories - Implications for Engineering Design and Safety Performance. Winnipeg, Canada.

Katsube, T.J. and Hume, J.P., 1987. Geotechnical Studies at Whiteshell Research Area (RA-3). CANMET Mining Research Laboratories Divisional Report MRL 87-52.

Kovari, D., Anstad, C.H. and Koppel, J., 1979. New developments in the Instrumentation of Underground Openings.Proceeding of the Rapid Excavation And Tunnelling Conference, Atlanta.

Lang, P.A., Everitt, R.A., Ng, L. and Thompson, P.M.,1986. Horizontal in situ stresses versus depth in the Canadian Shield at the Underground Research Laboratory. Proceedings of the International Symposium on Rock Stress and Rock Stress Measurements, Stockholm. pp 449-456.

Maury, V., 1987. Observations, researches and recent results about failure mechanisms around single galleries. Proceedings of the International Congress on Rock Mechanics, Montreal, Canada.

vol.2, pp. 1119-1128.

Santarelli, F.J. and Brown, E.T., 1987. Performance of deep wellbores in rock with a confining pressure-dependent elastic modulus. Proceedings of the International Congress on Rock Mechanics, Montreal, Canada. vol.2, pp. 1217-1222.

Thompson, P.M. and Lang, P.A. 1987. Geomechanical instrumentation applications at the Canadian Underground Research Laboratory. 2nd International Symposium on Field Measurements in Geomechanics, Kobe, Japan. vol. I.

```
* * * * * * * * * * * * * * * * * * * *
```

Séance II(b)

RAPPORT DE SYNTHESE DU PRESIDENT

A. BARBREAU (DPT/CEA, France)

```
* * * * * * * * * * * * * * * * * * * *
```

I.   INTRODUCTION

        La Séance II(b) qui comprenait huit communications, concernait des
études géotechniques spécifiques effectuées dans les roches cristallines, plus
particulièrement dans le laboratoire souterrain canadien (URL). C'est donc une
séance au cours de laquelle ont été présentés des travaux expérimentaux et au
cours de laquelle a été exposé la mise en oeuvre d'appareils de mesure avec,
comme principal objectif, la réponse aux deux questions suivantes :

        a) La nature et l'extension des dommages associés à l'excavation.

        b) Les conséquences sur la perméabilité du milieu.

        Je vais passer rapidement en revue les différentes communications,
soulignant quelques faits marquants.

II.   COMMUNICATIONS DE LA SEANCE II(b)

Communication n° 13 - N. Barton

        Cette communication concerne la modélisation du comportement historique
des fractures en relation avec les déformations qui se produisent autour d'un
tunnel et les conséquences sur la perméabilité du milieu. Une application à un
problème de tunnel est présentée. J'ai noté que cette étude montre la
difficulté d'obtenir des résultats précis avec les techniques conventionnelles
de mesure des déformations et la nécessité, pour modéliser les changements de
perméabilité, d'avoir recours à une représentation continue des phénomènes qui
affectent les fractures telles que fermeture, cisaillement, dilatation, etc.
D'importants phénomènes de cisaillement peuvent apparaître, entraînant une
augmentation de perméabilité. A noter que, de ce point de vue, la rugosité des
fractures peut jouer un rôle significatif. Les mesures de perméabilité
apparaissent donc indispensables pour étayer les modèles.

Communication n° 14 - W.F. Ubbes

        Cette communication est une présentation de la problèmatique de l'effet
des travaux et des solutions recherchées. Elle s'appuie sur l'expérience dite
"Mine-by" à CLIMAX (Nevada test site) et sur les essais effectués dans une
mine expérimentale par l'Ecole des Mines du Colorado. L'expérience de CLIMAX
se proposait de comparer les contraintes et les déplacements résultant de
travaux miniers avec les effets produits par une charge thermique.

Cette expérience a montré que les effets dus à l'excavation sont comparables à ceux résultant d'un échauffement, mais que tandis que ces derniers sont déterminés essentiellement par les propriétés thermiques de la roche, les effets dus à l'excavation dépendent davantage des conditions structurales. La complexité et l'hétérogénéité de ces conditions rendent plus difficile l'interprétation que dans le cas des effets thermiques.

L'expérience de l'Ecole des Mines du Colorado a notamment montré que la perméabilité de la roche dans la zone affectée dépend des contraintes autant que du nombre de fractures. Il est donc important de bien mesurer les contraintes et d'utiliser des modèles informatiques appropriés, prenant en compte les processus et les caractéristiques du milieu physique.

Communication n° 15 - P.A. Lang

Cette communication peut être considérée comme une présentation générale du programme d'études de l'effet des travaux réalisé à l'URL. Elle décrit les différentes études effectuées et les différentes techniques et appareillages mis en oeuvre. On a effectué des études sur la réponse de l'excavation, notamment son influence sur les variations de perméabilité.

Les mesures de perméabilité effectuées dans le puits ont révélé des divergences avec les prédictions des modèles ; elles donneraient une réduction de perméabilité au lieu d'un accroissement, indiquant que l'on ne maîtrise pas actuellement tous les phénomènes en cause, ce qui est un inconvénient du point de vue de la crédibilité des modèles (importance pour la sûreté). On a également rencontré des difficultés pour mesurer des contraintes avec des dispositifs à cordes vibrantes ; des capteurs de déformation ont fourni par contre des résultats intéressants.

En ce qui concerne l'expérience dans la chambre 209, les valeurs de perméabilité mesurées sont, comme pour le puits, opposées à celle résultant du modèle, dans le même sens (réduction de K) confirmant ainsi qu'il y a bien là une difficulté non résolue. Il apparaît que la zone endommagée s'étend jusqu'à environ 1 m, ce qui reste une valeur somme toute limitée. On peut aussi souligner le fait que, lorsque l'on est en présence de très faibles perméabilités, le temps nécessaire pour effectuer les mesures peut ne pas être compatible avec le programme d'avancement des travaux.

Une réflexion est en cours pour évaluer les techniques de mesures et notamment comment utiliser des traceurs. Le programme dit "Phase opérationnelle" qui concerne l'étude de l'évolution dans le temps de la zone endommagée en fonction de différents facteurs, tels que la lithologie, les contraintes, la forme des cavités est un programme fort intéressant, mais comme il est actuellement seulement en projet, on ne peut pas en tirer d'enseignements ni de conclusions aujourd'hui.

Les autres communications, c'est-à-dire les communications 16, 17, 18, 19 et 20 apportent des informations plus détaillées sur les différentes études, notamment sur des appareils de mesures.

La communication n° 16 présentée par P.M. Thompson, expose le BOF-EX, un extensomètre permettant l'observation du mouvement de fractures en forage (on mesure le déplacement relatif entre des ancres adjacentes). C'est un appareil simple, dont les utilisateurs paraissent satisfaits. Le PAC-EX, est un extensomètre à obturateurs qui a également donné de bons résultats.

La communication n° 17 de R.A. Everitt (présentée par P.A. Lang) concernait la cartographie de la zone endommagée autour d'un puits circulaire à la faveur de travaux de creusement, à 300 m de profondeur. Il est apparu que la répartition des dommages est en relation avec la texture de la roche et avec les principales orientations des contraintes.

La communication n° 18 de R. Koopmans concernait un dilatomètre destiné à évaluer le module de déformation par mesure des contraintes et déterminer ainsi l'extension de la zone perturbée. Les mesures effectuées dans la chambre 209 ont donné des résultats intéressants (notons toutefois que les mesures sur les fractures restent problématiques).

La communication n° 19 de P.A. Lang concernait les essais effectués dans la chambre 209. Il s'agit de l'étude des fractures intersectant la cavité. On a effectué des mesures au moyen de packers à cheval sur la fracture, à la faveur d'un trou de forage, et on a étudié les contraintes triaxiales à différentes distances de la fracture. Un réseau d'extensomètres à également permis des mesures de convergence des parois du tunnel. Les mesures de variation de contraintes sont en bon accord avec les prévisions, mais pas les mesures de perméabilité. Il est donc important de mesurer, au même endroit d'une fracture, les déplacements et les variations de perméabilité, afin de relier les phénomènes et de pouvoir faire une modélisation corecte. Il est nécessaire également d'améliorer les méthodes de mesures des débits d'eau dans les fractures. Les mesures soniques et les mesures de convergence sont satisfaisantes. Les modalités d'écoulement de l'eau dans les fractures sont modifiées par les mouvements de cisaillement qui peuvent affecter les relations entre fractures aquifères et réduire, par exemple, la perméabilité de l'ensemble, alors que l'on s'attendrait plutôt à une augmentation de perméabilité.

La communication n° 20 présentée par C.D. Martin, dernière communication de la séance II (b), concernait l'étude du comportement de roches soumises à de fortes contraintes postérieurement au creusement (observations de 255 à 352 m).

On a mesuré les contraintes in situ par fracturation hydraulique et par surcarottage, en fonction du temps. On a utilisé le dispositif TRIVEC qui donne les trois composantes x, y, z du déplacement. Il faut noter l'importance de la microfracturation dans le comportement de la roche en fonction du temps. Les phénomènes font appel à une combinaison de cisaillements et d'extensions. La zone perturbée est de l'ordre du mètre.

III.   CONCLUSIONS

Quelles conclusions générales peut-on tirer?

Je crois qu'il faut d'abord souligner l'importance et la qualité des programmes de recherches déjà réalisés ou en projet à l'URL. Certes, toutes les questions ne sont pas résolues, mais elles sont en bonne voie de l'être. De nombreuses techniques et de nombreux appareillages divers ont été utilisés et sont déjà opérationnels, permettant une meilleure caractérisation du milieu fissuré. On peut aussi souligner les points suivants :

a) L'une des difficultés rencontrées est l'évaluation de l'effet des excavations sur les fractures dans l'epace et dans le temps, son influence sur les variations de la perméabilité est par conséquent la difficulté de modéliser les phénomènes. Il est en effet, notamment, mal aisé de mesurer la perméabilité dans la zone de dommages. Il faut donc développer et perfectionner les procédures instrumentales dont l'expérience montre qu'elles sont souvent difficiles à mettre en oeuvre. Il y a donc là un effort de R-D à faire, notamment dans le domaine de la mesure des très faibles perméabilités.

b) Naturellement, il faut bien déterminer à l'avance ce qu'il faut mesurer et comment le faire et ne pas se perdre dans la mesure difficile des phénomènes non significatifs. De ce point de vue, les études de sensibilité paramétrique peuvent apporter des éléments de jugement utiles. Il peut être aussi nécessaire de faire certaines mesures préalablement aux travaux, car après il peut être trop tard, d'où la necessité d'une planification prévisionnelle.

c) Il convient d'approfondir notre connaissance des modalités de circulation d'eau dans les fractures sous l'effet des variations de contrainte, ce qui implique une bonne connaissance préalable tridimensionnelle des réseaux de fractures et il faut naturellement, en s'appuyant étroitement sur les données expérimentales, élaborer des modèles performants et réalistes. Cela implique une bonne compréhension des phénomènes conduisant à la création de la zone perturbée.

d) Ceci étant, il m'apparaît, à la lumière des résultats déjà obtenus, que les perturbations résultant de l'excavation dans les roches cristallines sont très limitées dans l'espace (quelques dizaines de centimètres à 1 m, rarement plus). Il semble donc que ce soit surtout au niveau des puits d'accès que ces phénomènes risquent d'avoir des conséquences sur la sûreté (solution de continuité avec la surface par l'intermédiaire de la zone fracturée), d'où l'importance des techniques de colmatage.

e) Pour terminer, je n'ai rien noté dans les communications présentées sur les effets "chimiques" d'altération ; ce n'est peut-être pas un problème important, mais il devrait sans doute être examiné.

* *

```
* * * * * * * * * * * * * * * * * *
```

Session II(b)

CHAIRMAN'S SUMMARY

A. BARBREAU (DPT/CEA, France)

```
* * * * * * * * * * * * * * * * * *
```

I.   INTRODUCTION

Session II(b) included eight papers dealing with specific geotechnical studies conducted in crystalline rock, particularly at the Canadian underground laboratory (URL). Experimental work was presented and the use of measurement equipment was described, the main objective being to answer the following two questions:

a) nature and extent of the damage associated with excavation; and

b) effects on the permeability of the rock mass.

I shall briefly review the papers, pointing out a few salient features.

II.   PAPERS PRESENTED IN SESSION II(b)

Paper No. 13 - N. Barton

This paper described the modelling of rock-joint behaviour in relation to the changes occurring in the rock mass surrounding a tunnel and the effects on permeability. An application concerning a specific tunnel problem was presented. The study showed that it is difficult to obtain accurate results with conventional techniques for measuring deformation and that, in order to obtain models of permeability changes, the phenomena affecting joints, such as closure, shear, dilatation, etc., should be represented by discontinuum analyses. Considerable shearing may occur, leading to increased permeability. In this connection, it should be noted that joint asperity may play a significant role. It is therefore essential to back up the models with permeability measurements.

Paper No. 14 - W.F. Ubbes

This paper presented the problems arising from the impact of the work and potential answers to them. It was based on the "Mine-by" experiment at CLIMAX (Nevada test site) and on the tests performed in an experimental mine by the Colorado School of Mines. The CLIMAX experiment proposed to compare stress and displacement effects from mining with those produced by a thermal load. The experiment showed that the effects of excavation were similar to those of heating, but whereas the latter were chiefly determined by the thermal properties of the rock, the effects due to excavation were more dependent on structural conditions. Since the latter are complex and heterogeneous, excavation effects are more difficult to interpret than thermal ones.

The experiment conducted by the Colorado School of Mines showed, among other things, that rock permeability in the zone affected depends as much on stress as on the number of fractures. It is therefore important to measure stress accurately and to use suitable computer models, taking into account the processes involved and the characteristics of the physical environment.

Paper No. 15 - P.A. Lang

This paper may be regarded as a general overview of the URL excavation response and excavation damage assessment programmes. It described the different studies and the different techniques and types of equipment used. Excavation response was studied, including its influence on permeability changes.

Permeability measurements made in the shaft diverged from the trend predicted by modelling. They indicated lower, not increased, permeability, suggesting that not all the mechanisms involved are fully understood at the moment, thereby reducing the credibility of the models (safety implications). Problems were also encountered when measuring stress with vibrating wire gauges. However, deformation sensors yielded useful data.

In the Room 209 experiment, as with the shaft, permeability data diverged in the same direction (lower K) from those predicted by the model, confirming that some mechanism is still not yet understood. The damage zone seems to extend to about 1 metre, which is after all very limited. It should also be noted that when the permeability levels are very low, the time required to make the measurements may be incompatible with the work schedule.

Work is being done on the assessment of measurement techniques, including the use of tracers. The "operating phase" programme investigating time-dependent behaviour of the damaged zone in relation to different factors such as lithology, rock stress and excavation geometry is of great interest, but since it is still at the project stage, no lessons or conclusions can be drawn at the moment.

The other papers, Nos. 16, 17, 18, 19 and 20, supplied more detailed information on several studies, in particular on the measurement apparatus.

Paper No. 16, presented by P.M. Thompson, described BOF-EX, an extensometer used for observing the movement of fractures during drilling (it measures relative displacement between adjacent anchors). This is a simple apparatus which seems to operate satisfactorily. PAC-EX is a packer extensometer showing equally good performance.

Paper No. 17 by R.A. Everitt (presented by P.A. Lang) dealt with mapping of the excavation damage zone around a circular access shaft during excavation at a depth of 300 metres. Damage distribution was shown to be correlated with rock fabric and with the principal stress orientations.

Paper No. 18, by R. Koopmans, described a dilatometer for assessing the deformation modulus by measuring stress and thereby evaluating the extent of excavation disturbance. The measurements made in Room 209 yielded useful data, although the fracture measurements were difficult to interpret.

Paper No. 19, by P.A. Lang, was about tests performed in Room 209. This was a study on the fractures intersecting the vault. Measurements were made by means of packers straddling the fracture through a borehole, and triaxial stress was observed at different distances from the fracture. A network of extensometers was also used to measure convergence of the tunnel walls. Stress variation measurements were in good agreement with forecasted values but permeability measurements were not. Consequently, it is important to measure displacements and permeability changes at the same point along a fracture so as to correlate the processes and build an accurate model. There is also a need to improve methods for measuring flow rates in the fractures. Sound and convergence measurements were satisfactory. The ways in which water flowed through the fractures were modified by shearing, which may affect relationships between water-bearing fractures and, for instance, reduce permeability in the system as a whole, whereas increased permeability would normally be expected.

Paper No. 20, presented by C.D. Martin, was the last paper in Session II(b), and described the behaviour of rock under the effect of high stresses following excavation (observations made from 255 to 352 metres). Stresses were measured in-situ by means of hydraulic fracturing and overcoring as a function of time. The TRIVEC system was used to measure three-dimensional displacement. Micro-cracking plays an important role in time-dependent rock behaviour. This occurs through a combination of shearing and splaying. The damaged zone extends to about one metre.

III. CONCLUSIONS

What general conclusions may be drawn?

First, the scale and high standard of the research programmes already conducted or planned at URL deserve mention. By no means have all the problems been solved, but success is around the corner. Many techniques and different kinds of equipment have been used and are already operational, improving our understanding of fractured rock. The following points are also worth mentioning:

a) One of the difficulties was how to assess the effect of excavation on fractures in space and time, and its impact on permeability changes. Consequently, it was difficult to model the processes involved. For instance, it is not easy to measure permeability within the damaged zone. Instrumental procedures must therefore be developed and improved, although experience shows that they are often difficult to apply. Research and development should therefore be conducted in this field, especially on the measurement of very low permeabilities.

b) Care should of course be taken to decide in advance what is to be measured and how, so as to avoid wasting time and effort on measuring non-significant processes. Parametric sensitivity studies may be useful at this stage. Certain measurements should also be made prior to excavation, since it might be too late afterwards; hence the need for forward planning.

c) Understanding of water flow modes in fractures under the effect of stress changes should be improved, requiring good prior knowledge of the three-dimensional features of fracture networks. In addition, reliable and realistic models must of course be designed which are closely based on experimental data. This means that the processes leading to the formation of the damaged zone must be well understood.

d) In the light of the results already obtained, however, it seems that the damage arising from excavation in crystalline rock is very limited in space (ranging from a few ten centimetres to seldom more than one metre). It is therefore in the area surrounding access shafts that such processes are likely to have safety implications (break with the surface through the fractured zone); hence the importance of sealing techniques.

e) Finally, I did not note any mention of the chemical effects of change in the papers presented. Perhaps this is not a major problem, but it ought to be examined.

SESSION III

EXCAVATION RESPONSE IN SALT

SEANCE III

EFFETS DE L'EXCAVATION SUR LES FORMATIONS SALINES

Chairman – Président

R.V. MATALUCCI
(SNL, United States)

LES EFFETS DES TRAVAUX D'EXCAVATION DANS LES
DEPOTS SOUTERRAINS DANS LES FORMATIONS SEL

EXCAVATION RESPONSES IN UNDERGROUND REPOSITORIES
IN SALT FORMATIONS

Prof. Dr. M. Langer, Prof. Dr. A. Pahl & Dr.-Ing. M. Wallner
Federal Institute for Geosciences and Natural Resources
Hannover, Federal Republic of Germany

ABSTRACT

The paper describes the work of the planning and construction of underground repositories for radioactive wastes in rock salt. The geotechnical
stability analysis is a critical part of the safety assessment. Engineering-
geological study of the site, laboratory and in-situ experiments, geomechanical modeling, and numerical static calculations comprise such an analysis.

With regard to excavation responses the principle features of the rheological processes in rock salt are plasticity, creep and fracturing. Treatment of these features involves geotechnical measurements (laboratory and in-
situ), theoretical investigations (continuum mechanics), and microphysical
considerations (e. g. dislocations, grain structure). A knowledge of the
thermo-mechanical behavior of rock salt is an essential part of any assessment
of the stability of underground openings for waste disposal purposes.

The integrity of the geological barrier can be assessed only by making
calculations with validated geomechanical and hydrogeological models. The
proper idealization of the host rock in a computational model is the basis of
realistic calculations of thermal-stress distribution and excavation damage.
Some results of these calculations for the preliminary repository design of
the Gorleben site and the ASSE II research mine are given.

RESUME

Le présent article expose les études pour la conception et la réalisation de stockages souterrains de déchets radioactifs dans le sel gemme. Pour
la démonstration de la sécurité, l'analyse de la stabilité geotechnique est un
point essentiel. Une telle analyse comprend l'étude géologique et géomécha-
nique du site, des essais en laboratoire et in situ, des modélisations et des
calculs statiques (recherches numériques). Les derniers résultats de la recherche sur le comportement méchanique des massifs de sel sont communiqués ici.

L'exemple de projet de stockage de Gorleben montre comment, par la recherche et les méthodes modernes de la géotechnique, il es possible de prouver
que les déchets peuvent être éliminés de la biosphère de manière sûre.

1.   INTRODUCTION

Permanent repositories for radioactive wastes in mines are meant to protect man and his environment from being harmed by the ionizing radiation from the wastes. To attain this objective, specific measures are to be carried out. The engineering concept for the repository and the requirements placed on the radioactive wastes (e. g. chemical and physical form, radionuclide content) are determined by the geological situation, which varies from site to site. Thus, the requirements cannot be put in the form of quantitative safety requirements valid for all cases. The necessary safety of permanent repository for radioactive wastes, therefore, can be demonstrated only by a site analysis, in which the entire system, "the geological situation, the repository, and the form of the wastes and the waste containers" and their interrelationships are taken into consideration. Therefore, technical, geotechnical and geological data are the basis of the site analysis.

The identification of the problems and their priorities requires a systematic examination of the measures necessary for planning, construction, and operation of the repository. This examination shows that the site analysis has three essential tasks:

-- assessment of the thermomechanical load capacity of the host rock (in this case, rock salt), so that deposition strategies can be determined for the site,
-- determination of the safe dimensions of the mine (e. g. stability of the caverns and safety of the operations), and
-- evaluation of the barriers and the long-term safety analysis for the authorization procedure.

Geotechnical and geological data are needed to answer the various questions associated with these three tasks. The following are examples of such questions:

-- Are the homogeneous parts of the rock salt large enough for the repository?
-- Which failure scenarios cannot be excluded on the basis of the structure of the salt dome?
-- Can the required width of the safety zones between the repository and unfavorable layers (e. g. Main Anhydrite and carnallitite) be observed?
-- What is the load capacity of the rock under the expected thermomechanical stress changes (e. g. rock burst and creep rupture)?
-- Will the caverns become unusable due to convergence, dilatancy, or subsidence?
-- Can the thermal load lead to decomposition of the minerals (e. g. loss of water of hydration) and hazardous migration of brines?
-- Can uncontrolled generation of gases or brines be avoided?
-- Will the integrity of the salt dome be influenced by long-term changes in the stress (e. g. fracture formation)?

To answer these and similar questions, an extensive and detailed characterization of the salt dome must be made, based partly on the underground exploration. This characterization, together with the requirements given in the repository concept, the design of the repository drifts, and the failure analysis, is used as a basis for deciding the suitability of the salt dome as

the host rock for a permanent repos.
umented for the authorization procedu.

oactive wastes. This is do.

2.   PLANNING AND DESIGN OF A REPOSITORY I.

2.1 Concept

The geoscientific conditions that must be tak.
planning a permanent repository result from the limiteu consideration in
capacity of the host rock, in this case the salt rock of omechanical load
(Fig. 1). The heat developed by high-energy radioactive w. alt dome
to the surrounding rocks. The increase in temperature cause. is transfered
changes in the properties of the rock, resulting in rock creep nificant
pressure-temperature conditions, this creep can lead to fracturing der certain
a negative effect on the stability of the repository caverns. Certa. which has
erals in salt deposits undergo chemical changes. For example, dehydrati.n
(loss of water molecules bound in the mineral) of carnallite or the migration
of brine towards parts of the rock at higher temperatures. Stress is also
produced by expansion of the rock resulting from the rising temperature. The
rising temperature causes expansion of the rock, which leads to stress, which
can concentrate in competent rock layers (e. g. Main Anhydrite). The deformation properties of the rock (creep) permits a certain amount of thermal expansion; however, owing to the stiffness of the rock, resistance to this expansion occurs, leading to thermally induced stress. If this is so great that
fractures appear in the anhydrite, the integrity of the salt dome in the case
of an unfavorable distribution of the anhydrite layers (i. e. providing migration pathways to the edge of the salt dome) would be placed in question.

Two geomechanical stability problems must be taken into account when
planning the repository: the stability of the underground caverns and the integrity of the salt dome (as barrier), as well as the chemical processes that
determine the limits of thermal loading. For the underground exploration,
this means that not only the geological situation, particularly the location
and thickness of the Main Anhydrite and carnallite (Stassfurt Potash Bed)
layers, but the parts of the salt dome containing mineral alterations and
pockets of brine or gas must be determined. Moreover, the large-scale thermo-
mechanical behavior of the salt dome must be determined for the homogeneous
parts of the dome for the analysis of the various repository concepts. There
are no guidelines for this. The technological state-of-the-art is determined
by the scientific state-of-the-art. Work is being done on this in the "System
Analysis, Mixed Concept" subproject of the direct deposition research project
of the Federal Ministry of Research and Technology.

2.2 Geotechnical Analysis of the Stability of Underground Openings

Although a permanent repository for radioactive wastes in the geological medium is an engineering project that has never been carried out before,
there are, however, many of its parts that are similar to previous underground
structures that have been safely and economically constructed. For example,
experience gained in the construction of underground storage for oil and gas
in salt domes and in the mining of salt can be utilized. Hence, there is extensive engineering information available for the rock mechanics aspects of
designing the mine. The process for designing the caverns is set out in the

...iation for Civil Engineering (DGEG). Rec-
...deration the special properties of salt rock
... fracturing) are being worked out by Working
...struction of caverns in salt rock, especially
...sitories for special wastes.

recommendations of the ...
ommendations that take ...
(e. g. creep, plastic ...
Group 4 of the DGEG ...
for the constructio ...

Central to ...culations for the safety analysis is the construction
of a model of th... that similates as close as possible to the actual conditions of the ...e. the geology, stress conditions, constitutive laws,
etc.) before a...he essential parameters (i. e. those decisive for the
engineer incl...ility) in his model, neglecting unimportant aspects, using
safety of th...for side effects, and investigating the decisive factors in de-
the ideal ...mining, aspects not included directly in the model must be sup-
tail. As ...y experience.
plement...

When all of the factors (e. g. primary stress, rock properties, construction processes, quality of the monitoring) have been assessed, the sum of
the certain factors must be so pre-dominant that failure is improbable for the
life of the repository. These criteria for a safety analysis are influenced
by various natural and technical factors (Fig. 3). Therefore, the analysis
can be carried out for the specific case only by a combination of engineering
geology, geotechnical, and rock mechanics investigations, monitoring and mining experience. The extent and exactness of the survey, as well as the safety
factors included in the calculations, are determined by the severity of the
damages that could occur. The different parts of the investigation may not be
considered separately, they are a functional unity and are interrelated (e. g.
the mechanical model of the rock, the parameter analysis, and the statics calculations). The status of the knowledge of the stability must be continually
checked for each phase of planning, construction, and operations. This is an
important part of the geotechnical safety concept described here.

These criteria for the safety analysis, which is the basis for the dimensioning of the mine, illustrate the interrelationships of the underground
survey with the other parts of the investigation, particularly with respect to
the consideration of the influence of natural factors (Fig. 3). These natural
factors include the geological situation (e. g. inhomogeneous bedding, fault
tectonics, petrography of the salt rock), the in situ stress conditions, presence of brine and/or gas, temperature-, and time-dependent fracture and deformation behavior of the salt rock. They must be determined with sufficient
accuracy that the rock mechanics simulation model reflects the actual conditions despite the use of ideal cases. Parameter analysis is very useful for
evaluating the importance of the data and determining whether further studies
are necessary.

2.3 Long-Term Safety Analysis and Barrier Assessment

From the formulation of the objectives of a permanent repository for
radioactive wastes can be concluded that failure of the system during the
post-operations phase must be impossible, i. e. the tranport of hazardous
amounts of radionuclides into the biosphere must be hindered. To guarantee
the health and safety of man during this long period, several independent
technological and geological barriers are used to prevent the release of hazardous substances. These barriers form a system consisting of the waste, the

repository proper, and the enclosing geological medium: the form of the waste, the container, backfill material in the boreholes and drifts, bulkheads, and finally, the host rock in the near field and far field. The development of a realistic and testable long-term repository concept to meet the requirements, taking into consideration all of the barriers, is a difficult task and is the subject of extensive research in many countries [1].

One possibility for carrying out a long-term safety analysis is shown in Figure 4. This concept includes the separate analysis of the individual barrier systems, both artificial and natural, the analysis of physical and chemical processes that occur in the near and far fields of the repository, and the assessment of failure scenarios. Risk analysis is used to assess the probability of failure of the container as a barrier. The other artificial barriers (back fill etc.) are assessed by a geotechnical stability analysis. The geological system is analyzed by forecasting future geochemical, hydrogeological, and tectonic occurrences. The analysis of the chemical and physical processes in the near field must include leaching, migration of brines and the associated changes in permeability. The convergence of the mine tunnels, resulting from creep of the salt rock, is one possible scenario for the release of stored wastes. Processes in the far field are uplift and subrosion of the salt dome, as well as fractures caused by thermally induced concentrations of stress. In the concluding failure analysis, the interaction of all the barriers during specific theoretically possible events that could release the stored wastes into the biosphere.

The probability for each failure scenario must be substantiated. This substantiation is provided by data from the underground, geological and geotechnical surveys. This data is also used for constructing models for the site-specific safety analysis. This data must permit a sufficiently precise estimate of the conservative assumptions that must be made. Hence, data must be obtained especially on the geological structures to identify failure cases, on the long-term geochemical processes of the salt dome, and on thermomechanical processes for the model calculations to simulate transport of wastes.

3.   RHEOLOGICAL PROCESSES IN ROCK SALT

Moreover, intensive basic research in the fields of geotechnical and rock mechanics has been and is being conducted in various countries involved in the permanent storage of radioactive wastes [2]. Thus, in principle, but not in every detail, the necessary tools (i. e. methodology, scientific principles, technical know-how, and experience in construction and operations) are available for reliably predicting the feasibility of various techniques for the permanent storage of radioactive wastes and the consequent safety aspects.

On the basis of the results of BGR studies, a review is given in the following sections on what has been achieved in the intensive research and development work conducted in the field of geotechnics to solve problems of rock mechanics involved with a permanent underground repository [3].

Pillars, caverns, and other mine constructions can be dimensioned on the basis of the above-described concept for a geotechnical safety analysis only if the mechanical behavior of the rock mass can be described with sufficient accuracy.

The foundations of the theoretical and experimental treatment of the deformation processes in salt rock are obtained from an analysis of the phenomena of creep, plasticity, and creep failure. Various reviews summarizing and assessing such analyses have been published and presented at special symposia in recent years.

3.1 Creep

Let us first deal with creep. Creep refers to time-dependent deformation under constant load. Creep is measured with devices in which a constant load can be maintained over months or even years. Since creep is a thermally activated process, it is important to study creep as a function of temperature. The data are usually presented in plots of creep rate versus applied compressive stress (Fig. 5).

The test results are supplemented by studies of microphysical deformation mechanisms in order to theoretically derive the equations for certain temperature and stress conditions. Only in this way can extrapolations be made from empirically based models to other conditions of stress, temperature, and deformation rates, as will prevail in a permanent repository for radioactive wastes. In the case of rock salt, the movement of dislocations (i. e. linear lattice defects) in the crystal lattice represent an important deformation mechanism. The movement of dislocations can be expressed as dislocation glide or dislocation climb. The creep laws describing the various deformation mechanisms can be represented in so-called deformation diagrams. An example of a deformation diagram for natural polycrystalline rock salt is shown in Figure 6.

Deformation diagrams are very useful for analyzing slow creep processes like those expected in the post-operational phase of a permanent repository for radioactive waste and those during the genesis of salt pillows and salt domes. These processes, called halokinesis, take place at such a slow rate, that laboratory measurements are not useful for a purely engineering interpretation. In the deformation diagram shown in Figure 6, it can be seen how far removed the conditions of halokinesis are from those of the laboratory tests. Therefore, the BGR developed a precision creep test rig with which the temperature can be maintained to within a thousandth of degree so that creep rates can be measured that are an order of magnitude slower than possible in normal tests. Only by combining geotechnical testing of materials and studies of microphysical deformation mechanisms is it possible to extrapolate from laboratory data to the deformation that takes place in nature over very long periods of time. It is often questioned whether laboratory data can be transferred to the much larger-scale in-situ situation. However, in-situ creep tests on rock salt pillars has demonstrated that results are equivalent to those of laboratory tests.

On the basis of the above-described considerations, a creep law for engineering purposes was derived for rock salt. The mathematical formulation of this law is shown in Figure 7. The values for the thermomechanical parameters are given in Figure 8; the rheologial model is illustrated in Figure 9.

The total strain rate is equal to the sum of the elastic strain rate, time-dependent reversible and irreversible strain rates (= creep deformation), and a failure deformation rate (= irreversible deformation with discretely distributed discontinuities and dilatation). Steady-state creep, the main component of long-term deformation, is formulated as the sum of two functions describing deformation climb and deformation glide.

Summarizing, one may conclude the following: Creep in rock salt can, both theoretically and empirically, be defined sufficiently for a perfectly acceptable engineering assessment of the phenomenon as a basis for a safety analysis. Moreover, test results establish that steady-state creep of natural, polycrystalline rock salt is governed less by impurities at grain boundaries than but the dislocation density and the pattern of the dislocations in individual crystals, together with the associated microphysical processes within the individual crystals. This certainly plays a decisive role in determining the boundaries of the homogeneous parts of the salt dome.

3.2 Creep Failure

The scientific elucidation of the phenomenon of creep failure in rock salt is more difficult. A very important task of geotechnics with respect to permanent repositories for radioactive waste is to ensure that the repository is so designed that neither fissures nor material failure that could lead to entry of water or brine into the repository occurs in either the near-field or far-field of the repository during the operations phase or during the post-operations phase. The major objective of creep failure studies, therefore, is the quantitative determination of the threshold conditions for such failures.

The tests for this are carried out for the most part in a triaxial test rig, in which lateral and axial pressures on the cylindrical samples can be varied with time independently of each other. The BGR has developed a special test rig (Fig. 10) for testing salt rocks that have a higher failure rate than other types of rocks. This rig can be used for tests up to 400 °C. The maximum force that can be applied by this apparatus is 2 500 kN with a maximum piston stroke of 17 cm. The maximum lateral pressure that can be applied to the sample is 1 000 bar. The temperature can be regulated with a precision of 1 °C. Cylindrical samples 25 cm high and 10 cm in diameter are tested. The hydraulic equipment occupies an entire basement room separate from the room containing the test rig owing to vibration. The rig is controlled and the data is recorded by computer.

The results of the stress or strain-controlled monoaxial and multi-axial tests for failure strength can be summarized as follows:

-- The strength of the rock depends to a greater extent than creep on the mineral composition of the rock, their distribution, and the texture.
-- Rock salt has a very low tensile strength.
-- The compressive strength of the rock increases with increasing isotropic pressure.
-- In the case of monoaxial loading, rock salt has a pronounced strength anisotropy, which decreases with increasing isotropic pressure.

Moreover, failure strength is also very dependent on the deformation rate. It can be seen in Figure 11 that fractures appear only when a certain deformation rate has been exceeded, which depends on the confining pressure. This limit is given by the steady-state creep equation obtained for disloca- tion glide. The horizontal straight lines in Figure 11 show the load-bearing capacity of the rock salt with respect to long-term stability with respect to fracturing at selected deformation rates. They serve as a basis for safety calculations for the operations phase of the underground repository. Thus, a reliable basis is available for calculating long-term safety of a permanent repository in terms of fracture mechanics.

4. MODELLING OF EXCAVATION RESPONSES

4.1 Principles of Modelling

Due to the multibarrier principle, the geological setting must be able to contribute significantly to the isolation over long periods of time. Al- though qualitative understanding, description models and expert judgement are important factors for a reasonable assurance that the integrity of the geo- logical barrier - within the system "waste product, disposal facility, geo- logical situation" will be valid, the assessment of the integrity of the geo- logical barrier should be performed by calculations with validated geomechan- ical and hydrogeological models. The proper idealization of the host rock and the surrounding geological formations into a computation model is the basis of a realistic calculation. The natural system has to be considered with given properties as permeability, thermo-mechanical behaviour, tectonic fractures as well as in-situ stress. Initial conditions given by the operation of the re- pository have to be taken into account, whereby the repository as a geotech- nical system can be optimized to the existing geological situation. This op- timization must regard systemimmanent conditions, e. g. geological and geo- chemical longterm processes, limiting values for temperature, rock failure modes. The adequacy of the computation model with respect to predict the bar- rier efficiency has to be proved by a validation procedure.

It is obvious that modelling can only reach a certain level of accu- racy, since the actual behaviour of a complex geological structure will always remain unknown up to a certain extend.

The geoscientific approach to overcome this general difficulty is a continuous improvement if the model, appropriate to the improved knowledge of the input data. The main features of this approach are the etablishment of a consistent constitutive relationship for the mechanical and hydrological be- haviour, validation of this model, and quantification of site-relevant input data.

Herein, model validation has to follow a strict scientific procedure:

-- prior to validation, the numerical code used for computation has to be verified. This means, it has to prove that the code gives mathematically correct answers,
-- model validation is achieved through successful predictions for laboratory tests or in-situ tests, taking into account a consistent constitutive model, proper boundary conditions, and initial condi- tions. Model validation in this sense is not curve fitting by

back-analysis but a demonstration to what extent a particular consistent model is able to describe the response of the host rock,
although the constitutive model perhaps does not take into account
the entire mechanical and hydrological behaviour, and
-- a validated constitutive relationship for the rock mass is then the
proper basis for modelling of the site-specific geological situation. However, site-specific units of equal behaviour and related
parameters have to be determined and are to be confirmed by field
tests.

The numerical models used in the assessment of the efficiency of the
geological barriers have to represent the physical system of the geological
medium. The degree of representation (simulation) is significant because the
misapplication of mathematical models may produce erroneous results. Anyway,
the use of mathematical models must be justified by the site data. Even if
these data were based on a comprehensive site investigation, there is always a
certain amount of uncertainities modelling a geological system. Therefore,
boundary approaches must be used in a conservative manner. The extent to
which boundary approaches can be justified and demonstrated as being conservative will affect the reliance of the estimation of the barrier efficiency.

Nevertheless, numerical calculations based on models are of particular
significance in the case of the final storage of wastes since licensing procedures require the prior reliable and convincing demonstration of safety.

4.2 Geomechanical Models and Calculations

The main aim of geomechanical modelling is directed to stability calculations in order to demonstrate that stress redistribution due to mining operations and possible thermally induced stresses do not endanger the failure-
free state of equilibrium in the rock, do not cause any inadmissable convergence or support damage during the operative period, and maintain the long-
term integrity of the rock formations. It is necessary to calculate the distribution of stress and deformation in the formations surrounding the mine,
under consideration of the temperature-dependent rheological properties of the
rock and to compare them with the limiting load-bearing capacity of the rock
mass. Above all, this requires the formulation of the geomechanical model and
the associated calculation models, the study of parameters, and the definition
of failure criteria.

To build up a calculating model, certain idealizations are unavoidable.
In the terms of this model one can distinguish between:

-- rock-mechanical models, and
-- static models.

The rock-mechanical model encompasses:

-- the geological structure of the rock mass,
-- material laws to describe the time, temperature- and stress-dependent deformation and strength behaviour of the rock mass, and
-- the primary state of the rock mass (stresses, temperature).

In the static model the following conditions have to be determined:

-- geometry (establishing the load-bearing capacity of the rock mass
   by spatial, planar or rotationsymmetric substitute systems),
-- stress states and their changes over time,
-- thermal effects (e. g. storage and heat-generating wastes, ventilation), and
-- mechanical effects (e. g. leaching, backfilling, static loads, etc.

A calculating method which is particularly suitable for the solution of
such geomechanical problems is the method of the finite elements (FEM). This
numerical calculation method which is specifically tailored to the requirements of automatic data processing even allows consideration of many important
factors, e. g. tectonic features, operating conditions, mine geometry, construction methods etc. to an accuracy closely approaching the real situation.
The FEM method is also suitable for the calculation of non-linear limiting
values. For the specific purpose of the analysis of waste disposal facilities, the Federal Institute of Geosciences and Natural Resources (BGR),
F.R.G., has developed the ANSALT computer program (analysis of non-linear
thermomechanical analysis of rock salt) [4].

The development focused on the following:

-- Implementation of a material law formulation for salt rock which
   includes temperature-dependent rheological deformation behaviour
   according to the current state of know how, as described in the
   preceding, and which can be expanded in a simple, modular fashion.
-- Development and integration of computer techniques for the simulation of rock mechanical processes and mining measures.
-- Optimization of computing techniques in regard to providing a user
   friendly program.

The capacity of the ANSALT program system was proven in participation
at the 2nd WIPP Benchmark Program in the USA. ANSALT contains in its element
bibliography all single, double and threedimensional isoparametric elements
for the discretization of underground cavity structures which are necessary to
process rock mechanical problems. Furthermore, certain parameters and initial
conditions are simulated in a suitable fashion, as can arise due to the peculiar tasks involved in mining engineering underground. The program system
consists of four selfcontained but communicating subroutines (Fig. 12):

-- ANSPRE, a pre-processor for simple and reliable preparation of the
   input data,
-- ANTEMP, a finite element program to calculate the heat transport
   problem,
-- ANSALT, a finite element program to solve thermo-mechanical problems, and
-- ANSPOST, a post-processor for graphical analysis and presentation
   of the results.

The long-term assessment on the geological barrier integrity cannot be
evaluated from experiments alone but is only possible by computations. In addition geomechanical computations on the rock behaviour of a waste mine have
the following objectives:

-- analysing thermo-mechanical processes by calculations shall lead to a proper assessment of consequences,
-- experience-based conclusions can be extended by computations,
-- rock-mechanical criteria for a stable mine design can be developed from computational parametric studies,
-- such criteria are necessary to adapt a preliminary mine design to the real geological situation.

4.3 In-situ Stress and Permeability

The initial state of the rock mass, e. g. the in-situ stress, has a particularly decisive influence on the calculations, and thus on the engineering safety analysis. Determination of the natural stress in a rock mass is one of the most difficult problems in rock mechanics. For this reason, great efforts are being made throughout the world to develop more effective methods of measuring stress.

The following methods and equipment will be used for the project Gorleben:

-- the BGR overvoring method for measuring the initial stresses,
-- the BGR dilatometer test for measuring borehole deformation under defined loads,
-- mechanical frac tests to measure stress and stress changes,
-- slot tests and compensation tests using large area flat-jacks (BGR system),
-- hydraulic displacement sensors (Gloetzl type) at various depths and orientation for long-term recording of stress,
-- vacuum probe for measuring permeability of the rock and back-fill, and
-- biaxial tests on hollow, cylindrical test bodies in laboratory experiments.

Stress sensors and dilatometers in boreholes are used to determine in-situ stress by the over-coring method. This well-known method could previously be used only in boreholes less than 100 m deep. Further development of the method and development of special equipment now makes it possible to measure stress and deformation in boreholes 300 m deep and more. Especially the use of a down-hole computer (to avoid the necessity of a cable in the drilling rod) makes such depths possible.

The overcoring method for measuring in-situ stress is shown schematically in Figure 13. The probe, which functions inductively, is lowered together with the downhole computer into a pilot borehole with a diameter of 4.6 cm. The borehole is then overcored with a diameter of 14.6 cm; during this process the borehole diameter is continually measured as a measure of stress release deformation. Direction-dependent stress values are then calculated from this data. This calculation can be made only if the stres and deformation behavior of the rock mass is known. For this reason, dilatometer measurements are made at the same depth. For calculating the in-situ stresses in salt rock non-linear deformation as a function of time (e. g. creep law, Fig. 7) must be taken into account [5].

In general, the salt rock retains its impermeability if the average stress is sufficiently high. This is not the case, however, in the immediate vicinity of the tunnel. The salt rock becomes brittle and begins to fissure, forming a zone of permeability that cannot be neglected. This permeability decreases, owing to the viscoplasticity of the salt, as soon as the creep of the salt rock is stopped by the backfill and the container, which causes the average stress to rise again. After the tunnel has been filled with crushed salt and the backfill is compacted by the surrounding salt rock and with time looses its porosity and permeability. This process is accelerated by elevated temperatures.

To determine the permeability of the salt rock in the loosening zone of a tunnel, measurements are made with a newly developed vacuum measuring equipment. This equipment consists of a measuring and processing control system and several double packers with an integrated sensor housing for pressure and temperature measurements and a variable packer interval of 1 - 5 m. The double packer can be used in boreholes with diameters of 46 - 66 mm and 76 - 116 mm.

5.   EXAMPLES OF THE CALCULATION OF LARGE-SCALE DEFORMATION IN PERMANENT UNDERGROUND REPOSITORIES

   5.1 Asse II Mine

The former Asse II salt mine is currently being used for research on permanent repository for radioactive wastes. During the 1970s, low-level radioactive wastes were stored in barrels in the mine. The Federal Government has left itself the option of being able to use the mine again for storing low-level radioactive wastes at a later date. For this reason, the Clausthal mining authority requested the BGR to assess the stability of the Asse Mine.

In terms of rock mechanics, the large workings at several levels, chiefly on the southwest flank of the salt dome, have an influence on the stability of the rock mass. The stability is also influenced by the continuous, steady movement of the salt in the pillars between the galleries demonstrated by a comprehensive geodesic survey. These were the decisive characteristics that had to be included in the statics model (Fig. 17). A cross section through the salt dome was covered with a net of finite elements for the numerical calculations (Fig. 15). The stress, deformation, and creep rate was then calculated for each grid point. These were then compared with the in-situ measurements. The results can be presented on the finite element grid itself. Deformation is shown in Figure 16, zones of tensile stress that lead to fracturing are shown in Figure 17. Summarizing, the stability of the Asse Mine may be assessed as follows:

The calculations have shown that with time the stress distribution around the workings on the southwest flank approaches steady-state conditions, i. e. with time the salt rock mass reaches an equilibrium in which stress is redistributed as a result of deformation caused by creep. Owing to the creep capability of the salt rock, a sudden collapse of the pillar system is not to be expected. In the long term, however, gradual convergence of the mine cavities with progressive defromations and slow destruction of the pillar system cannot be excluded.

5.2 Far-field Analysis Gorleben

In the preceding example, thermomechanical processes did not play a decisive role in the stability analyses. In the case of the permanent repository planned at Gorleben, however, it is intended to store high-level radioactive wastes and heat-generating intermediate-level radioactive wastes. For this reason, the thermomechanical properties of rock salt are of particular importance for the numerical calculations for the safety analysis for the permanent repository at Gorleben. Therefore, these properties definitely have to be included in the geomechanical models used for the planning of the repository [6]. Even though the details of the geological structure of the Gorleben salt dome will be determined during underground exploration, some information is already available about the gross structure of the salt dome from deep boreholes at Gorleben. This information is the basis of the model shown in Figure 18.

5.2.1 The Computational Model

The objective of the far-field analysis is the determination of the thermal and thermomechanical effects of the deposition of heat-producing radioactive wastes on the salt as a geological barrier. In terms of rock mechanics, the following aspects must be taken into consideration for the determination of the thermal load capacity of a salt structure for which the values for the thermal and mechanical parameters must be known:

-- maximum permissible temperature:
    -- to prevent the release of water of hydration, e. g. from carnallite, and
    -- at the margins of the salt dome to prevent mobilization of aqueous solutions and the associated leaching.
-- integrity of the salt rock as geological barrier to hinder the movement of fluids in the salt dome:
    -- fracture formation due to thermomechanical reduction of stress, with a tendency to tensile stress, particularly near to the top of the salt dome, and
    -- opening of paths in competent layers of the salt, e. g. anhydrite and clay layers.

Only aspect of fracture formation is taken into consideration for this initial far-field analysis, i. e. the heat generated by the radioactive wastes is being evaluated in terms of fracture formation in the salt at the top of the salt dome. The basis for this failure scenario (the formation of fractures) is illustrated in Figure 19 on the basis of a highly simplified model of the salt dome. The figure shows (a) temperature, (b) horizontal stress for the case of a high creep capacity of the salt rock, and (c) horizontal stress for the case of a low creep capacity of the salt rock as a function of depth (from ground level to below the repository) at four times after deposition of the radioactive wastes.

The increase in temperature in the repository results in a confined thermal expansion of the salt. This leads to higher stress in the heated part of the salt dome and reduction of stress, with a tendency to tensile stress, at the top of the salt dome. Owing to the kinematic boundary conditions of a half-space and in order to maintain equilibrium, a characteristic thermally induced stress distribution is produced.

The finite element grid shown in Figure 20 was used for the initial thermomechanical far-field analysis. It represents a plane NW-SE section through the Gorleben salt dome. The section extends to a depth of 4 000 m over a distance of 9 000 m. To simplify the model, only four homogeneous areas were considered: the cap rock (Quaternary/Tertiary), the country rock (Cretaceous/Bunter), the basement (Rotliegende), and the salt rock. The values for the thermal and mechanical parameters are given in Table 1.

5.2.2 Deposition Schemes considered in the Calculations

The following deposition schemes have been considered in the first phase of planning a mixed wastes repository:

-- deposition only in boreholes,
-- deposition in drifts at three levels,
-- mixed wastes deposition A, and
-- mixed wastes deposition D.

The calculations for all of these schemes are based on a IE/DE ratio of 500 t/200 t. The following cooling times of the wastes are assumed:

-- spent fuel in casks: 30 years,
-- HLW canisters: 40 years,
-- HTR: 10 years, and
-- MLW(Q): 7 years.

Each of the four deposition schemes contains several sections in which different containers can be stored separately or in specified combinations. The far-field analysis of each of the four schemes is based on the section in which the greatest amount of heat is generated. The data for this are given in the form of mean loading values for volumes or areas in Fig. 19.

In addition, a reference thermal output is given based on earlier calculations for the integrated disposal concept. For an analysis of thermomechanical fissure formation at the top of the salt dome, the following parameters were varied for the parameter study: (a) reference thermal source: 30 %, 50 %, 70 %, and 100 % of the values given in Table 2; (b) section width: 200 m and 300 m; and (c) low and high creep capacity. The direct consideration of different thermal outputs, different decay rates, and combined loading for volumes and areas will be treated in a second step.

5.2.3 Results

The results of six simulations carried out during the report period are compiled in Table 3. The mean maximum temperature in the center of the repository and the maximum tensile stress at the top of the salt dome are given.

Temperature isolines are shown in Figure 21 for (a) before deposition and (b) about 110 years and (c) about 810 years after deposition. It can be seen that relatively large increases in temperature occur only in the immediate vicinity of the repository.

The temperature at three points is shown as a function of time in Figure 22: the center of the repository (line 1, node 253), the level of the repository (line 2, node 257), and the top of the salt dome above the center of the repository (line 3, node 267). The maximum mean temperature of 159 °C in the center of the repository occurs after about 90 years. The maximum temperature of about 100 °C at the level of the repository is reached after about 125 years. The temperature increase at the top of the salt dome occurs very slowly, reaching only low values.

The thermomechanical effects of heat generation in the repository at the times 0, 50, and 100 years after deposition are shown in Figure 23 in the form of the stress distribution from ground level to a depth of 2 000 m in a vertical profile through the center of the repository (line 1 = horizontal stress, line 2 = vertical stress, line 3 = von Mises effective stress). It can be seen that in the reposistory near-field an increase in stress will occur as a result of confined thermal expansion and later a reduction of stress will occur as a result of thermal contraction. The changes in stress in the heated repository will cause a reduction of compressive stress at the top of the salt dome, resulting in the production of tensile stress there. This stress becomes compressive again when the repository cools.

The stress at the top of the salt dome above the center of the repository is shown in Figure 24. The reduction of the horizontal primary stress (line 1) occurs very rapidly. The tensile stress reaches a maximum of 2.5 MPa after 42 years. The tensile stress remains the same for almost 150 years and becomes compres-sive after about 170 years.

The influence of the heat output on the increase in temperature and the reduction of stress at the top of the salt dome is shown in Figure 25. A 100 % of the heat output given in Table 2, column 5 results in a temperature increase from 41 °C to 159 °C and a reduction of the primary stress from -6.0 MPa to +2.5 MPa.

The normalized values for ΔT_0 and $\Delta\sigma_0$ therefore are 118 °C and 8.5 MPa. Corresponding values for 30 %, 50 %, and 70 % of the heat outputs given in Table 3 lead to the curves shown in Figure 25. Whereas the maximum mean temperature in the repository is very nearly a linear function of the heat output, the thermally induced tensile stress is distinctly nonlinear. Only heat output below 40 % of the values given Table 2 does not lead to tensile stress at the top of the salt dome.

The occurrence of tensile stress also depends on the width of the repository: The narrower the repository, the lower the stress. If the creep capacity of the salt rock is high, tensile stress is not produced even at high heat output, thus reducing the hazard of fissure formation.

6.   REFERENCES

[1]   Langer, M. et al.: Engineering-geological methods for proving the barrier efficiency and stability of the host rock of a radioactive waste repository. Proc. IAEA Int. Symp., Hannover, SM 289/23, 1986, p. 20.

[2]   Hardy, H. R. jr. & Langer, M. (ed.): The mechanical behavior of salt. Proc. 1. Conf., Pennslv. State Univers., 1981, Trans. Tech. Publ., Clausthal, p. 901.

[3]   Langer, M.: Rheology of rock salt and its application for radioactive waste disposal purposes. Proc. Int. Symp. on Engineering in Complex Rock Formations, pp. 1 - 19, Peking, 1986.

[4]   Wallner, M. & Wulf, A.: Thermo-mechanical calculations concerning the design of a radioactive waste repository in rock salt. Proc. ISRM Symp. Rock Mech. Cavern and Pressure Shafts. Aachen, Vol. 2, 1982, pp. 1003 - 1012.

[5]   Pahl, A. & Heusermann, St.: Der Einfluß des zeitabhängigen Stoffverhaltens auf die Bestimmung gebirgsmechanischer Parameter. FELSBAU, 6 (1988), Essen, z. Z. in Druck.

[6]   Wallner, M.: Stability demonstration concept and preliminary design calculations for the Gorleben repository. Proc. Conf. Waste Management '86, Tuscon, Vol. 2, 1986, pp. 145 - 151.

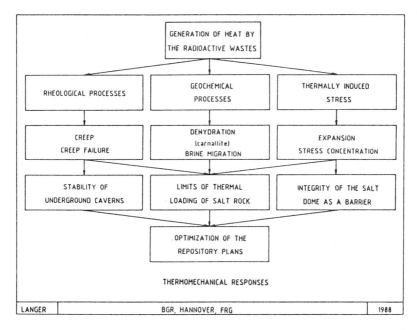

Fig. 1: Thermomechanical responses of an underground repository in salt media

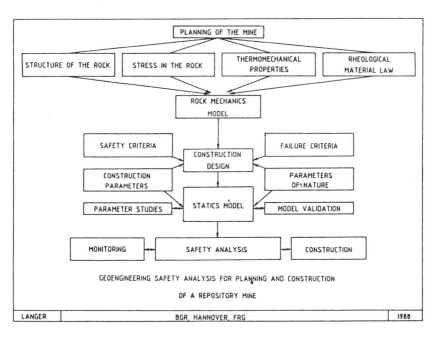

Fig. 2: Geoengineering Safety Analysis for planning and construction of a repository mine

SAFETY CRITERIA	NATURAL INFLUENCES	TECHNICAL INFLUENCES	MEASURES
Deformations	Geological conditions	Cavity geometry	Geological exploration
Stresses	Tectonics	Building processes	Geotechnical investigation
Failure mode	Primary stress	Method of utilization	Static design
Bearing capacity	Mechanical rock characteristics	Conditions of operation	Control test
Brine incursion	Gas and brine deposits	Temperature	Mining measures

Fig. 3: Safety criteria concerning the stability of underground openings.

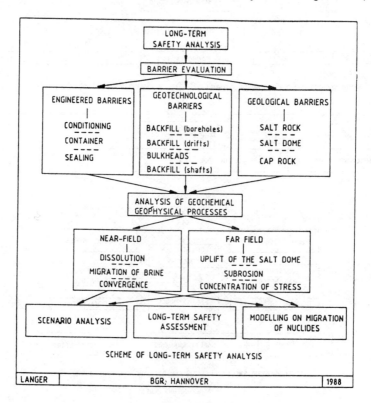

Fig. 4: Scheme of long-term safety analysis.

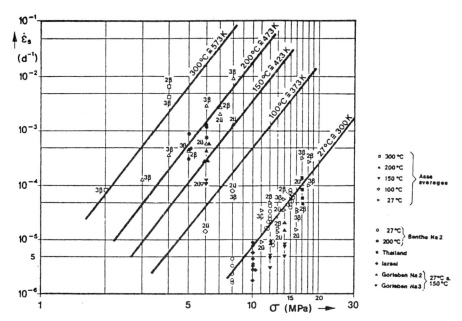

Fig. 5: Creep of rock salt. Creep rate versus applied compressive stress (examples)

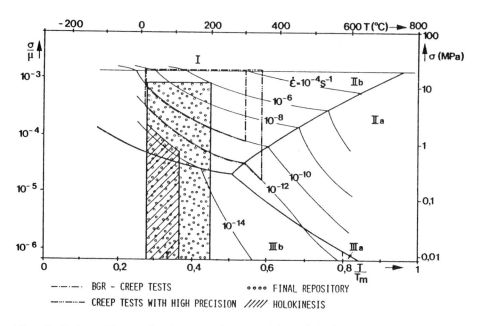

Fig. 6: Deformation-mechanism map for natural rock salt

$$\dot{\varepsilon}_{ij} = \dot{\varepsilon}_{ij}^{el} + \dot{\varepsilon}_{ij}^{th} + \dot{\varepsilon}_{ij}^{cr} + \dot{\varepsilon}_{ij}^{f} \qquad (1)$$

$$\dot{\varepsilon}_{ij}^{el} = -\frac{v}{E}\,\dot{\sigma}_{kk}\,\delta_{ij} + \frac{1+v}{E}\,\dot{\sigma}_{ij} \qquad (2)$$

$$\dot{\varepsilon}_{ij}^{th} = \alpha_t\,\dot{T}\,\delta_{ij} \qquad (3)$$

$$\dot{\varepsilon}_{ij}^{cr} = \frac{3}{2}\frac{\dot{\varepsilon}_{eff}^{cr}}{\sigma_{eff}}\,s_{ij}\,, \qquad \dot{\varepsilon}_{eff}^{cr} = \sum_{i=1}^{3}{}^{i}\dot{\varepsilon}_{eff}^{cr}\,(S,\sigma_{eff},T)$$

$${}^{1}\dot{\varepsilon}_{eff}^{cr} = A_1\exp(-Q_1/RT)(\sigma_{eff}/\sigma^{*})^{n_1}$$

$${}^{2}\dot{\varepsilon}_{eff}^{cr} = A_2\exp(-Q_2/RT)(\sigma_{eff}/\sigma^{*})^{n_2}$$

$${}^{3}\dot{\varepsilon}_{eff}^{cr} = 2[B_1\exp(-Q_1/RT) + B_2\exp(-Q_2/RT)]\,\times$$

$$\sinh\left(D < \frac{\sigma_{eff}-\sigma_{eff}^{0}}{\sigma^{*}} > \right) \qquad (4)$$

$$\dot{\varepsilon}_{ij}^{f} = \frac{1}{\eta} < F > \frac{\partial F}{\partial \sigma_{ij}}$$

$$F = \alpha\left(\frac{|I_\sigma|}{\sigma^{*}}\right)^{m-1} I_\sigma + \sqrt{\mathrm{II}_s} - k \qquad (5)$$

Fig. 7: Mathematical formulation of a material law for rock salt

elastic parameters   failure parameters

$E = 25\,000 \pm 5000$ MPa $\alpha = 1.37$

$v = 0.25 \pm 0.05$ $n = 0.72$

$\alpha_t = 0.45 \cdot 10^{-4}\,k^{-1}$ $k = 0$

steady state creep parameters

$A_1 = 1.2 \cdot 10^{22}\,s^{-1}$ $B_1 = 1.5 \cdot 10^{7}\,s^{-1}$

$A_2 = 1.7 \cdot 10^{14}\,s^{-1}$ $B_2 = 3.6 \cdot 10^{-2}\,s^{-1}$

$Q_1 = 27\,kcal\,mol^{-1}$ $n_1 = 5.5$

$Q_2 = 12.9\,kcal\,mol^{-1}$ $n_2 = 5.0$

$D = 3.5 \cdot 10^{3}$ $\sigma_{eff}^{0} = 20$ MPa

Fig. 8: Thermomechanical parameters for rock salt

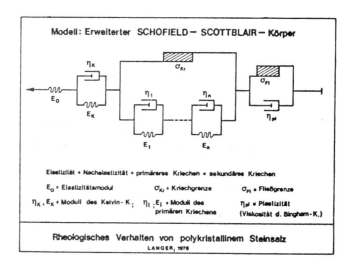

Fig. 9: Rheological model for rock salt

Fig. 10: Triaxial test apparatus for testing rock salt (BGR)

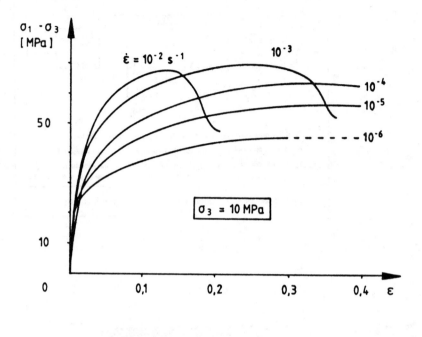

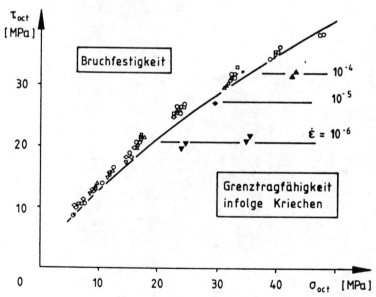

Fig. 11: Long-term strength of rock salt

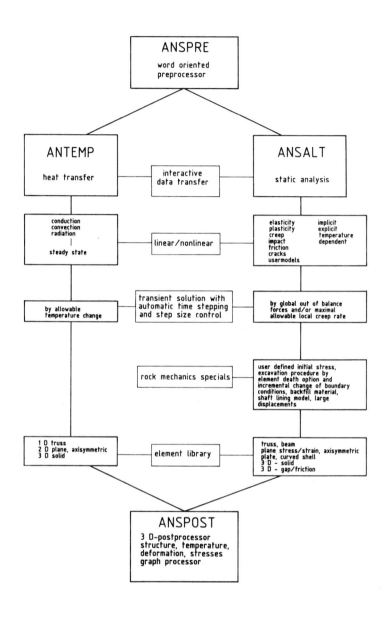

Fig. 12: Computer code ANSALT (BGR)

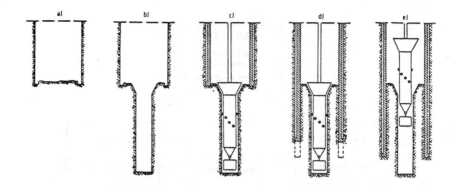

Fig. 13a: The BGR-overcoring method
- Coring a borehole with a diameter of 146 mm a) with a pilot borehole diameter of 46 mm b)
- Installation of the BGR probe in the pilot hole c)
- Overcoring of the probe and measurement in the release of stress in the core during the overcoring process d)
- Removal of the probe after overcoring e)

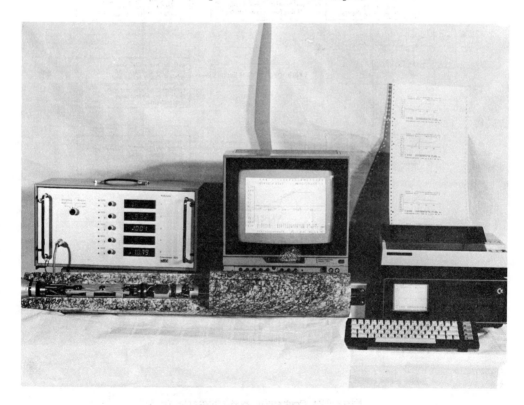

Fig. 13b: Measurement device

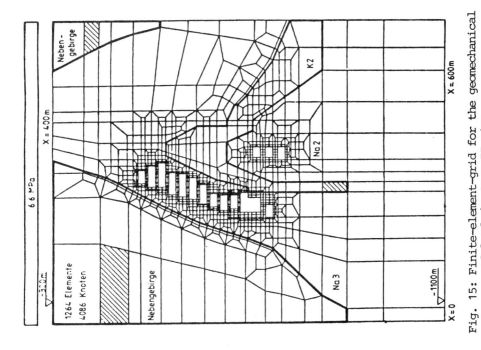

Fig. 15: Finite-element-grid for the geomechanical model of the Asse II mine

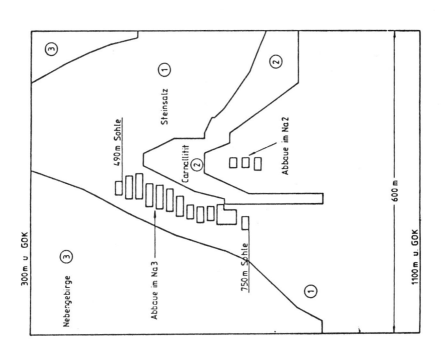

Fig. 14: Geomechanical model of the Asse II mine

Fig. 17: Zones of tensile stress around mine workenings of Asse II (tensive cut off)

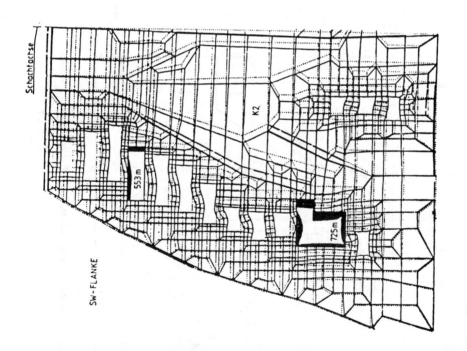

Fig. 16: Deformational structure around the workenings of Asse II

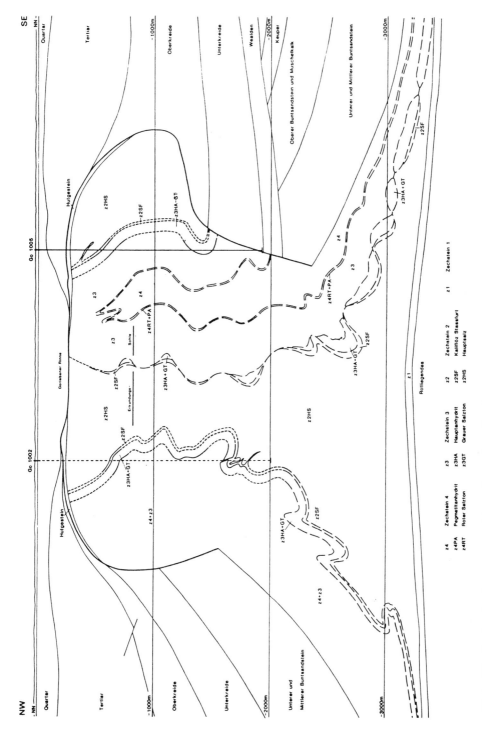

Fig. 18: Geological cross section through salt dome GORLEBEN

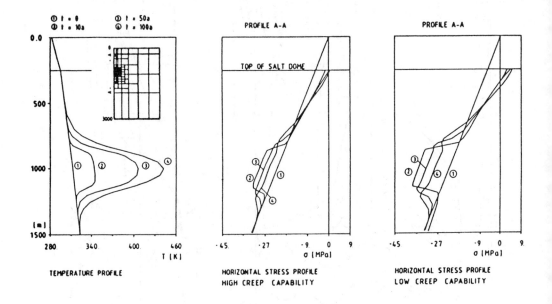

Fig. 19: Thermal induced stress

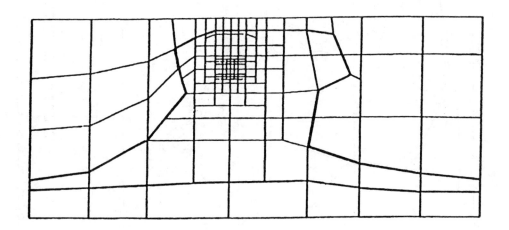

Fig. 20: Finite-element-mask

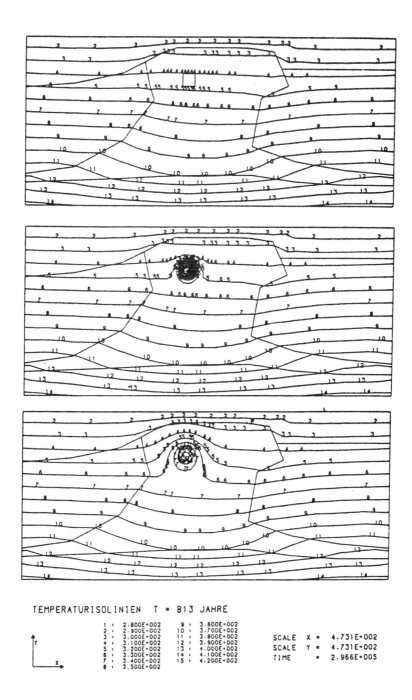

TEMPERATURISOLINIEN T = 813 JAHRE

1 :	2.800E+002	9 :	3.600E+002
2 :	2.900E+002	10 :	3.700E+002
3 :	3.000E+002	11 :	3.800E+002
4 :	3.100E+002	12 :	3.900E+002
5 :	3.200E+002	13 :	4.000E+002
6 :	3.300E+002	14 :	4.100E+002
7 :	3.400E+002	15 :	4.200E+002
8 :	3.500E+002		

SCALE X = 4.731E+002
SCALE Y = 4.731E+002
TIME   = 2.966E+005

Fig. 21: Temperature distribution

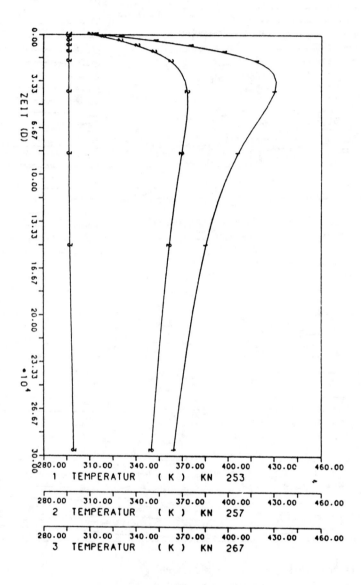

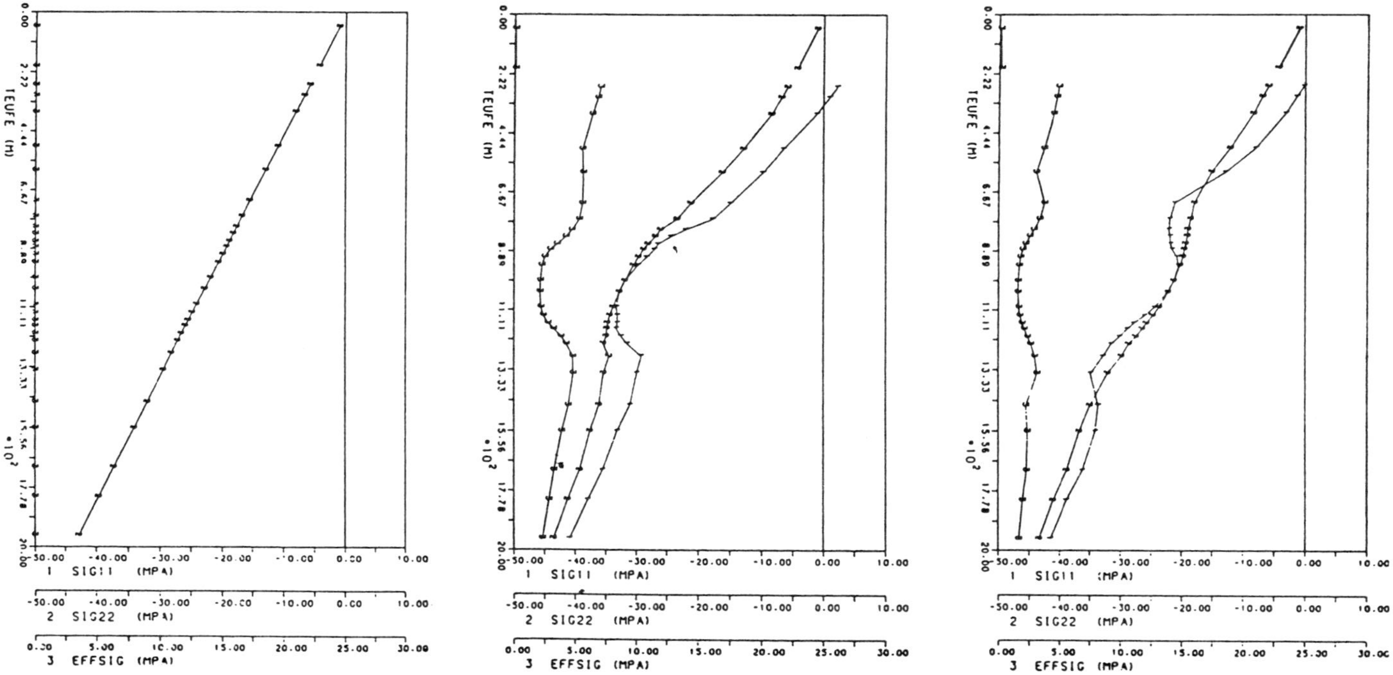

Fig. 23: Thermal induced stress redistribution

- 381 -

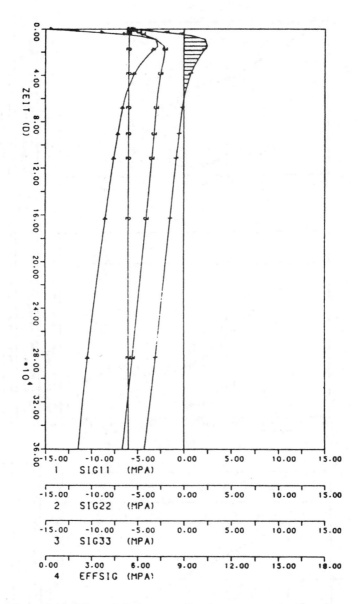

Fig. 24: Time history of stresses at the top
of the salt dome

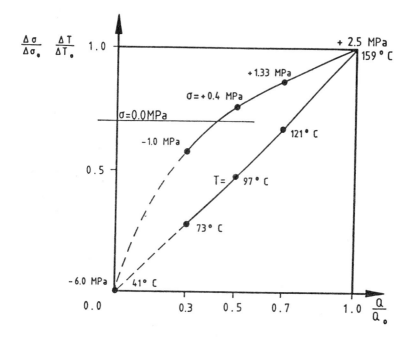

Fig. 25: Dependence of thermal induced effects due to heat output

Table 1: Parameters used for the far-field analysis

| | thermal parameters | | | | elasticity parameters | | | mechanical parameters |
| | T | α_t | λ | $c \cdot \rho$ | ρ | E | ν | creep parameters |
	K	K^{-1}	$\dfrac{W}{m \cdot K}$	$\dfrac{W \cdot d}{m^3 \cdot K}$	$\dfrac{kg}{m^3}$	MPa	-	steady-state creep: $\dot\epsilon_{eff} = A_1 \cdot \exp(-Q_1/RT) \cdot (\sigma_{eff}/\sigma^*)^{n_1} + A_2 \cdot \exp(-Q_2/RT) \cdot (\sigma_{eff}/\sigma^*)^{n_2}$
cap rock (Quaternary/ Tertiary)		1.0×10^{-5}	2.15	22.0	2,200	100	0.33	–
country rock (Cretaceous/ Bunter)		1.0×10^{-5}	2.0	22.0	2,200	100	0,33	–
basement (Rotliegende)		1.0×10^{-5}	2.0	22.0	2,200	17,000	0.33	–
salt rock	300 350 400 500 600	4.0×10^{-5}	5.44 4.53 3.88 3.02 2.47	22.0	2,200	25,000	0.25	1. high creep capacity (upper limit) $A_1 = 0.18$ (d^{-1}) $Q_1 = 54.0$ $(kJ \cdot mol^{-1})$ $n_1 = 5.0$ 2. low creep capacity (lower limit) $A_1 = 0.05$ (d^{-1}); $A_2 = 2.1 \times 10^6$ (d^{-1}) $Q_1 = 58.0$ $(kJ \cdot mol^{-1})$; $Q_2 = 113.0$ $(kJ \cdot mol^{-1})$ $n_1 = 5.0$; $n_2 = 5.0$

Table 2: Thermal output for the cases studied

time since deposition (years)	thermal output					
	1. (W/m³)	2. (W/m²)	3. (W/m³)	4. (W/m²) +	(W/m²)	5. (W/m³)
0	0.168	16.94	0.185	13.1	0.118	0.2186
10	0.134	13.49	0.142	11.0	0.091	0.1725
20	0.111	11.17	0.116	9.32	0.074	0.1375
30	0.093	9.38	0.096	7.93	0.061	0.1146
60	0.057	5.75	0.055	5.35	0.035	0.0617
160	0.025	2.49	0.019	2.78	0.012	0.0190
260	0.018	1.82	0.014	2.09	0.009	0.0131
360	0.015	1.52	0.012	1.74	0.007	0.0108
460	0.013	1.31	0.010	1.51	0.006	0.0093
710	0.0095	0.941	0.007	1.11	0.004	0.0065
960	0.007	0.703	0.005	0.86	0.003	0.0045
1,960	0.003	0.318	0.0015	0.46	0.001	
2,960	0.002	0.222	0.0008	0.36	0.0005	
3,960	0.002	0.189	0.0006	0.31	0.0004	
4,960	0.002	0.175	0.0005	0.29	0.0003	
7,460	0.001	0.145	0.0004	0.24	0.0003	
9,960	0.001	0.124	0.0004	0.21	0.0003	

Legend:

1. = deposition only in boreholes, reference field:
   3,654 HLW canisters + 6,258 spent fuel in canisters;
   286.5 x 300 x 300 m

2. = deposition in drifts at three levels, reference field:
   112 casks - 8 LWR-spent fuel elements + 175 casks
   - 6 HLW canisters + 288 casks - 8,400 HTR spent fuel
   elements

3. = concept A, reference field: 4,248 HLW canisters;
   184.2 x 300 x 300 m

4. = concept D, reference field: 210 casks - 8 LWR spent
   fuel elements + 3,450 HLW canisters in 300-m boreholes

5. = reference thermal output

Table 3: Results of the thermomechanical calculations

no.	job no.	initial conditions			results		remarks
		width of drift (m)	creep law (see Table 1)	thermal output (W/m³)	maximum temperature* (°C)	stress at the top of the salt dome (MPa)	
1	ASABS4	300	1.	0.2186	159	+ 2.5	maximum stress after about 42 years
2	ASABS5	300	2.	0.2186	159	- 1.0	maximum stress after about 18 years
3	ASADS1	300	1.	0.1530	121	+ 1.3	maximum stress after about 42 years
4	ASADS2	300	1.	0.1093	97	+ 0.4	maximum stress after about 42 years
5	ASADS3	300	1.	0.0656	73	- 1.0	maximum stress after about 85 years
6	ASAFS2	200	1.	0.2186	140	+ 1.2	maximum stress after about 45 years

Legend:

1. = low creep capacity
2. = high creep capacity
* = maximum temperature after 85 - 95 years

DEFORMATIONAL ROCK MASS RESPONSE DURING A NUCLEAR WASTE REPOSITORY SIMULATION EXPERIMENT AT THE ASSE SALT MINE

K. Wieczorek, G. Staupendahl, T. Rothfuchs
Gesellschaft für Strahlen- und Umweltforschung mbH München
Institut für Tieflagerung, D-3300 Braunschweig (FRG)

ABSTRACT

A joint US/FRG nuclear waste repository simulation experiment was performed at the Asse Salt Mine from 1983 through 1985.

For the installation of four individual test sites a special 60 m long, 7.5 m high, and 10 m wide underground test room was excavated at the 800 m-level. At two of the test sites the HLW disposal was simulated by the emplacement of nine heaters with a total power of 10 kW in 6 m deep boreholes. At the two remaining test sites, cobalt-60-sources of 18,000 Curies each were additionally emplaced in the boreholes to simulate Photon-emitting HLW.

In order to observe the rock mass response, room closure (convergence) measurements at the different test sites were started already during the excavation period. Lateron, horizontal extensometers were installed at the midheight of the gallery wall to monitor the rock mass disintegration in the vicinity of the test room. Vertical room closure measurements were only started after equipment installation, simultaneously with start-up of heating. Heating simulated the heat release from HLW and lasted for about 2 1/2 years.

This paper presents the measurement results together with a comparison to numerically calculated data.

DEFORMATION DE LA MASSE ROCHEUSE PENDANT UN ESSAI DE SIMULATION DE DEPOT DE DECHETS RADIOACTIFS MENE DANS LA MINE DE SEL D'ASSE

RESUME

Une expérience de simulation de dépôt des déchets radioactifs a été menée conjointement par les Etats-Unis et la République fédérale d'Allemagne dans la Mine de Sel d'Asse de 1983 à 1985.

Une chambre d'essai souterraine de 60 mètres de long, 7.5 mètres de haut et 10 mètres de large a été creusée à la cote 800 mètres pour aménager quatre sites d'essai individuels. Dans deux de ces sites, on a simulé le stockage définitif de déchets de faible activité en mettant en place neuf radiateurs ayant un pouvoir calorifique total de 10 kW dans 6 trous de forage de 6 mètres de profondeur. Dans les deux derniers sites, on a rajouté des sources de cobalt 60 de 18 000 curies chacune dans les trous de forage pour simuler des déchets de faible activité émetteurs de photons.

Pour observer les effets sur la masse rocheuse, on a commencé à procéder à des mesures de fermeture de la chambre (convergence) à divers sites d'essai dès la période d'excavation. Plus tard, des extensomètres horizontaux ont été installés à mi-hauteur de la paroi de la galerie pour contrôler la désagrégation de la masse rocheuse à proximité de la chambre d'essai. Les mesures de la fermeture de la chambre dans le plan vertical n'ont commencé qu'après la mise en place de l'équipement, parallèlement au démarrage du chauffage. Le chauffage qui simulait le dégagement de chaleur de déchets de haute activité, a été maintenu pendant deux ans et demi.

Les auteurs exposent les résultats des mesures et les comparent aux données numériques calculées.

From 1983 through 1985 the "Brine Migration Test" (BMT), an HLW simulation experiment involving cobalt-60-sources, was performed at the Asse Salt Mine near Braunschweig (FRG) as a US/FRG co-operative effort of the Office of Nuclear Waste Isolation and the Institut für Tieflagerung (IfT) of the Gesellschaft für Strahlen- und Umweltforschung.

The BMT consisted of four individual test sites. At two of them the emplacement of HLW was simulated by electrical heaters installed in boreholes, while at the other two test sites both electrical heaters and cobalt-60-sources were used to investigate the effects of heating and gamma-irradiation. In order to resolve the issues of rock mass/waste package interaction the temperature field, brine inflow into the heater boreholes, borehole gas pressure and composition, and rock mass stress changes and displacements were monitored during the experiment. In order to detect structural changes in the rock salt samples were taken and lab-tested before and after the experiment. Corrosion specimens remained in the heater boreholes during the BMT and were afterwards examined. The final report of the project will soon be published [1].

Figure 1 shows a cross section of the Asse Salt Mine. The BMT was performed on the 800 m-level of the mine, in the area marked "HAW/MAW Experimental Area". A new test gallery (see Figure 2) was mined for the experiment by means of a part-face heading machine. The four test sites were arranged in a line, with a distance of 15 m between two sites. Test Site 1 and 2 were the non-radioactive ones, while the cobalt-60-sources were installed at Test Site 3 and 4. Apart from the radioactive sources all test sites were equipped identically. Figure 3 shows a typical test site. Around the central borehole containing the main heater (and - at Test Site 3 and 4 - the cobalt-60-sources) eight guard heaters were installed at 1.5 m distance to the central borehole axis. The heater midplane was at 4.57 m below floor, the length of the central heater and the cobalt-sources was 2 m. The total electrical power at each site was 10 kW to obtain a maximum salt temperature of 210 °C and a temperature gradient of 3 °C/cm on the central borehole wall. The cobalt-60-sources had an activity of 18,000 Curies each.

In order to observe the rock mass response, horizontal room closure measurements at all test sites were started already during mining of the test gallery. Two horizontal extensometers were installed at each test site, and a vertical extensometer was placed in the floor at 1.5 m distance to the central borehole axis of Test Site 2 (see Figure 4). All extensometers were triple steel rod extensometers with wedge-type anchors fixed at depths of 2.7 m, 7.4 m, and 20.0 m. The time-dependent change in distance between each anchor and the reference plate was monitored by LVDTs. Extensometer measurements as well as vertical room closure measurements could only be started after equipment installation, simultaneously with start-up of heating. Floor heave measurements by mine-surveying methods were performed to obtain absolute displacements of the test gallery floor.

Besides the displacement measurements thermomechanical Finite-Element calculations were performed by the Rheinisch-Westfälische Technische Hochschule (RWTH) Aachen as a subcontractor of the IfT [2]. To simplify the problem and reduce the expenditure only Test Site 2 was considered and a two-dimensional axisymmetric approach was chosen. Thus, the guard heaters had to be approximated by a heater ring and the test gallery by a cylindrical shape. It is understood that this simplification may have an effect especially on the horizontal displacements. Figure 5 shows the mesh layout of the FE model. The mesh is made up of isoparametric 8-node elements. The initial stress is 15 MPa inside the rock mass and zero on the central borehole wall.

The following Figures 6 through 12 show the measured data in comparison to the calculational results. The figures show only the measured curves of Test Site 2 in order to maintain some clearness. The other measurements showed similar results. In all figures Test Day 0 denotes the start-up of heating at the Test Sites 1 and 2. At the Test Sites 3 and 4 heating started with a delay of 204 days. Heating lasted until Test Day 863, then the heater power was reduced to zero in seven identical steps each taking four days.

The Figures 6 and 7 show the measured and calculated curves for the horizontal and vertical closure, respectively. In both figures an increase of the closure rate with start-up of heating can be observed. The effect of the heater power reductions is also evident. After 863 days of heating the horizontal closure is about 45 mm and the vertical closure about 100 mm. The closure rates for different time intervals are shown in Table I.

Table I: Measured and Calculated Room Closure Rates for Different Time Intervals

| | Room Closure Rate in 10^{-2} mm/day | | | |
| | Horizontal Closure | | Vertical Closure | |
	Measured	Calculated	Measured	Calculated
Prior to Heating (before Test Day 0)	3.4	2.1	5.1	2.3
Start-up of Heating (Test Day 1 - 35)	5.3	1.2	47.1	58.3
Heating Phase (Test Day 104 - 197)	7.1	5.2	14.5	20.2
Constant Temperature (Test Day 509 - 863)	4.6	4.5	7.9	8.6

There is a good agreement between measured and calculated curves for both the horizontal and the vertical closure. The measured horizontal closure is, however, higher than the calculated one, while the vertical closure is lower. The same can be stated for the closure rates. This effect is interpreted to be due to the simplified FE model, which becomes more obvious when we compare Figure 6 to the figures showing the horizontal extensometer readings (Figures 8 and 9). Although the horizontal closure is higher than calculated, the extensometer readings are lower. So to speak the closure seems to result from the rock mass being pushed in as a block of large extent. It is obvious that this could not happen in a cylindrical cavity, which explains the difference between measurement and calculation. A three-dimensional modelling of the test gallery would be likely to result in a higher horizontal closure.

Figure 10 shows the measured and calculated floor heave at Test Site 2. This figure is quite similar to Figure 7 showing the vertical closure. This is due to the nearly neglectable roof sag which can be taken from the comparison of floor heave and vertical closure (see Figure 11).

The readings of the vertical extensometer are shown in Figure 12. The change of distance between the reference plate and the anchor at 20.0 m depth below floor is nearly the same as the vertical closure (or the floor heave, respectively) for both the measurement and the calculation. Therefore, the anchor can be considered as nearly immovable. The measured changes of distance between the reference plate and the other anchors, however, are much lower than calculated. While after about 200 days of heating the highest dilatations take place in the region between 7.4 m and 20.0 m below floor, the calculation postulates them in the region above 7.4 m below floor. By analogy to the discussion of horizontal displacements this can be explained by the difference between reality and FE model.

As a summary, it can be stated that the calculation results could be improved by using a three-dimensional model. The two-dimensional approach, however, produced good results for the closure calculation at a comparatively low expenditure.

REFERENCES

[1] T. Rothfuchs, K. Wieczorek, H.-K. Feddersen, G. Staupendahl, A.J. Coyle, H. Kalia, J. Eckert, "Nuclear Waste Repository Simulation Experiments - Asse Salt Mine, Federal Republic of Germany: Final Report", Office of Nuclear Waste Isolation and Gesellschaft für Strahlen- und Umweltforschung mbH München, to be published in 1988

[2] G. Albers, R. Elsen, K. Hahne, "Nachrechnungen zum HAW-Simulationsversuch mit Kobalt-60-Quellen im Salzbergwerk Asse", Rheinisch-Westfälische Technische Hochschule Aachen for Gesellschaft für Strahlen- und Umweltforschung mbH München under Contract No. 31/139919/86, February 1987

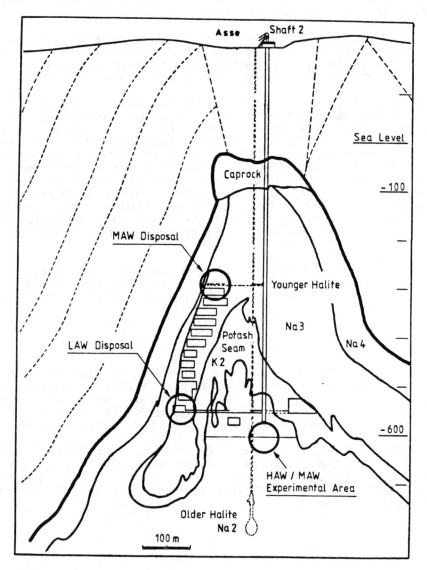

Figure 1: Asse Mine Cross Section

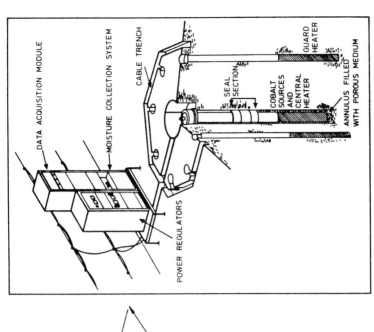

Figure 3: Typical Test Site

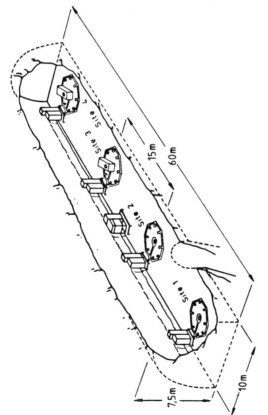

Figure 2: Test Gallery

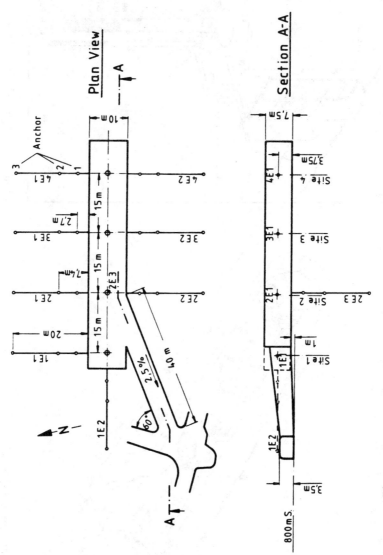

Figure 4: Arrangement of Extensometers

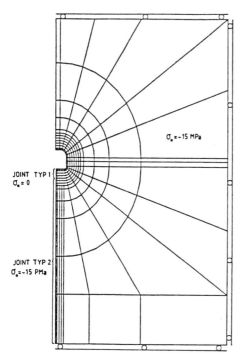

JOINT TYP 1
$\sigma_\bullet = 0$

$\sigma_\bullet = -15$ MPa

JOINT TYP 2
$\sigma_\bullet = -15$ PMa

Figure 5: Mesh Layout and Initial Boundary Conditions
of the Thermomechanical Model

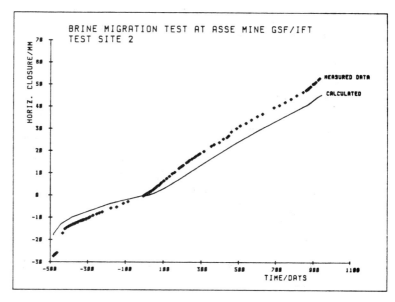

Figure 6: Measured and Calculated Horizontal Closure at
Test Site 2

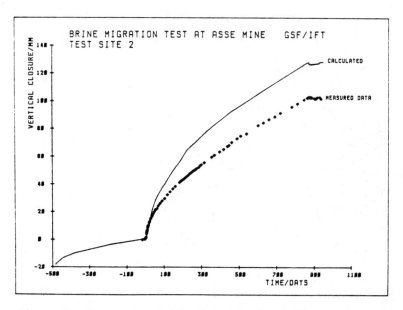

Figure 7: Measured and Calculated Vertical Closure at
Test Site 2

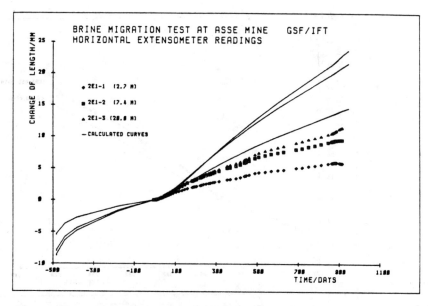

Figure 8: Measured and Calculated Horizontal Extensometer
Readings of Test Site 2 (North)

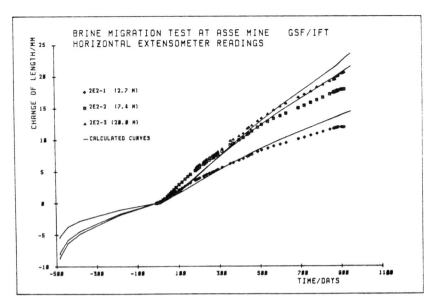

Figure 9: Measured and Calculated Horizontal Extensometer
Readings of Test Site 2 (South)

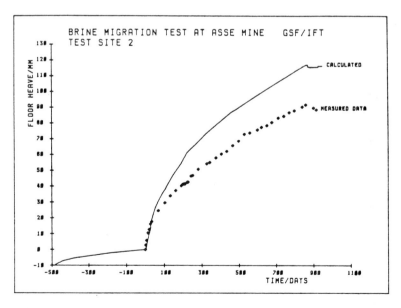

Figure 10: Measured and Calculated Floor Heave at Test
Site 2

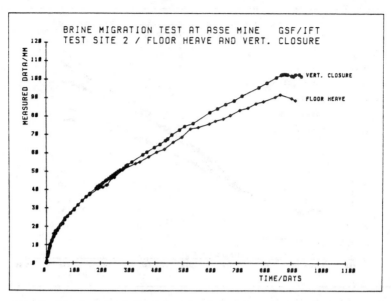

Figure 11: Comparison of Floor Heave and Vertical Closure
at Test Site 2

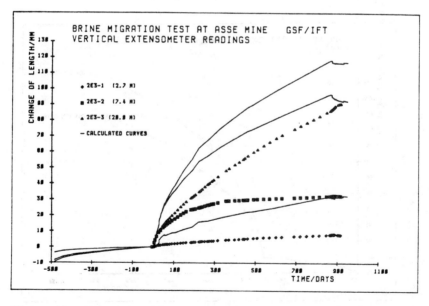

Figure 12: Measured and Calculated Vertical Extensometer
Readings of Test Site 2

INVESTIGATION OF CAVITY RESPONSES BY MICROSEISMIC METHODS

C. Heick   Kavernen Bau- und Betriebs-GmbH

D. Flach   Gesellschaft für Strahlen- und
                    Umweltforschung mbH
                    Institut für Tieflagerung

Abstract

Microseismic measurements in rock salt have been taken in both
the Asse mine and the abandoned Hope potash mine, both situated
in the State of Lower Saxony of the Federal Republic of Germany.

In addition to the location of seismic events, microseismic
measurements also provide information on the physical parameters
of these processes. Both projects show that a majority of the
microseismic sources located are situated very close to galleries and workings. Processing and interpretation techniques
taken from seismology indicate that small source radii are
formed when the events occur near galleries. Bigger fault planes
are observed in the vicinity of more extensive cavities. In both
cases the radii are of the same scale as those of the diameters
of the galleries and workings.

Because all microseismic events, provided that they are not
blast induced, take place only in high stress fracture zones,
these processes have to be considered in light of weakness
areas. Microseismic data acquisition lends itself well to the
location and monitoring of these areas.

ETUDE DES EFFETS DES TRAVAUX D'EXCAVATION PAR LES METHODES MICROSISMIQUES

RESUME

On a procédé à des mesures microsismiques sur des roches salines, dans la mine d'Asse et dans la mine de potasse désaffecté de Hope qui se trouvent l'une et l'autre dans le Land de Basse-Saxe (République fédérale d'Allemagne).

Outre la localisation des événements sismiques, les mesures microsismiques ont fourni des informations sur les paramètres physiques qui interviennent dans ces processus. Ces deux projets ont montré que les sources microsismiques repérées se trouvent en majorité à très courte distance des galeries et des chantiers. Les techniques de traitement et d'interprétation empruntées à la sismologie indiquent que lorsque ces événements se produisent près des galeries, il se forme à partir de chaque source des zones de déformation de faible rayon. On observe des plans de faille plus importants à proximité des cavités plus grandes. Dans les deux cas, les rayons sont du même ordre de grandeur que le diamètre des galeries et des chantiers.

Etant donné que tous les événements microsismiques, à condition qu'ils ne soient pas provoqués par une explosion, se produisent dans des zones de fissures soumises à de fortes contraintes, il convient d'étudier ces processus dans le contexte des zones de moindre résistance. L'acquisition de données microsismiques se prête bien à la localisation et à la surveillance de ces zones.

1   Introduction

Observations of passive seismic processes have been underway at the Asse mine near Wolfenbüttel, FRG, since 1979 and at the former potash mine of Hope, north of Hannover, FRG, since 1986.

Passive seismics includes all events which result in the emission of seismic wave energy without being artificially excited, e.g. explosive shots or falling weights. Such events are predominantly movements of neighbouring rock blocks relative to one another. A prerequisite for this is the presence of discontinuities along which movements can take place, i.e. laying within the material or fracture fissures in an otherwise homogeneous medium.

In mines isolated cases of spalling from roof sections can occur along roads and in chambers. Upon impact these blocks also transmit a fraction of their kinetic energy as seismic energy. These events are not intended for inclusion in this study.

Microseismic activity refers to all events of low energy requiring registration. An unequivocal definition is not available, the term usually covers microquakes with source radii of up to a few meters.

2   Measuring Microseismic Activity in Mines

Seismic events in mines result from equalization movements on existing or new discontinuities in the rock mass. The cause is stress relocation frequently as a result of mass relocations. These include cutting off roads and, as described later, the flooding of underground cavities.

Thermally induced fracturing, e.g. caused during heater tests, can also be detected. Whereas these are generally of limited spatial extension, stress redistribution because of load changes can effect large areas of a mine. It is possible to configure data acquisition systems to match various objectives.

From March 1984 onwards the abandoned potash mine of Hope was filled with saturated brine from the solution mining of subsurface storage caverns. The flooding was accompanied by a concurrent research program involving not only geophysical measurements but also measurement programs for geochemistry and rock mechanics [1].

The microseismic monitoring for the Hope mine covered the entire northern section, known as Deposit 1/2 (Fig. 1). Observations prior to the flooding showed that Deposit 1/2 was relatively inactive, in a one month period six events were located [2].

Immediate to the start of flooding there was a several fold increase in event frequency (Fig. 2).

Location of the sources, which was achieved using first onset times derived from the seismograms of at least 4 geophone sites, showed that these events were mainly concentrated along the roads and chambers (Figs. 3 + 4).

As described above, every recorded microseismic event should be regarded as analogous to seismological observations on movements on a fault plane.

The frequency content of the microseismic signals enables calculation of the surface of the seismic source [3]. To this end the seismograms are frequency analyzed and represented in the form of a displacement spectrum (Fig. 5). The corner frequency f_c is entered directly into the equation for calculating focus radius r.

$$r = \frac{2.34}{2\pi} \cdot \frac{V_s}{f_c} \qquad\qquad [3]$$

where V_s = shear wave velocity.

Other source parameters as well as the seismic wave energy and magnitude can be calculated.

The calculated source values show a relationship between source radii and the diameter of the roads and chambers. Source radii and cavity diameters are directly proportional and of approximately the same scale.

The origin for the occurrence of seismic activity could be traced to the stress relocations caused by loading as a result of inflooded brine. Thermal influences were discounted since, amongst others, no events occurred either along the inflow paths or in the deeper section of the mine where relatively cold brine collected.

From these observations it can be concluded that the vicinity of roads and chambers should not be regarded a homogeneous composite rock structure. Indeed in stressed rock zones discontinuity surfaces are formed where neighbouring blocks move relative to each other.

It is suspected that these discontinuities are not newly formed fault planes but movements on existing fissures. This is corroborated by the observation that no activity is recorded in mine sections already flooded. Presumably the brine penetrated the fissures thereby reducing the frictional forces and causing aseismic movements.

This provokes the question of when the disturbance to the rock formation occurs, i.e. when the fracture planes are formed. It is suspected that this occurs during or shortly after the excavation phase. To answer this question it is planned to perform appropriate tests in the Asse mine. During excavation the immediate vicinity of a newly driven road will be monitored seismically.

Observations performed to date in the Asse mine have shown that driving new cavities produces activities in other sections of the mine. That is, as in flooding the Hope mine, mass displacements caused seismic movements in Asse. The immediate vicinity of the excavations work could not linked to the instrumentation systems and could therefore not be monitored. It was only possible to pinpoint sources with sufficient accuracy in the rock zone enclosed by the geophone array.

The data acquisition system in the Asse mine is technically identical to the system in Hope [4]. This facilitates direct comparison of data of the two measurement series. It was shown that in Asse too the sources were found close to the existing roads and chambers. Moreover, here too calculations showed that the fault planes were closely related to the diameters of the neighbouring cavities.

Considerable activity was registered in the vicinity of chambers in the south flank of the mine (Figs. 6 + 7).

Further evidence of equilization movements in the zones close to the roads was given by observations during a heating up test in 1985 in the Asse mine [5]. In addition to a large number of events in the immediate vicinity of the heater a further 8 foci at distances of up to 100 m from the heater were located (Fig. 8).

Whereas the three events to the north of the heater occurred within only 5 seconds, the others occurred at relatively regular intervals during 3 month test period.

The relationship to the roads is clear. The changes in the stress field here also apparently led to movements close to the roads.

3   Microseismic Monitoring of Subsurface Structures

Data acquisition systems for measuring microseismic activities are suitable for localizing weak zones near subsurface cavities in salt and for recognizing changes in the vicinity of roads and workings. These are generally the zones where critical stress states are located, leading to movements on boundary planes or existing fracture planes or even breaking up of the rock. Stress balance is reintroduced with the completion of these movements.

Microseismic arrays consist of geophone stations arranged around the zone to be investigated. The data from at least four geophones are required to locate an event. Installing three-component transducers at the measurement cross sections is necessary to be able to calculate the source parameters. The best frequency range for measurement data acquisition must be selected depending on the spatial extent of the zone to be investigated. Extended foci produce low frequency signals, foci in the centimeter range on the other hand produce higher frequencies of up to a number of kHz. Figure 9 shows a typical data acquisition layout.

A great advantage of microseismic monitoring is that the zone to be investigated does not itself have to be fitted with sensors. Movements can be observed from the outside and quantified.

Because fault planes must generally be regarded as potential brine intrusion sites, knowledge of their spatial location and extent is of great importance.

Literature

[1]   Untersuchungen endlagerrelevanter Vorgänge vor, während und nach der Flutung des Kalibergwerkes Hope. 1. Bericht, 1985; 2. Bericht, 1986.

[2]   Heick, C., Hente, B.: "Accompanying Geophysical Observations During the Flooding of a Salt Mine", Second Conference on the Mechanical Behaviour of Salt, Hannover (FRG), 1984, in press.

[3]   Brune, I.N.: "Tectonic Stresses and the Spectra of Seismic Shear Waves from Earthquakes", I. Geoph. Res., 75, 1970, 4997-5009.

[4]   Hente, B., Gommlich, G., Flach, D.: "Microseismic Monitoring of Candidate Waste Disposal Sites", Proc. of the Third Conference on Acoustic Emission/Microseismic Activity in Geologic Structures and Materials; The Pennsylvania State University, Trans. Tech. Publications, Clausthal, 1981, 393-401.

[5]   Hente, B. et al.: "Microseismic Activity and Seismic Travel Time Measurements", Nuclear Science and Technology, Report EVR 10827 ENII, 85-97.

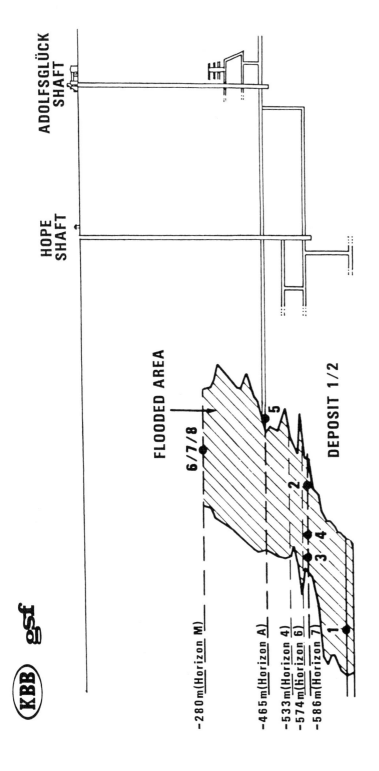

THE HOPE POTASH MINE / DEPOSIT 1/2

Figure 1 - The Hope potash mine with Deposit 1/2

THE HOPE POTASH MINE
FLOODING RATE AND EVENT RATE

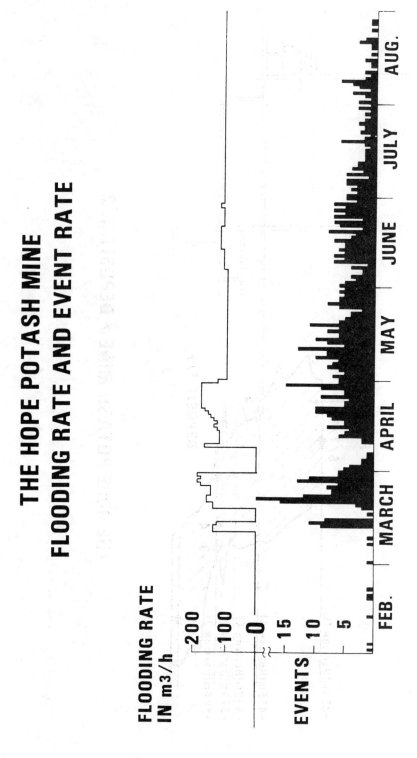

Figure 2 - Flooding rate and numbers of events

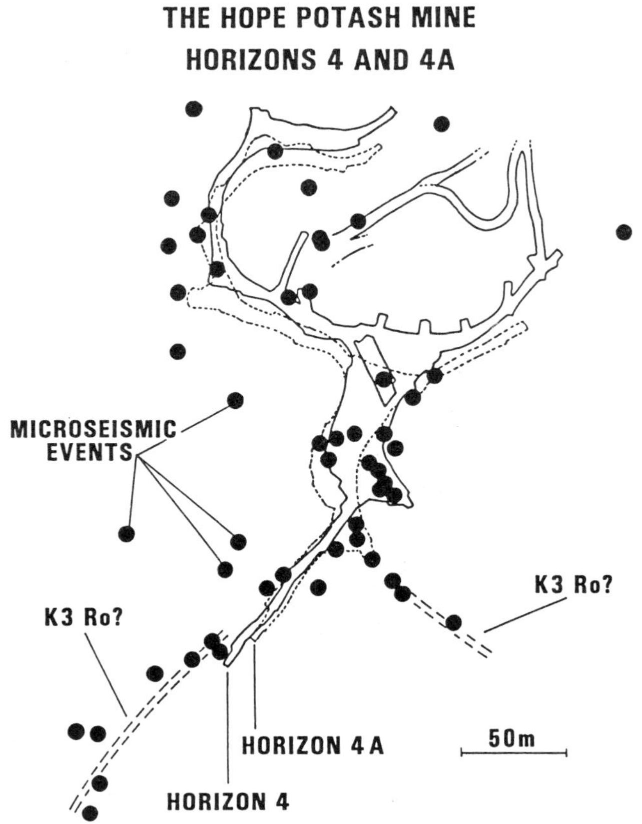

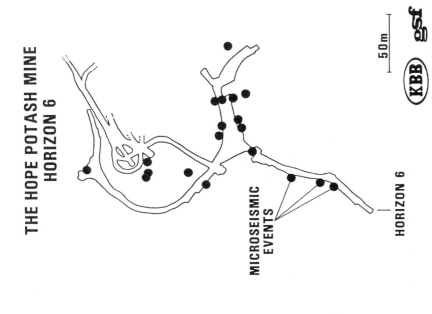

THE HOPE POTASH MINE
HORIZON 6

MICROSEISMIC
EVENTS

HORIZON 6

50m

KBB gsf

Figure 4 - Horizon 6 with microseismic events

THE HOPE POTASH MINE
HORIZONS 4 AND 4A

K3 Ro?

50m

HORIZON 4 A

MICROSEISMIC
EVENTS

HORIZON 4

K3 Ro?

Figure 3 - Horizons 4 and 4a with microseismic events

- 407 -

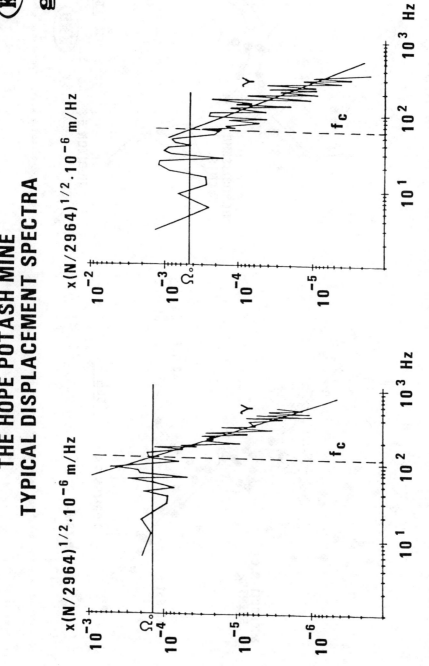

Figure 5 – Displacement spectra

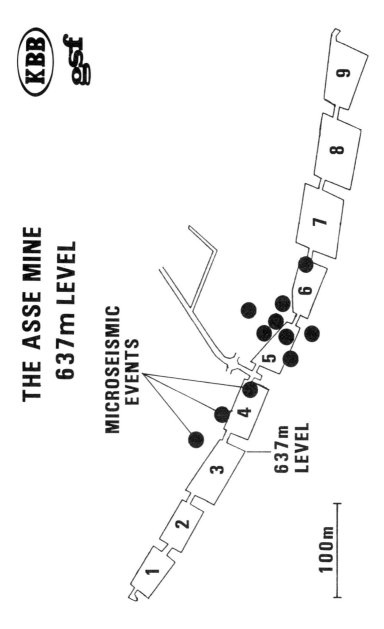

THE ASSE MINE
637m LEVEL

MICROSEISMIC
EVENTS

637m
LEVEL

100m

Figure 6 - 637-m-level with microseismic events

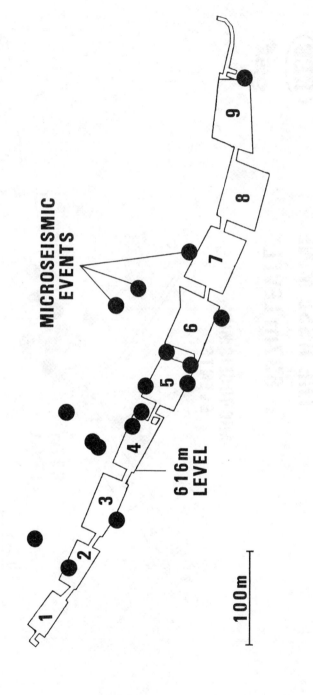

THE ASSE MINE
616m LEVEL

MICROSEISMIC
EVENTS

616m
LEVEL

100m

Figure 7 - 616-m-level with microseismic events

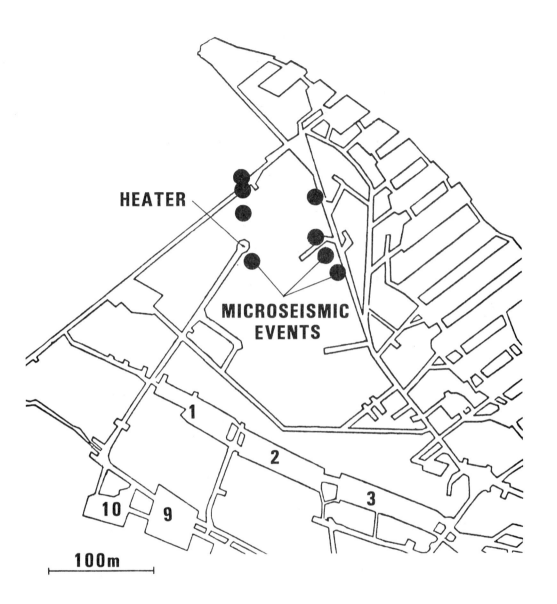

HEATER

MICROSEISMIC
EVENTS

1

2

3

10   9

100m

THE ASSE MINE
750m LEVEL

Figure 8 - 750-m-level with microseismic events

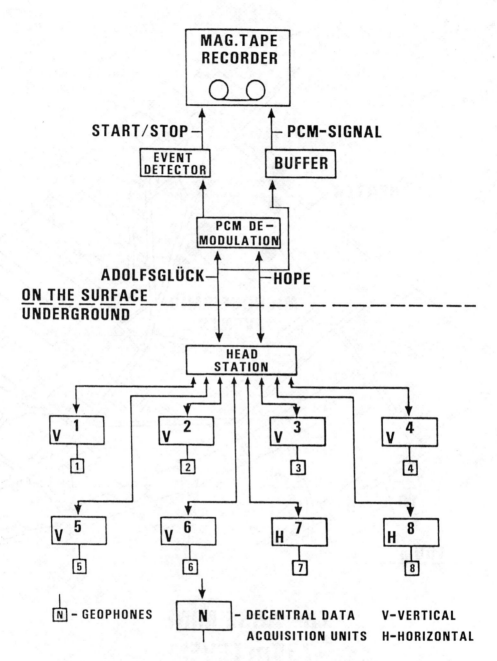

Figure 9 - Block diagram of a data
acquisition system

AN INTERIM REPORT ON EXCAVATION EFFECT STUDIES AT THE WASTE ISOLATION PILOT PLANT: THE DELINEATION OF THE DISTURBED ROCK ZONE

David J. Borns
Division 6331
and
John C. Stormont
Division 6332
Sandia National Laboratories
Albuquerque, NM 87185

ABSTRACT

For nuclear waste repositories with both long operational periods (50 yr) and long performance assessment periods (10000 yr), the Disturbed Rock Zone (the zone of rock in which the mechanical and hydrologic properties have changed in response to excavation; abbreviated as DRZ) is important to both operational (e.g. slab or fracture failure of the excavation) and long term performance (e.g. seal system performance and fluid transport). At WIPP, the DRZ has been characterized with three approaches: visual observation; geophysical methods; and permeability measurements. Visual observations in drillholes indicate that fluids and fractures are common in the host rock of the underground facility. Geophysical studies have utilized radar, transient (TEM) and direct current (DC) electromagnetic methods. Radar has been useful, but the penetration of radar is limited by the water content and bedded nature of the host rock. The TEM method was able to detect a four-fold increase in resistivity from 1 to 5 m into the rock. This trend reflects a four-fold increase in the moisture content from near the excavation (0.5 to 1% by weight) to 5 m into the host rock (2 to 3% by weight). The DC method has been able to detect zones of moisture around the excavation. Numerous gas permeability measurements indicate that beyond 2 m from an excavation halite and interbeds (anhydrite and clay) allow very low gas flow (calculated permeabilities < 1 microdarcy for gas flow tests and < 0.01 microdarcy for brine-based permeability tests). Within 2 m of the excavation, very high flow rates (10^4 SCCM) were measured. All three approaches have defined a DRZ at WIPP extending laterally throughout the excavation and varying in depth from 1 to 5 m, according to the size and age of the opening.

RAPPORT INTERIMAIRE SUR LES ETUDES DES EFFETS DES TRAVAUX
D'EXCAVATION AU SITE DU PROJET PILOTE DE CONFINEMENT DES DECHETS :
DELIMITATION DE LA ZONE ROCHEUSE PERTURBEE

RESUME

Dans le cas des dépôts de déchets nucléaires qui devraient être
exploités pendant une longue durée (50 ans) et dont les performances devraient
être évaluées à long terme (10 000 ans), la zone de roches perturbées (zone de
roches dans laquelle les propriétés mécaniques et hydrologiques ont été
modifiées sous l'effet des travaux d'excavation) revêt de l'importance, tant
du point de vue de l'exploitation (éboulement par décollement ou fissure de
l'excavation) que de celui des performances à long terme (tenue du dispositif
de scellement et transport des fluides par exemple). Au site du projet pilote
de confinement des déchets, la zone de roches perturbées a été caractérisée au
moyen de trois méthodes différentes : observation visuelle ; méthodes
géophysiques et mesures de perméabilité. Les observations visuelles faites
dans les trous de forage ont permis de constater que les fluides et les
fissures étaient fréquentes dans la roche réceptrice de l'installation
souterraine. Les études géophysiques ont été réalisées au moyen de méthodes
électro-magnétiques par radars alimentés par courant transitoire et courant
continu. Ces radars se sont avérés utiles, mais leur pénétration est limitée
par la teneur en eau et la stratification de la roche réceptrice. La méthode
TEM a permis de déceler une multiplication par un facteur 4 de la résistivité
de la roche entre 1 et 5 m de profondeur. Cette tendance traduit une
multiplication équivalente de la teneur en humidité qui varie de 0.5 à 1 pour
cent en poids à proximité de l'excavation et passe à 2 à 3 pour cent en poids
à 5 m à l'intérieur de la roche réceptrice. La méthode fondée sur le courant
continu a permis de déceler des zones d'humidité autour de l'excavation. De
nombreuses mesures de perméabilité au gaz ont montré qu'au-delà de 2 m à
l'intérieur d'une formation d'halite ou d'une formation interstratifiée
(anhydrite et argile) excavée, les possibilités d'écoulement de gaz étaient
très faibles (perméabilité calculée < 1 microdarcy pour les essais
d'écoulement gazeux et < 0.01 microdarcy pour les essais de perméabilité à
partir de saumure. Dans les deux premiers mètres d'épaisseur, on a mesuré des
débits très élevés (10^4 cm^3 normalisés). Les trois méthodes utilisées au site
du projet pilote de confinement des déchets ont permis de définir une zone de
roches perturbée entourant toute l'excavation et dont l'épaisseur varie de 1 à
5 m, selon la taille et à l'ancienneté de l'ouverture.

INTRODUCTION

Following excavation of underground openings, a Disturbed Rock Zone (abbreviated as DRZ) forms in the wall rock and is defined as the zone of rock in which mechanical properties (e.g., elastic modulus) and hydrologic properties (e.g. permeability and fluid inflow) have changed in response to the excavation. The present extent of the DRZ around workings at Waste Isolation Pilot Plant (WIPP) is delineated in this paper based on the measurement of rock properties in field studies. The measurement of these rock properties is relatively straightforward, but the processes involved in the development of a DRZ are complex although basically related to stress relief and/or rapid strain rates. The formation of the DRZ drives coupled processes such as changes in permeability in response to fracture growth. This report will delineate the DRZ in the underground facility at WIPP based on field observations and the available measurements of rock properties and provide a basis for further studies on the processes that result in the development of a DRZ. Since nuclear waste repositories have both long periods of operation (50 yr) and performance assessment (10000 yr), characteriza-

tion and understanding of the DRZ is important to the evaluation of operational (e.g. slab or fracture failure of the excavation) and long term performance (e.g. seal system performance and fluid transport).

The WIPP underground facility lies 653 m below the surface (Fig.1). The underground areas are divided basically between three tasks: site and preliminary design validation (SPDV); technology experiments; and demonstration of full-scale TRU waste disposal. Mining of the SPDV areas started in 1982, the experiment areas in 1984, and the waste disposal panels in 1986.

At WIPP, the DRZ has been characterized using three approaches: visual observation; geophysical methods; and measurement of hydraulic properties. All three approaches have defined a DRZ at WIPP extending laterally throughout the excavation and varying in depth from 1 to 5 m, according to the size and age of the opening. This report contains some initial results of an ongoing experimental program to develop a more detailed 3D definition of the DRZ

VISUAL OBSERVATIONS IN BOREHOLES

During the development of the underground workings at the WIPP site, numerous drillholes were made made for stratigraphic studies, test areas for experiments, and construction foundations. As the presence of a DRZ became apparent, Bechtel National (1986) drilled numerous holes to specifically investigate the DRZ. Therefore, we have a data base of direct observations of the DRZ spread both spatially and in time. These visual observations using drillholes indicate that fractures (*with apertures greater than 2 mm and visible without enhancement to the naked eye*) and fluids are common in the wall rock of the underground facility. The observations from these boreholes suggest the development of an elliptical pattern of fractures around an excavation as summarized in an idealized crosssection of a WIPP Room (Fig. 2). The basic features of these observations are as follows:

° An arcuate fracture system, concave towards the opening, develops in the floor and the back crosscutting the stratigraphy.

° Separations may develop along stratigraphic markers such as clay seams.

° Shear displacements are observed along some fractures and separations.

° Vertical fractures and spalling are observed within the ribs.

A reexamination of existing boreholes was conducted in 1987 [2]. This reexamination showed that the extent of observed fracturing in borehole arrays increased from 48% of the array locations in 1986 to 73% in 1987. The locations without fractures are largely restricted to drifts with narrow spans (4x4 m). In the oldest 11x4 m test rooms (Room 1 through 4), 100% of the locations exhibited fractures of 2mm or greater.

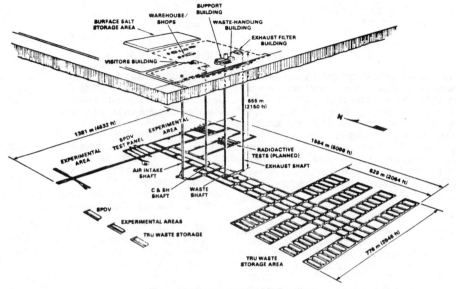

Fig. 1 Underground WIPP Facility

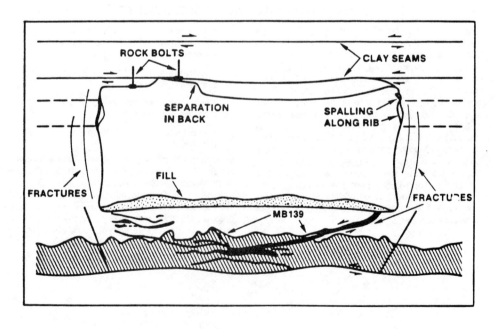

Fig. 2 Observed Fracture Pattern for 4x11 m Room

GEOPHYSICAL OBSERVATIONS

In-mine Electromagnetic Surveys

Electromagnetic methods measure the apparent resistivity of the host rock. Properties, such as permeability and fluid content, can be inferred or calculated from the resistivities. The initial phase of this study was the measurement of the electrical conductivity of the wall rock using conventional electromagnetic coupling equipment. Two systems were used: the EM-31 and EM-34 systems manufactured by Geonics, Ltd., of Toronto, Canada [3] [4]. In both of these the mutual coupling between two induction coils is measured and converted to apparent resistivity. With the EM-31 system, the two coils are separated by a distance of 3 m along the excavation surface. In the EM-34 system, the two induction coils are separated by 20 m. These configurations result in a distance of search (penetration) of 1 to 2 m (for the EM-31) and 10 to 20 m (for the EM-34). Measurements were made at 3.2 m (EM-31) and 7.2 m (EM-34) intervals along the same traverses in selected drifts at WIPP (Fig. 3). In essence, this survey con-

figuration measures the electrical conductivity (or resistivity-the reciprocal of conductivity) adjacent (EM-31) to the mined opening and 10 m (EM-34) away from the opening.

For the EM-31 system, the measured conductivities for the wall rock up to 2 m from the excavation range from 1 to 5 millisiemans per meter (from 200 to 1000 ohm-meters). For the EM-34 system, the deeper conductivity measurements, up to 20 m from the excavation, range from 7 to 10 millisiemans (100 to 140 ohm-meters). The EM-31 and EM34 measurements are compared with resistivities measured in salt mines in Germany [5]. Based on this comparison, we expect the free water content of the salt around the mine opening to increase from 0.5 to 1.0% (by weight) at the excavation surface to between 2.0 to 3.0% at several meters depth (Fig. 4). This observation may reflect an alteration of the wall rocks of the tunnels due to drying by the ventilation system.

The second phase was the measurement of the

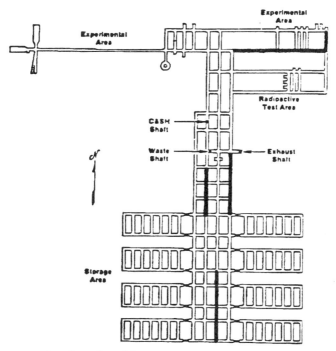

Fig. 3 Location of EM-31 and EM-34 Traverses in Black

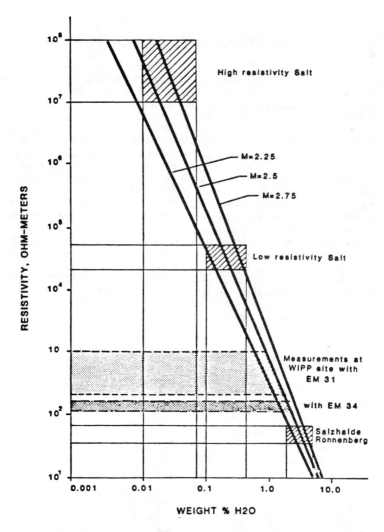

Fig. 4 Apparent Resistivity versus Water Content

Explanation of Figure 4: This figure displays the relationship between apparent resistivity and water content for different factors of cementation or consolidation (m) for Archies Law (m = 2.5 to 2.75 for Asse salt [Kessels et al., 1985]), crosshatched ranges are for resistivities of both Asse salt (high and low resistivity salt) and salt taolings pile (Salzhunde Ronnenberg) in which the water content was determined indepently (Kessels et al., 1985); the stippled ranges are apparent resistivities of salt at the WIPP facility horizon; the water content of WIPP salt can be extrapolated from the intersection of the WIPP resistivities with the lines for the different consolidation factors.

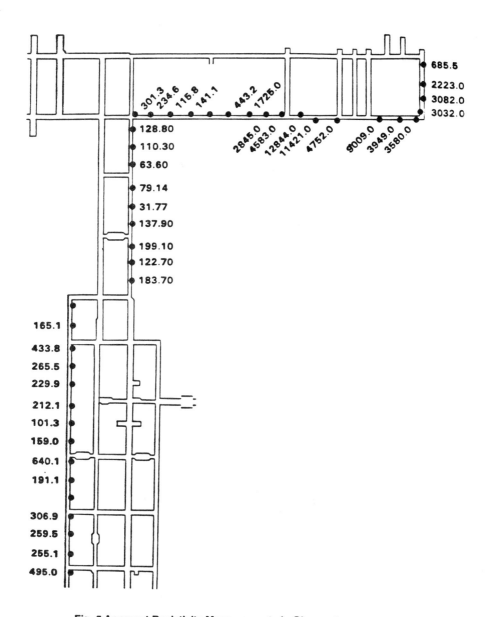

Fig. 5 Apparent Resistivity Measurements in Ohm-meters

electric field and electrical potential in the mine openings with a source of direct current sited on the surface. The rocks around the mine workings were energized using a fixed dipole source located on the Earth's surface. Two wells (located 1.0 km apart and with 300 m deep casings) were used as electrodes. From the determination of the electric field at selected stations in the underground workings, apparent resistivities were calculated for the rock mass near the excavations.

The preliminary survey in the underground facility shows a broad range of lateral variation in resistivity (30 to 10,000 ohm-m, Fig. 6). Some of this variation can be attributed to dehydration of the host salt adjacent to heated rooms and to zones where the wallrock is rich in brine and seen to visibly weep to the surface of the excavation. The sensitivity of this method is approximately proportional to the size of the conducting unit being detected [4]. Roughly, a unit 1 m in length is detectable at a distance of 1 m and a body 30 m in length is detectable at a distance of 30 m.

Borehole Moisture and Density Measurements

The USGS has conducted a series of trial moisture and density measurements in boreholes under-ground at WIPP [6]. These measurements are based on standard geophysical logging methods, such as neutron-epithermal neutron, neutron-gamma, and the Troxler-moisture meter. The specific probes for each method were calibrated using mined material from WIPP to account for the effects specifically associated with the brine at the WIPP site, such as the high-chloride background. The moisture and density measurements were made in an array of drillholes (9 m deep) in the roof, back and floor of a drift used in the gas-flow testing described below. The moisture content determined from these methods ranges from 3 to 5 weight % for the halite and as high as 9 % in the anhydrite and clay marker beds above and below the excavations and within the DRZ. Although the application of these methods at WIPP is still developmental and the calibration may require more examination, the halite moisture values from the logging probes are in the same range as the values determined by the TEM method described above. These moisture contents are 5 to 10 times larger than the moisture contents assumed earlier for WIPP salt [7]. The moisture content of halitic interbeds cannot be attributed to accessory minerals: for example, polyhalite, which contains approximately 6% by weight water and comprises less than 5% of rocksalt, can only contribute 0.3% of the water present in the rocksalt.

HYDRAULIC TESTING

Brine Inflow

During the excavation of the underground WIPP facility, brine inflow was observed in shallow drillholes drilled from the repository [8] [9] or as weeps on the ribs and back of the facility [10]. The occurrence of brine inflow into the facility up to 1987 has been documented by Deal and Case [11]. The migration of brine into controlled boreholes in heated and unheated salt was described and modelled by Nowak [12] and Nowak and McTigue [13].

Gas-Flow Testing

Gas flow tests at the WIPP site were conducted from horizontal, vertical (both up and down), and angled boreholes drilled from the WIPP drifts [14]. Test fluid (nitrogen) was injected into the test interval (a portion of the borehole isolated by the packer system). The majority of the tests were either constant pressure flow tests or pressure decay tests conducted from single boreholes. The principle data from both constant pressure flow tests and pressure decay tests are flow rates of gas from the test interval into the sur-rounding formation. To facilitate comparisons of flow rates measures during different conditions (test pressure and test interval size), flow rates have been normalized to 0.07 MPa (10psig) working pressure and a test interval of 1 m length and 13 cm diameter [14]. Permeabilities can be calculated by assuming the rock is a homogeneous, isotropic porous medium and the flow obeys Darcy's Law.

Results of these gas flow tests in test intervals composed of rocksalt are given in Figure 6 as the logarithm of the normalized flow rate versus the distance of test interval from the excavation. Beyond 1 m the flow rates are consistently small (< 1 SCCM) and within 1 m of the excavation, flow rates vary by many orders of magnitude (1 to 10^3 SCCM).

When the test interval containing an interbed layer was distant from the excavation, the measured flow rates were low (< 1 SCCM). Relatively high flow rates occur when the interbed is within about 2 m of the excavation and the measurement has been made near the center of a drift or intersection. As illustrated in Figure 7, measured flow rates are always less than 1SCCM (one standard cubic centimeter per minute) when the measurement is made from near the rib, and the measurements made from near the center are greater than 1 SCCM. As shown in Figure 8, the wider the drift, the more flow is measured in the interbed.

Gas-flow measurements were conducted in the entryways to the first storage panel during May and June 1986 and repeated about one year later. The S1600 entryway is 4.3 m high by 4.6 m wide, and the S1950 entryway is 4.3 m high by 6.7 m wide. At these locations, the tests were performed in test intervals which contain the first significant interbeds above (Seam B) and below (MB139) the excavation (Fig. 2). The tests were made near the center of the excavation. Initial measurements were made in these locations about 1 month after the excavation of the drifts. The result of the measurements are given in Table 1. Initially, during five of the six tests conducted in the rock above the drift, the formation produced gas rather than accepting gas. One year later, the gas from the formation had dissipated, and flows less than 10 SCCM were measured. In tests conducted in the rock below the drift, the initial gas injected flow rates were considerably greater in the wider drift. One year later, the flow rates nominally increased in the narrow drift but increased substantially in the wider drift.

In the 3.5x6 m N1100 drift an array of holes were drilled radially to a depth of about 10 m. Gas flow tests were conducted in numerous intervals along each hole. Figure 9 presents the results as contours of the normalized flowrate magnitude surrounding the drift. These results clearly demonstrate increased flow rates in the flow rates in the immediate vicinity of WIPP facility excavations. The contours in the halite suggest an circular or elliptical pattern centered on drift with flow-rates decreasing radially outward from the excavation. Anomalous zones of high flow were measured in MB139.

Tracer Gas Studies

Tracer gas studies have been conducted from the N1420 drift in 1986 [14] and the N1100 drift in 1987 [15]. The tests involve injecting a diluted tracer gas into a packed off region of the test borehole and sampling the gas in surrounding boreholes for its arrival. In most instances, plastic tarps were fixed to the drift floor to catch gas moving from the test interval into the drift. The gas samples were analyzed for evidence of tracer using electron capture gas chromatography. The purpose of the N1420 testing was to provide estimates of fracture continuity and apertures in MB139. The tracer arrivals indicate that the flow path width in MB139 increases as the span of the drift increases - from an estimated 0.002 cm in a 6.6 m wide drift to 0.04 cm at the intersection of two 11 m wide drift. It was also discovered during these tests that tracer gas was being transmitted from MB139 to the drift via the approximately 1 m thick layer of salt between MB139 and the floor of the drifts. Consistent with the inferences about MB139, it was found that apparent vertical flow paths in the salt increase with drift dimension - from 0.002 cm in a 6.6 m wide drift to 0.02 cm at the intersection of two 11 m wide drifts.

In the 6.6 m wide N11000 drift, tracer gas studies were conducted in three boreholes: one vertically up; one vertically down; and one horizontal. The test intervals were all in the salt within 1 m of the drift face. The tests in the vertically up and vertically down boreholes both indicate larger flow paths in the vertical direction (between the test interval and the drift). The test in the horizontal borehole indicated flow paths parallel to the drift face appreciably larger than those between the test interval and the drift. The inferred aperture of the flow path in all tests was small, about 1×10^{-6} m.

SUMMARY

Possible Mechanical Processes Active in the DRZ

o Strain-rate dependent brittle failure of an elliptical zone of hostrock immediately around the opening, in which the brittle failure envelope based on a strain-rate criterion is exceeded by the accelerated strain-rate adjacent to the opening [16].

o A volume of disturbed rock develops bounded by the excavation face and the elliptical surface of "the Active Opening" [17]. This volume of rock can separate (decouple) from the host rock along a shear zone that follows the elliptical surface of "the Active Opening."

o Shear induced by the opening along planes of weakness such as clay seams, which are close to the excavation but do not intersect the excavation [18].

o A pressure arch develops symmetrically above and below the opening resulting in the redistribution of stresses and the development of stress concentration about the opening [19] [200].

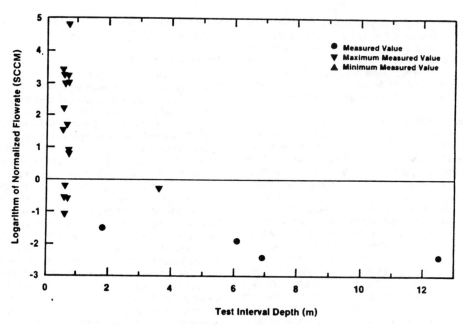

Fig. 6 Gas Flow Rates in Halite Test Intervals

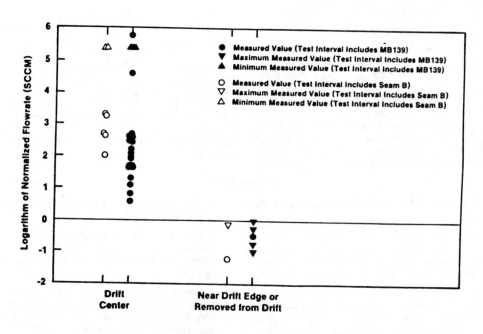

Fig. 7 Flow Rates in Interbed Layers within 2 m of Drifts

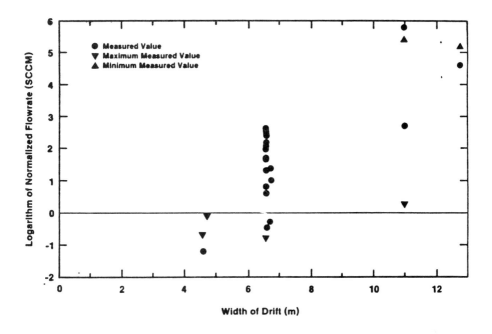

Fig. 8 Width of Drift vs. Gas Flow Rate in MB139

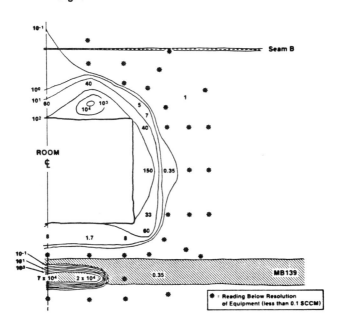

Fig. 9 Contours of Gas Flow Rate around N1100 Drift @ 4yr

TABLE 1

NORMALIZED FLOW RATE (SCCM)

	1986 TESTS	1987 TESTS
S1600 Drift (4.6 m wide)		
Seam B (up)	produced gas	0.3 *
	produced gas	0.0 *
	produced gas	0.0
MB139 (down)	0.06	0.3 *
	<0.23	0.0
	0.11	0.3
S1950 Drift (11 m wide)		
Seam B (up)	0.01	0.0 *
	produced gas	0.0
	produced gas	2.0
MB139 (down)	0.57	1300
	24.0	>10^5
	11.8	>10^5

* below detection limits of instrumentation

Within the pressure arch, zones that are in tension develop within the hostrock.

These processes give us a framework to understand the development of the DRZ at WIPP. But at WIPP, it has not been demonstrated rigorously, as yet, that these processes are active.

Effects of the DRZ on Rock Properties and Behavior

Rocksalt, which experiences excessive strain or is decoupled from the far-field host rock, will no longer behave like intact rocksalt. Such changes in behavior and properties are observed in hydraulic tests and geophysical measurements within the DRZ that has developed around the underground WIPP facility. The DRZ is characterized by distinct changes in permeability and water content of the host rock from the excavation surface out to 1 to 2 meter from the excavation. Geophysical studies of resistivities suggest that the permeability and/or fluid content of the DRZ also varies laterally.

Character of the Disturbed Rock Zone

Underground excavations at the WIPP site result in the development of a DRZ. This DRZ can be envisioned as a series of elliptical structures centered on the excavation (Fig. 11, modified from Gramberg

and Roest, 1984). The structures developed within the DRZ are characterized by mesoscopic and microscopic fracturing in both the halite and anhydrite interbeds at the facility horizon in response to stresses developed during excavation or excessive strain induced by creep. . The rocksalt in the ribs develops nearly vertical fractures parallel to the drift due to the low radial stresses near the ribs. These fractures may become extensive enough to result in spalling.

Effects of the Age and Size of the Excavation

The magnitude of the structures develop within the DRZ appears to be a function of both the size and the age of the opening. In the wider openings within a 5 yr period at WIPP, fractures are observed at all locations. Around these openings, a zone of the host rock has separated (decoupled) from the surrounding rock and has become stress relieved. In this case, the effective size of the opening , which includes the excavation and the decoupled rock around it (the *Active Opening* of Mraz [1980]), is not defined by the excavation but by the zone of the fractures that separate the DRZ from the intact host rock. In the narrower openings at WIPP within the same 5 yr period, fracturing is not observed at all locations. However, with time the narrow rooms show an increasing distribution of fractures (50% of areas after 3 years and 70% after 5 years).

REFERENCES

[1]   Bechtel National, Inc., *Interim Geotechnical Field Data Report*, DOE-WIPP-221, prepared for the U. S. Department of Energy by Bechtel National, Inc., San Francisco, California, (1986)

[2]   Francke, C.," Excavation Effects and Fracture Mapping, input to 1987 GFDAR," memorandum to R. Mckinney, IT Corporation, WIPP site, August 18, 1987, (1987)

[3]   Keller, G. V., Skokan, C. K., Andersen, H. T., Pfeifer, M. C., Keller, S. D. Keller, and Kim, K. D," Final Report: Studies of Electrical and Electromagnetic Methods for Characterizing Salt Properties at the WIPP Site, New Mexico," Contractor report to Sandia National Laboratories (Contract No. 04-1295), Department of Geophysics, Colorado School of Mines., (1987)

[4]   Pfeifer, M. C., *Multicomponent Underground DC Resistivity Study at the Waste Isolation Pilot Plant, Southeast New Mexico*, M. S. Thesis, Colorado School of Mines, T-3372, (1987)

[5]   Kessels, W., Flentge, I.,and Kolditz, H, " DC geoelectric sounding to determine water content in the salt mine ASSE (FRG)": *Geophysical Prospecting*, 33., (1985), 436-446

[6]   Hudson, J., personal communication to J. S. Stormont on USGS moisture and density measurements in bedded rock salt at the WIPP site, (1987

[7]   Powers, D. W., Lambert, S. J., Shaffer, S.-E., Hill, L. R., and Weart, W. D.," Geological Characterization Report, Waste Isolation Pilot Plant (WIPP) Site, Southeastern New Mexico," *Sandia Report*, SAND78-1596, (1978)

[8]   Popielak, R. S., Olsen, R. L., and Smith, Z. A., 1983, Results of Gas Testing in WIPP Underground Drifts, March 6-23, 1983, Memo to D. K. Shukla, dated April 13, 1983, WIPP Support Contractor Project No. NM78-648-090B

[9]   Popielak, R. S., 1983, Gas Testing Program: Phase II, Task 2, memo to D. K. Shukla et al., dated May 9, 1983, WIPP Support Contractor Project No. NM78-648-080B

[10]   Alcorn, S. R., Occurrence of Brine "Weeps" in the WIPP Facility, Memo to D. K. Shukla, dated October 3, 1983, WIPP Support Contractor Project No. NM78-648-080A, (1983)

[11]   Deal, D. E., and Case, J. B., "Brine Sampling and Evaluation: Phase I Report," Report No. DOE-WIPP-87-008, U. S. Department of Energy (Waste Isolation Pilot Plant), Carlsbad, NM, (1987)

[12]   Nowak, E. J., "Preliminary results of brine migration studies in the Waste Isolation Pilot Plant (WIPP)," *Sandia Report*, SAND86-0720, (1986)

[13]   Nowak, E. J., and McTique, D. F., "Interim results of brine transport studies in the Waste Isolation Pilot Plant (WIPP)," *Sandia Report*, SAND87-0880 , (1987)

14]   Stormont, J. C., Peterson, E. W., and Lagus, P. L., "Summary of and Observations about WIPP facility horizon flow measurements through 1986," *Sandia Report*, SAND87-0176,(1987)

[15]   Peterson, E. W., Summary of N1100 Drift Tracer Gas Travel Time Tests, personal communication (23 October 1987) to J. C. Stormont (SNL), (1987)

[16]   Dusseault, M. B., Rothenburg, L., and D. Z. Mraz, " The design of openings using multiple mechanism viscoplastic law", *28th US Symposium on Rock Mechanics,*(1987) 633-642

[17]   Mraz, D., "Plastic behavior of salt rock utilized in designing a mining method," CIM Bulletin, 73, (1980),11-123,

[18]   Brady, B. H. G., and Brown, E. T., *Rock Mechanics for Underground Mining*, George Allen and Unwin, London, 527p, (1985)

[19]   Peng, S. S. *Coal Mine Ground Control*, John Wiley and Sons, New York, 450p ,(1978)

[20]   Coates, D. F., *Rock Mechanics Principles*, CANMET, Energy Mines and Resources Canada, Ottawa, Monograph 874, (1981)

EVALUATION OF EXCAVATION EFFECTS ON ROCK-MASS PERMEABILITY
AROUND THE WASTE-HANDLING SHAFT AT THE WIPP SITE[*]

Richard L. Beauheim   George J. Saulnier, Jr.
Sandia National Laboratories[+]   INTERA Technologies, Inc.
Earth Sciences Division 6331   6850 Austin Center Blvd.
P.O. Box 5800   Suite 300
Albuquerque, NM, USA 87185   Austin, TX, USA 78731

ABSTRACT

Pulse-injection permeability tests were performed in six subhorizontal
boreholes drilled at four horizons in the Waste-Handling Shaft at the Waste
Isolation Pilot Plant (WIPP) site to evaluate potential changes in permeability caused by stress relief around the shaft. The tests were performed using
a triple-packer test tool capable of simultaneous isolation and pressure and
temperature monitoring of three intervals within a borehole. The tests were
designed to provide a "profile" of permeability with increasing distance from
the shaft wall. All lithologies tested had very low hydraulic conductivities,
with values ranging from 6 x 10^{-15} to 1 x 10^{-13} m/s. No significant changes
in permeability as a function of distance from the shaft wall were observed.
Thus, if a zone of enhanced permeability caused by stress relief exists around
the shaft, it is probably within the first 1-2 m of rock around the shaft that
could not be tested using the available test-tool configuration. The testing
revealed the presence of a depressurization zone around the shaft, apparently
caused by drainage of fluid from the formations into the shaft.

[*]This work was supported by the U.S. Department of Energy (U.S. DOE) under
 contract DE-AC04-76DP00789.
[+]A U.S. DOE facility.

EVALUATION DES EFFETS DES TRAVAUX D'EXCAVATION SUR LA PERMEABILITE
DE LA MASSE ROCHEUSE AUTOUR DU PUITS DE MANOEUVRE DE DECHETS AU SITE
DU PROJET PILOTE DE CONFINEMENT DES DECHETS

RESUME

Des essais de perméabilité par injection pulsée ont été réalisés dans
six trous de forage sub-horizontaux pratiqués dans quatre horizons du puits de
manoeuvres des déchets au site du projet pilote de confinement des déchets
(WIPP) pour évaluer les modifications éventuelles de la perméabilité
provoquées par le relâchement des contraintes autour du puits. Les essais ont
été réalisés au moyen d'un instrument d'essai à triple obturateur capable de
mesurer simultanément l'isolement, la pression et la température dans trois
intervalles à l'intérieur d'un trou de forage. Les essais ont été conçus pour
obtenir un "profil" de perméabilité à une distance croissante de la paroi du
puits. Toutes les formations testées avaient des conductivités hydrauliques
très faibles d'une valeur comprise entre 6×10^{-15} et 1×10^{-13} m/s. On n'a
observé aucune modification significative de la perméabilité en fonction de la
distance par rapport à la paroi du puits. Ainsi, s'il existe autour du puits
une zone de perméabilité plus forte provoquée par un relâchement des
contraintes, elle se trouve probablement dans les 1 ou 2 prémiers mètres de
roche autour du puits, lesquels n'ont pas pu être testés compte tenu de la
configuration de l'instrument disponible pour l'essai. Ces essais ont révélé
la présence d'une zone de dépressurisation autour du puits, apparemment
provoquée par le drainage dans le puits de l'eau contenu dans les formations.

1. INTRODUCTION

Creation of an underground opening in rock causes a redistribution of the stress field existing in the surrounding rock mass. This redistribution may extend to a distance of 5 to 6 times the radius of the opening [1]. Close to the opening, the rock commonly responds to the new stress field by fracturing and converging on the opening. This creates a disturbed rock zone (DRZ) with permeability enhanced relative to that of the intact rock mass. The hydraulic properties of this DRZ must be taken into consideration when planning the sealing of the opening.

The Waste Isolation Pilot Plant (WIPP) is a U.S. Department of Energy research and development facility designed to demonstrate safe disposal of transuranic radioactive wastes resulting from the nation's defense programs. The WIPP facility lies in bedded halite in the lower Salado Formation in southeastern New Mexico. The Plugging and Sealing Program at the WIPP is responsible for the design of plugging and sealing materials and procedures to be used in sealing the WIPP shafts. As part of this program, a field study was initiated in 1987 to measure rock-mass permeability as a function of distance from the shaft wall in the WIPP Waste-Handling Shaft in an effort to delineate the effective extent of the DRZ assumed to exist around the shaft [2].

2. WASTE-HANDLING SHAFT

The WIPP Waste-Handling Shaft (WHS) was originally blind bored as a 1.8-m diameter ventilation shaft in 1981 and 1982. In 1983 and 1984, the ventilation shaft was expanded into the present WHS by drill-and-blast slashing. The WHS is 697 m deep, and is lined with concrete through the upper 274 m (Figure 1). The finished diameter of the lined portion is 5.8 m and the diameter of the unlined portion ranges from 6.1 to 7.0 m. The concrete liner in the upper portion of the shaft is intended to seal off the water-bearing units present above the Salado Formation. These include dolomite, anhydrite/gypsum, claystone, mudstone, and siltstone. The unlined portion of the shaft is entirely within the Salado Formation, and penetrates predominantly halite, with minor anhydrite, polyhalite, and claystone interbeds.

3. TRIPLE-PACKER TEST TOOL

The tool used for the permeability testing was a three-packer assembly capable of isolating the bottom of a borehole and two intermediate zones (Figure 2). Pressure transducers and thermocouples were located in each isolated interval, and each interval could be individually pressurized. A fourth transducer monitored the pressure of the packer-inflation system. The transducers and thermocouples were monitored with a data-acquisition system controlled by a Hewlett Packard 85 computer. The triple-packer tool was installed in each test hole a single time, and the three isolated intervals were tested sequentially without resetting the test tool. The length of the packer elements and an operational constraint that no equipment could protrude from the boreholes into the shaft meant that testing was not possible within 1.64 m of the wall of the shaft. The bottom-hole test intervals ranged in length from 2.26 to 12.75 m, depending on borehole length, while the intermediate and near-shaft test intervals had fixed lengths of 1.11 and 1.26 m, respectively.

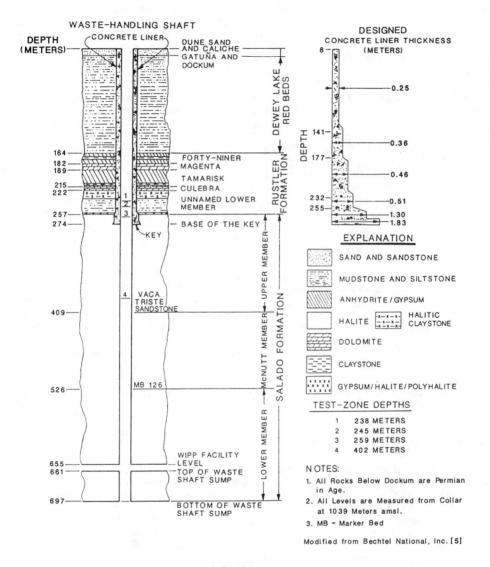

WASTE-HANDLING SHAFT

DEPTH (METERS)

CONCRETE LINER

DUNE SAND AND CALICHE
GATUÑA AND DOCKUM

164 — FORTY-NINER
182 — MAGENTA
189
215 — TAMARISK
222 — CULEBRA
UNNAMED LOWER MEMBER
257
274 — BASE OF THE KEY

KEY

VACA TRISTE SANDSTONE
409

MB 126
526

WIPP FACILITY LEVEL
655 — TOP OF WASTE SHAFT SUMP
661
697 — BOTTOM OF WASTE SHAFT SUMP

DEWEY LAKE RED BEDS
RUSTLER FORMATION
SALADO FORMATION
UPPER MEMBER — McNUTT MEMBER — LOWER MEMBER

DESIGNED CONCRETE LINER THICKNESS (METERS)

DEPTH
8 —

0.25

141 —
177 — 0.36

0.46

232 — 0.51
255 — 1.30
1.83

EXPLANATION

SAND AND SANDSTONE

MUDSTONE AND SILTSTONE

ANHYDRITE / GYPSUM

HALITE   HALITIC CLAYSTONE

DOLOMITE

CLAYSTONE

GYPSUM / HALITE / POLYHALITE

TEST-ZONE DEPTHS

1	238 METERS
2	245 METERS
3	259 METERS
4	402 METERS

NOTES:

1. All Rocks Below Dockum are Permian in Age.

2. All Levels are Measured from Collar at 1039 Meters amsl.

3. MB = Marker Bed

Modified from Bechtel National, Inc. [5]

FIGURE 1

Generalized Stratigraphy in the Waste-Handling Shaft Showing the Test Zones and the Concrete-Liner Thickness Profile

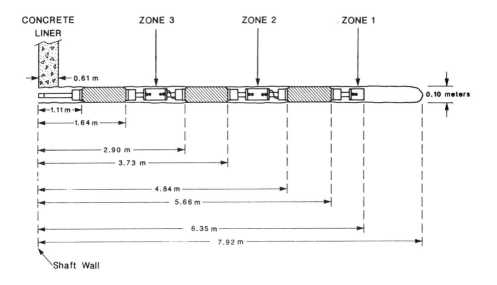

FIGURE 2

Configuration of the Multipacker Test Tool in Borehole W245E

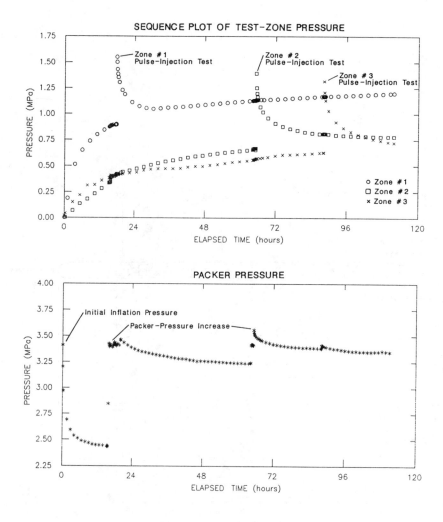

FIGURE 3

Linear–Linear Sequence Plot of the Test–Zone and Packer–Inflation Pressures
During Testing in Borehole W245E

4. PERMEABILITY-TESTING PROGRAM

Because of testing-time and accessibility limitations in the WHS, the permeability testing was intended more as reconnaissance testing to provide a relatively quick indication of the range and variability of permeability than as a carefully controlled program to obtain data for precise permeability estimation. Quantitative estimates of permeability were obtained, but the resolution of the estimates was not as great as could have been obtained had more time been available for testing.

Four geologic horizons intersected by the shaft which are considered as candidate locations for seal emplacement were selected for testing. Three of these horizons are within the lined portion of the shaft: a mudstone at a depth of 238 m; a claystone at a depth of 245 m; and halite at a depth of 259 m. Halite at a depth of 402 m was the fourth test horizon. At each of these depths, one or two 0.10-m diameter holes were drilled or cored to depths ranging from 7.9 to 12.7 m. The holes were inclined 6° to 7° below the horizontal to allow filling with brine. No fracturing was evident from examination of the core from the holes.

The general testing procedure was as follows. One to 14 days after a hole was drilled, the hole was filled with brine and the triple-packer tool was set in the hole at the desired depth. After a temperature-equilibration period of about 2 hr, the packers were inflated to about 3.45 MPa, and after another 0.5- to 2-hr equilibration period, the test intervals were shut in (isolated). A 1-day "packer-compliance" and test-zone equilibration period followed, during which the packer-inflation pressure typically decreased as the packer elements stretched, and the fluid pressures in the test intervals rose toward the formation fluid pressures existing in the adjacent rock. Following the packer-compliance period, the packer-inflation pressure was increased back to about 3.45 MPa one to two hours before the beginning of the testing.

Testing consisted of applying a pressure pulse of about 0.7 MPa to each test interval as nearly instantaneously as possible and monitoring the decay of the pulse. The bottom-hole interval (zone #1) was usually tested first, followed by the middle interval (zone #2), and finally the interval closest to the shaft (zone #3). Each pulse was allowed to decay for a minimum of one day before testing began in the adjacent interval. The pressure history of a typical test sequence from initial packer inflation to the end of testing is shown in Figure 3. Pressures monitored in the isolated zones adjacent to the test intervals responded to the test-zone pressure pulses in the majority of tests conducted. However, the responses indicated that the pressures were probably caused by packer movement due to the pressure changes rather than to direct pressure (fluid) communication between isolated zones.

The primary limitation of this testing procedure was that insufficient time was available between the initial isolation of each test zone and the subsequent testing for the test-zone pressure to equilibrate completely with that in the surrounding rock. This limitation was due to a restriction on the overall operational time available for testing in the shaft. Because of the inadequacy of the pretest pressure-equilibration periods, the pulse tests did not start from ideal static-pressure conditions. Instead, the pressure was rising at the beginning of each test, and the continuation of this rise was superimposed on the subsequent test-induced pressure response. This phenomenon is shown clearly in the zone #1 pressure response shown on Figure

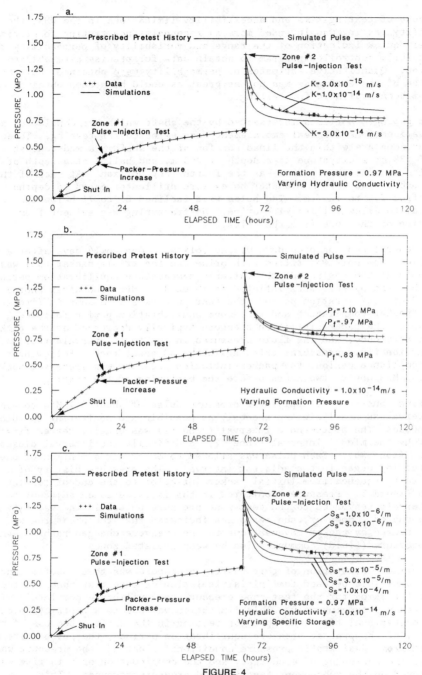

FIGURE 4

Simulation of the Borehole W245E Zone #2 Pulse-Injection Test

3, which shows that the pressure was rising before the pulse injection, and after an initial pressure decrease following the pulse injection, the pressure increased for the duration of the monitoring period. These nonideal antecedent conditions made the test data difficult to interpret analytically. Therefore, a numerical model was used for data interpretation.

5. TEST-ANALYSIS METHODS

The well-test simulator GTFM [3] was selected to interpret the pulse-test data because of its ability to include the effects of a complex pretest bore-hole pressure history as a time-variant prescribed-pressure boundary condition on the borehole. GTFM was developed using graph-theoretic-field-modeling techniques [4]. These techniques constitute a generalized methodology for modeling the behavior of continuum-type problems based upon linear graph theory, continuum mechanics, and a spatial-discretization procedure. The GTFM methodology results in a set of algebraic equations identical to those derived using finite-element or finite-difference methods. Full details of the theoretical basis and numerical implementation of the GTFM approach for the analysis of borehole-test results are presented by Pickens et al. [3].

GTFM was used to simulate the pulse-test data in an iterative fashion. Discrete values of the two hydraulic parameters considered to have the highest uncertainty, hydraulic conductivity and formation fluid pressure, were selected from ranges anticipated to be appropriate, and suites of simulations were performed using different combinations of these parameters. The simulations were compared to the observed pressure data and the parameter values were varied until an adequate match between the simulation and the data was judged to have been obtained. Figure 4a shows a typical suite of test simulations using a constant formation fluid pressure and varying hydraulic conductivity. Figure 4b shows a complementary suite of simulations using a constant hydraulic conductivity and varying formation fluid pressure. In this instance, the best match between observed data and a simulation was obtained using a hydraulic-conductivity estimate of 1×10^{-14} m/s and a formation-fluid-pressure estimate of 0.97 MPa.

A third hydraulic parameter, specific storage (which incorporates both formation compressibility and porosity), was considered sufficiently uncertain to warrant a sensitivity study of the effects of varying its value on the best-fit hydraulic-conductivity and formation-fluid-pressure estimates. Figure 4c shows a suite of simulations holding hydraulic conductivity and formation fluid pressure constant and varying specific storage through a two-orders-of-magnitude envelope of uncertainty. The specific-storage sensitivity study showed that hydraulic-conductivity estimates could be in error by about ± one-half order of magnitude and that formation-fluid-pressure estimates could be in error by about ±15% because of uncertainty in the precise value of specific storage.

The GTFM simulations shown in Figure 4 are based on an assumption that flow to/from the slightly inclined test holes was radial and isotropic. This assumption, although widely accepted in the analysis of tests performed in vertical boreholes, may not be entirely justified in the case of inclined test holes, particularly in bedded strata. If the vertical permeability of the rock is significantly lower than the horizontal permeability, flow may not enter or leave the borehole radially, which implies both vertical and

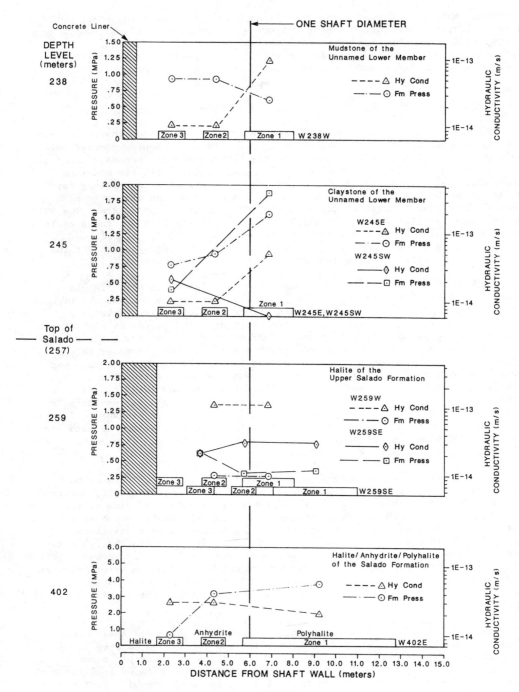

FIGURE 5
Estimated Formation Pressure and Hydraulic Conductivity vs. Distance From the Shaft Wall

horizontal components, but may be effectively restricted to horizontal only. In the case of the WHS testing, if flow to/from the slightly inclined holes was considered to be horizontal only, the interpreted hydraulic conductivities would be an order of magnitude higher.

Little is known about hydraulic anisotropy in evaporites; flow to/from the holes in the Salado Formation may well have been radial. In the mudstones and claystones of the lower Rustler Formation, horizontal flow may be more significant than vertical flow. The actual hydraulic behavior during the WHS testing was probably somewhere between the two extremes of purely radial flow and purely horizontal flow. Thus, the actual hydraulic-conductivity values lie within an envelope defined by the two end members which spans one order of magnitude. This uncertainty applies only to absolute values of hydraulic conductivity, however, and not to interpretations of the presence or absence of a DRZ. The differences in hydraulic conductivity between test zones in a borehole, from which conclusions about disturbance are drawn, are independent of the uncertainty in the flow conceptualization. The hydraulic-conductivity estimates presented below are from the simulations performed assuming radial flow to/from the inclined boreholes.

6. TEST RESULTS

Mudstone of the unnamed lower member of the Rustler Formation was tested in one borehole at the 238-m depth level. The intervals tested extended from 5.66 to 7.92 m, 3.73 to 4.84 m, and 1.64 to 2.90 m from the collar of the borehole. The concrete liner is 0.84 m thick at this location, and therefore the zone #3 test interval began 0.80 m into the rock. No increase in hydraulic conductivity with increasing proximity to the shaft is apparent from the tests at this depth. The highest hydraulic conductivity, 1×10^{-13} m/s, was found in the interval farthest from the shaft (zone #1). The hydraulic conductivity of the two intervals closest to the shaft is 1×10^{-14} m/s (Figure 5), an order of magnitude lower. Zone #1 had the lowest formation fluid pressure, about 0.62 MPa, compared to 0.97 MPa in the other two test zones (Figure 5). Fluid-pressure gradients were expected to be toward the shaft. These observations of higher hydraulic conductivity and lower pressure in the most-distant test interval may indicate that the mudstone is vertically heterogeneous. The dip of the hole (7°) caused a difference in elevation of about 1 m over the length of the hole. The bottom of the hole may have been in a layer of the mudstone more permeable than the layers slightly higher that were tested in the other two intervals. The higher permeability of the lower layer could allow more fluid drainage to the shaft, thus accounting for the lower apparent formation fluid pressure.

Claystone of the unnamed lower member of the Rustler Formation was tested in two boreholes at the 245-m depth level. The intervals tested in borehole W245E extended from 5.66 to 7.92 m, 3.73 to 4.84 m, and 1.64 to 2.90 m from the collar of the hole, with a concrete-liner thickness of 0.63 m. The intervals tested in borehole W245SW extended from 5.66 to 8.08 m and from 1.64 to 2.90 m (the intermediate zone was not tested in this hole), with a concrete liner thickness of 0.71 m. In W245E, the highest hydraulic conductivity, 5×10^{-14} m/s, was found in the zone farthest from the shaft. The other two zones have apparent hydraulic conductivities of 1×10^{-14} m/s. A fluid-pressure gradient toward the shaft was indicated by the tests in W245E (Figure 5). In W245SW, hydraulic conductivity is higher in the zone nearest the shaft (2 x

10^{-14} m/s) than in the zone farthest from the shaft (6 x 10^{-15} m/s), but the difference is not great and both values are quite low. As in W245E, the fluid-pressure gradient in W245SW is toward the shaft (Figure 5).

Halite of the upper Salado Formation was tested in two boreholes at the 259-m depth level. The intervals tested in borehole W259W extended from 5.66 to 7.42 m, 3.73 to 4.84 m, and 1.64 to 2.90 m from the collar of the hole, with a concrete-liner thickness of 1.65 m. The intervals tested in borehole W259SE extended from 7.06 to 10.97 m, 5.13 to 6.24 m, and 3.04 to 4.29 m from the collar of the hole, with a concrete-liner thickness of 1.62 m. In hole W259W, the zone closest to the shaft could not be tested because an apparently open and fluid-pressurized microannulus between the rock and the concrete liner was within the test zone. Repeated attempts to impose a pressure pulse on this zone failed, as the microannulus rapidly bled the pressure away. The tests of the other two zones in W259W were successful, however, and produced identical hydraulic-conductivity estimates of 1 x 10^{-13} m/s and identical formation-fluid-pressure estimates of 0.28 MPa for the two zones (Figure 5). Tests of all three zones were successful in W259SE. Interpreted hydraulic conductivities of the three zones are similar (3 x 10^{-14} m/s for zones #1 and #2, 2 x 10^{-14} m/s for zone #3). The formation-fluid-pressure estimates for zones #1 and #2 indicate a gradient toward the shaft, but the apparent formation fluid pressure of zone #3 is considerably higher (Figure 5), perhaps reflecting the fluid pressure in the microannulus between the concrete liner and the rock. No indication of the presence of a disturbed zone having relatively high permeability was observed in the testing at the 259-m depth level.

At the 402-m depth level, one borehole was drilled with a 6° dip. This hole intersected beds of polyhalite, anhydrite, and halite of the Salado Formation. Polyhalite was tested in an interval extending from 5.66 to 12.75 m into the hole, anhydrite was tested from 3.73 to 4.84 m, and halite was tested from 1.64 to 2.90 m; no concrete liner is present at this depth. The hydraulic conductivities of all three zones are similar, 2 x 10^{-14} m/s for zone #1 and 3 x 10^{-14} m/s for zones #2 and #3. The formation fluid pressures of the three zones indicate a gradient toward the shaft (Figure 5). Inasmuch as three different rock types were tested in this hole, the hydraulic conductivities of the three zones cannot be compared directly to provide evidence for the existence or absence of a DRZ. However, all three hydraulic conductivities are much lower than would be expected if hydraulically significant fractures were present, providing at least a qualitative indication of little disturbance of the rock mass at this depth.

7. CONCLUSIONS

Permeability testing at four horizons in the WIPP Waste-Handling Shaft produced no evidence for the existence of a disturbed rock zone around the shaft. Tests in most of the holes showed no increase in permeability with increasing proximity to the shaft, and those increases that were observed are not considered significant. The lithologies tested included mudstone, claystone, halite, anhydrite, and polyhalite. Hydraulic conductivities of all units were similar and quite low; assuming radial flow to the boreholes, the overall range was from 6 x 10^{-15} to 1 x 10^{-13} m/s. If horizontal flow predominated over radial flow in any of the observed responses, the interpreted hydraulic-conductivity values could be up to an order of magnitude higher. The maximum permeability variation observed in a single hole was one order of magnitude, which may be attributable to natural vertical heterogeneity within that mudstone unit. Hydraulic conductivities would be expected to be several orders of magnitude higher if open fractures were present in the rock. Formation-fluid-pressure estimates indicate the presence of a zone of depressurization around the shaft caused by drainage of fluid into the shaft.

Because of tool constraints, the 0.80 to 1.64 m of rock closest to the shaft could not be tested. Thus, the possibility remains that a DRZ does exist, but was restricted, at the time of testing, to a thin envelope around the shaft. Future testing might reveal whether or not the DRZ propagates outward from the shaft with time. Considering that only six boreholes were tested at only four horizons, the possibility also exists that tests at other horizons might have shown significant disturbance. Nevertheless, these tests show that permeability enhancement (fracturing) in response to stress-field redistribution around a shaft is not a pervasive process.

A more extensive testing program is planned for the WIPP Air-Intake Shaft. This shaft, currently under construction, will be raise-bored to a 6.2-m diameter. As soon after shaft construction is complete as possible, three holes will be tested at each of 11 horizons: six horizons in the water-bearing units above the Salado, and five horizons in the Salado. One to two years after initial testing, selected holes will be retested to investigate time-dependent changes in the rock-mass properties around the shaft. Additional retesting will be performed five years after initial testing. More time will be allotted for testing of each hole in this program than was available during the WHS testing to reduce the uncertainty in the test interpretations. A different tool configuration will also be used to allow testing of the rock closer to the shaft than was possible during the WHS testing. The results of the Air-Intake Shaft testing are expected to provide much more conclusive information on the development of a disturbed rock zone around a shaft than could be obtained from the reconnaissance testing performed in the WHS.

8. REFERENCES

[1] Brady, B.H.G., and E.T. Brown. 1985. Rock Mechanics for Underground Mining. George Allen and Unwin, London, 527 p.

[2] Saulnier, G.J., Jr., and J.D. Avis. 1988. Interpretation of Hydraulic Tests Conducted in the Waste-Handling Shaft at the Waste Isolation Pilot Plant (WIPP) Site. Sandia National Laboratories, Contractor Report SAND88-7001.

[3] Pickens, J.F., G.E. Grisak, J.D. Avis, D.W. Belanger, and M. Thury. 1987. "Analysis and Interpretation of Borehole Hydraulic Tests in Deep Boreholes: Principles, Model Development, and Applications", Water Resources Research, v. 23, no. 7, pp. 1341-1375.

[4] Savage, G.J., and H.K. Kesavan. 1979. "The Graphic Theoretic Field Model, I, Modelling and Formulations", Journal of the Franklin Institute, v. 307, pp. 107-147.

[5] Bechtel National, Inc. 1985. Quarterly Geotechnical Field Data Report. U.S. DOE, Waste Isolation Pilot Plant Office, Carlsbad, NM, DOE-WIPP-218.

RAPPORT D'AVANCEMENT SUR LE PROJET COSA

A PROGRESS REPORT ON PROJECT COSA

N.C. Knowles — WS Atkins Engineering Sciences, Epsom, U.K.
B. Come — Commission of the European Communities, Brussels

ABSTRACT

The COSA project was set up in 1984, under the aegis of the CEC's research programme in the management and storage of radioactive waste, with the broad objective of assessing the current capability to predict the geomechanical response of rock salt to thermal loading. The project has progressed from code verification studies in the early stages to more recent comparison exercises based on in-situ behaviour. In the latest phase a series of experimental heat and pressure tests carried out at the Asse mine in West Germany are being modelled and predictions of the thermomechanical behaviour are being compared. Ten European organisations are participating. A key feature of this exercise is that the calculations are performed 'blind', i.e. without any prior knowledge of the observed behaviour. Interest centres around the various constitutive models for rock salt and the assumptions as to the initial in-situ state of stress.

The paper presents an overview of the project to date and attempts some broad conclusions.

RESUME

Le project COSA a été lancé en 1984 dans le cadre du programme communautaire de R&D en matière de gestion et stockage des déchets radioactifs. Son objectif général est d'évaluer les possibilités actuelles quant à la prédiction du comportement du sel soumis à un dégagement de chaleur. Les étapes initiales de vérification des codes ont progressivement fait place a des exercises de comparaison a des expériences in-situ. C'est ainsi que la phase actuelle s'attache a modéliser un ensemble d'essais de chauffage et de mise en pression, réalisés dans la mine d' Asse (RFA), et a comparer les prédictions du comportement thermomécanique du sel. L'exercice regroupe dix organismes de la Communauté. Sa caracteristique fondamentale est que les calculs sont réalisés par les participants sans que ceux-ci connaissent à l'avance les résultats effectivement mesurés. Les divers modèles envisagés pour le comportement du sel, et les hypothèses sur l'etat initial des contraintes en place dans le massif, sont les principaux centres d'intérêt de cette phase.

On présente ici un panorama général de la deuxieme étape du projet COSA, ainsi que quelques conclusions préliminaires.

1. INTRODUCTION

Research studies into the geomechanical behaviour of repositories for radioactive waste in rock salt have now been pursued for nearly two decades. Such studies include fairly sophisticated computer calculations to predict the long term stress and deformation history within the host strata and a number of computer programs are now in use [1]. Project COSA (Comparison of computer codes for salt) was organised as a benchmark exercise with the broad objective of comparing the reliability of predictions made with these codes.

The project was set up by the Commission of the European Communities as part of its research programme in the Management and Storage of Radioactive Waste, following a meeting of CEC research contractors in December 1983 [2]. The first phase (COSA I) was completed in July 1986, the second phase (COSA II) commenced in November 1986 and is now approximately 2/3 complete. Twelve organisations have participated in COSA II, with the U.K. engineering consultants WS Atkins Engineering Sciences (formerly Atkins R & D) acting as the non-participating (and hopefully impartial) co-ordinator and scientific secretary.

Participants in COSA II are listed in Table 1, together with the codes in use. Ten 'mechanical' codes have been used, together with complementary or associated codes for thermal calculation. All but one (CYSIF) are based on the finite element method. They range from relatively small special purpose programs specifically developed for this generic problem (e.g. GOLIA, FLORA) to large-scale general purpose codes which are commercially available (e.g. ADINA, ANSYS). All participants have considerable experience of geomechanical analysis and are regarded as 'expert users' of their particular codes.

Table 1 Participants and Codes

Participant	Country	Thermal Code	Mechanical Code
Foraky, in association with Centre d'Etude de L'Energie Nucleaire (CEN/SCK)	Belgium	HEAT	FLORA
Laboratoire de Genie Civil of the University of Louvain-la-Neuve (LGC)	Belgium	SOLVIA-TEMP	PLACRE
Rheinisch-Westfalische Technische Hochschule (RWTH) - Aachen	Germany	FAST-RZ	MAUS
Kernforschungszentrum, Karlsruhe (KfK)	Germany	ASYTE-KA	ADINA
Laboratoire de Mecanique des Solides, Ecole Polytechnique Palaiseau (LMS)	France	THERM	GEOMEC
Departement des Etudes Mecaniques et Thermiques, CEA Saclay (CEA)	France	DELFINE	INCA
Centre de Mecanique des Roches, Ecole des Mines, Fontainebleau (EMP)	France	CHEF	VIPLEF CYSIF
Istituto Sperimentale Modelli e Strutture, Bergamo (ISMES)	Italy	GAMBLE*	GAMBLE*
Energieonderzoek Centrum Netherlands, Petten (ECN)	Netherlands	ANSYS	ANSYS GOLIA
Empresa Nacional de Residuos Radiactivos/School of Mines, Polytechnic University of Madrid (ENRESA)	Spain	ADINA-T	ADINA
Gesellschaft fus Strahlen und Unveltforschung, Braunsweig (GSF)	Germany	)	) rock salt specialists
Technical University, Delft	Holland	)	)

* GAMBLE is a pseudonym for a commercially available general purpose code

2. GENERAL ORGANISATION OF THE EXERCISE

A key feature of most benchmark exercises, and one which has, as far as possible, been scrupulously observed in Project COSA is that calculations are performed blind (i.e. without any prior knowledge of the behaviour to be predicted). In the present instance candidate benchmark problems have been discussed with participants in plenary session and a draft specification for each was drawn up by the co-ordinators and circulated to all participants. The specification was revised in the light of comments received, finalised and re-issued. Participants then performed their calculations and submitted the results to the co-ordinators, for collation and comparison with others. Direct communication amongst participants about the exercise was discouraged; the co-ordinators acting as a focal point for all matters. In this way any latent subjectivity, subconscious prejudice or bias stemming from past experience was minimised. The co-ordinators, quite deliberately, did not participate, in order to maintain maximum impartiality.

3. PROJECT COSA - PHASE I

The first phase of Project COSA was essentially a code verification exercise, with two problems being benchmarked [3]. This phase has been described elsewhere [4], [5], [6] and hence only the salient features are briefly mentioned here.

Benchmark 1 was a hypothetical problem, complex enough to demonstrate the strengths and weaknesses of the participating codes, yet sufficiently well-posed for there to be semi-analytical solutions available for comparison. It was intended principally to assess the numerical performance of the codes, particularly in respect of their temporal integration performance over the long timescales involved for repository performance.

Benchmark 2 was based on a laboratory experiment, the intention being that the codes would attempt to replicate the observed behaviour. A mutually agreeable "common" model was drawn up in discussion with all participants: geometry, boundary conditions, loading history and material law were all defined. It was left to each participant to decide on the most appropriate spatial and temporal discretisation (i.e. number, type and arrangement of elements, time step and solution method) and other aspects peculiar to his code. In short all participants were to solve the same mathematical model in a manner most appropriate to the codes available to them.

A 'common' mathematical model means, of course, that only those capabilities that are available in all the participating codes can be used. The "quality" of the model is thus limited to that described by the "lowest common factor". For this reason, attention focussed more on comparisons of the code predictions with each other, rather than with the experimental behaviour. All predictions were plausible in isolation, but wide variations were evident between the results as initially reported to the co-ordinators. Following discussion of the initial results at a plenary meeting, several participants submitted revised sets of results which reduced the variation appreciably. This benchmark thus highlighted the difficulty of obtaining good predictions of non-linear behaviour with even the most experienced teams.

4.   PROJECT COSA - PHASE 2

In the second phase of the COSA project, attention is directed to the reliability of predictions of real, in-situ behaviour. At the same time participants are allowed greater freedom to model the problem according to the dictates of their experience and to permit interpretation and characterisation of the material constitutive behaviour in a form suitable for their own computer codes. Suitable sources of in-situ behaviour for comparison, are however, not abundant. In practice they are confined to the relatively few experiments carried out in the USA and Germany in domal salt. After a brief search (the search criteria are listed in Table II) it was decided to use the experiments carried out by ECN from 1979 onwards in the 300m dry drilled borehole in the Asse mine, W. Germany [7]. This test was considered optimum despite the fact that it was familiar, in part, to certain participants and also that the measured behaviour was relatively limited.

Table II - Desirable Attributes for COSA II Benchmarks

    o   In-situ tests
    o   realistic timescale
    o   adequate thermal loading
    o   well-established behaviour
         - i.e. adequately instrumented
    o   well-posed problem - in terms of geometry,
         material and loading data
    o   unfamiliar to participants
    o   relevant to European needs

The three experiments are indicated schematically in Figure 1. The first, the so called Isothermal Free Convergence test (IFC) comprises monitoring the inward creep convergence under the lithostatic stress close to bottom of the borehole over a period of some 800 days. The second involved monitoring the pressure build up on close fitting heated probe ("Heated Pressure Probe 1" Test - HPP1). In the third test a slightly smaller diameter probe was similarly heated and the thermally induced free convergence was monitored ("Heat Free Convergence Probe" - HFCP).

4.1   I.F.C. Benchmark

    4.1.1 Experimental Details

         Figure 1 gives details of the time and location of the

experiment. Convergence measurements were begun 3 days after completion of the borehole, which was considered to have been drilled "instantaneously" compared to the time span of the experiments. Measurements were taken, on average every 2-3 days and continued for a period in excess of 800 days. In addition, the convergence at varying depths was measured on 4 occasions by raising and lowering the probe. The overall behaviour is indicated by the dotted line in Figure 2. (The two dicontinuities occur when the probe was raised for other measurements and are attributed to imperfections in the borehole wall). No other measurements were taken.

4.1.2 Model Details

Notwithstanding the limited number of measurement points, participants were asked to calculate displacement histories of the borehole wall at several levels and also time histories of both stress and displacement at a number of points at the 292m level out to 20m radius. All participants opted for an axisymmetric discretisation with the outer boundary at between 20m and 40m radius. The vertical extent of the models varied considerably: several participants opted for a thin slice (typically 1m thick) with plain strain conditions imposed on the top and bottom horizontal boundaries and repeated the analysis for each required level. One (CYSIF) used a semi-analytical method based on a 1-dimensional formulation assuming axisymmetry in the vertical plane and plain strain conditions in the horizontal plane. Others modelled the depth range between 200m and 320m directly.

4.1.3 Boundary Conditions - Lithostatic Stress

Quite deliberately participants were allowed to exercise their own judgement in defining the lithostatic stress state. Assuming a homogeneous overburden of 1040m, the theoretical (hydrostatic) value at the 292m level is approximately 22.3 MPa. However, Asse has been extensively mined over some 70 years and there is empirical evidence that the true value is somewhat less. By backfitting the observed behaviour to his earlier estimates Prij [7] showed by analytical calculations that an appropriate value was 17 MPa i.e. a reduction of 5 MPa due to mining. In contrast, Pudewills [8] estimated the reduction to be only 2 MPa by using a 2-d finite element macro-scale model of the salt formation and surrounding lithology. Not all participants were privy to this work and accordingly wide variations in the assumed lithostatic stress are apparent (Table III).

4.1.4 Material Data

It was originally intended that participants would be supplied with "raw" material test data for Asse Na2 salt. This would have allowed each team to make its own interpretations of the data and derive an appropriate constitutive law. Unfortunately such data were not available at the time required and participants selected from the following:

i.   a limited number of experimental curves showing results of isothermal tests at constant load on Na2 salt by GSF.

ii.   results of two short term creep tests carried out at the Ecole des Mines Fontainebleau [9], to which a Lemaitre law was fitted [10].

iii. existing Norton type creep equations in popular usage by researchers dealing with Asse Na2 [3] (see Table III).

The majority of participants opted for the latter.

4.1.5 Results

Principal interest is in the borehole convergence at the 292m level, for which results are illustrated in Figures 2, 3 and 4. (The latter results were submitted after the initial ones shown in Figures 2 and 3 had been discussed and assessed). Several teams submitted a number of results which were clearly intended to bound the experimental behaviour, and accordingly the label 'variant' in Figure 3 is somewhat subjective.

The dominating influence of the lithostatic stress and constitutive equations is evident. Indeed if a Norton type constitutive law is used it is possible to 'normalise' the behaviour with respect to the elastic solution for a cylindrical cavity, and the interdependence of elastic modulus, creep modulus and lithostatic stress is then apparent. For such results there is remarkable consistency in the predictions of 'normalised' convergence (Figure 5). This can be viewed as an additional verification of the individual codes capabilities. Differences in discretisation, boundary conditions, temporal integration schemes etc. appear to have a minor influence.

The CYSIF predictions are worthy of note. This team used the Lemaitre strain hardening creep law described in 4.1.4 above and a lithostatic stress corresponding to the theoretical overburden.

4.2   HPP Benchmark

4.2.1 Experimental Details

The HPP experiment followed chronologically the IFC test (figure 1). A close-fitting heated probe was lowered down the borehole (nominal diameter 315mm) until it 'stuck' at the 262m level. A heater power of 4.7kw was switched on and the subsequent pressure build up on the probe was measured together with the probe temperatures over a period of 60 days.

4.2.2 Model Details

Participants were asked to predict the temperature and stress fields in the salt mass over the duration of the experiment. For the purpose of the benchmark it was assumed that there was perfect contact between the probe and the salt from the beginning. However, as a parametric variant the KfK team modelled the problem with a 1.5mm gap. For the mechanical (i.e. stress) calculations, there were noticeable variations in the external boundary conditions, the representation of the probe, the overall size of model and in the spatial discretisation (number size and type of element used). There is no evidence that the results are sensitive to these variations however.

4.2.3 Lithostatic stress

The nominal lithostatic stress for an overburden of 1010m is 21.7 MPa. Following the experience of the IFC several participants reduced this by up to 2 MPa.

4.2.4 Material Data

As in the IFC benchmark, participants were allowed latitude in their choice of material model. In addition to the material data described in 4.1.4 the results of selected uniaxial creep test performed by BGR/GSF on Asse Na2 salt at constant temperature and stress were made available. The EMP team used this data and fitted to it a Lemaitre law modified to account for the apparent temperature dependence in the activation energy [11]. Other participants used the same creep models as for the IFC benchmark (Table IV). Worthy of note is the variation in the values of elastic modulus, since this directly influences the initial thermo-elastic response.

4.2.5 Results

Figure 6 presents the calculated temperature histories of the borehole wall at the mid-plane of the probe (i.e. where the thermal response is maximum). The majority of codes are reasonably consistent in their predictions [the generally lower response predicted by FLORA is attributed to a lower than specified heater power and a data error is suspected in the SOLVIA-TEMP calculations but has yet to be confirmed], but they are markedly higher than measured. Possible reasons for the difference include: errors in the thermal properties of the salt, heat/power loss up the borehole/cables etc. and measurement error.

The predictions of pressure build up on the probe (figure 7) vary widely; this is characteristic of all the stress predictions. It is clear that the pressure is dominated by the initial elastic response to the spatial thermal gradient and thus sensitive to variations in temperature and elastic modulus. The generally higher than-measured predictions of pressure are thus consistent with the higher-than-measured values of temperature. The effect of an initial gap between salt and probe (as in the KfKNA calculations) is both to delay the build up of pressure until the gap closes and to attenuate the peak value (due to creep relaxation).

Following the initial thermo-elastic response some relaxation due to creep is predicted by most codes. The effect is generally small however (because the relaxation is mitigated by the tendency for the pressure to rise in order to accommodate the increasing thermal strain with time). Thus in contrast to the IFC benchmark, variations in the values assumed for lithostatic stress and creep properties do not have a marked influence on the behaviour at the borehole wall at least for the duration of the test. (This is not the case however within the salt mass).

4.3   HFCP Benchmark

Calculations on this benchmark are still in progress at the time of writing this paper, so no results are available. They will be reported in due course.

Table IV - HPP Benchmark
Material Data Used

| Code label | Elastic Modulus Gpa | Creep law constants in $\dot{\varepsilon} = A\sigma^n\exp^{-Q/RT}$ | | Notes |
		A	n	
ANSYS	7.6	22.9	5.5	
FLORA	12	0.36	5.0	
GAMBLE	12	0.36	5.0	
ADINA(1)	12	0.36	5.0	
ADINA(2)	12	0.18	5.0	
PLACRE	24	0.36	5.0	
GEOMEC	25	0.36	5.0	
MAUS (1)	25	–	–	Strain hardening law
MAUS (2)	25	1.65	5.0	
VIPLEF	25	–	–	Lemaitre-Tijani Law

5. CONCLUDING REMARKS

Benchmark exercises, such as COSA, inevitably stimulate controversy in that they highlight the inadequacy (or otherwise) of computer-based predictions. In COSA I a remarkable number of "human errors" tended to obscure the difficulty of obtaining suitable data to describe the material response, but simultaneously revealed an apparent weakness in Q.A. procedures for performing sophisticated calculations.

The choice of the Dutch experiments at Asse for the in-situ benchmarks of COSA II has proved to be very satisfactory, despite initial reservation that objectivity would be compromised by some participants' familiarity with the experiments. In fact the benchmark has allowed the original analytical work to be re-assessed and to benefit from comparison with the work of a broader group of experts.

In the IFC benchmark the difficulties in defining both the ambient lithostatic stress and suitable parameters to fit to the creep model are apparent.

In the HPP benchmark the difference between the measured and predicted temperature fields (which have yet to be fully reconciled) account in large part for the differences between measured and predicted values of stress. Nevertheless substantial differences clearly arise from the variations in data values characterising the material behaviour. On the context of this benchmark there appears to be considerable benefit to be obtained from further rheological testing at Asse with the ultimate aim of characterising the constitutive properties into a form where they can be used for 'blind' calculations.

Table III

IFC Benchmark

Lithostatic Stress and Material Data Used

Code Label	Figure	Lithostatic Stress MPa	Creep Constant*	Elastic Modulus GPa
ADINA (1)	2	20.3	0.36	12
ADINA (2)	3	22.3	0.36	12
ADINA (3)	3	22.3	0.18	12
MAUS	2	17	0.18	24
MAUS (A)[1]	4	17	slow strain hardening law	
MAUS (B)[1]	4	17	fast strain hardening law	
MAUS (C)[1]	4	22	slow strain hardening law	
MAUS (D)[1]	4	22	fast strain hardening law	
ASTREA	2	22	0.36	24
GAMBLE 1	2	22	0.36	24
GAMBLE 2[1]	4	20	0.36	24
GOLIA-FAME	2	17	0.36	24
INCA (1)	2	17	0.36	24
INCA (2)[1]	4	20.3	0.36	24
CYSIF	2	22	Lemaitre law	25
FLORA (1)	2	16	0.36	24
FLORA (2)	3	20(H) 16(v)	0.36	24
FLORA (3)	3	25(H) 20(v)	0.36	24
FLORA (4)[1]	4	21.9	0.36	24

* $\dot{A}$ in $= A \sigma^n \exp(-Q/RT)$ except where noted, in which:-

σ =   equivalent stress MPa
n   =   5.0
Q   = 5.4×10^4 J/mol
R   =   8.314 J/mol/^{o}1C
T   = ^{o}K
$\dot{\varepsilon}$ =   strain rate/day)

(1)   later variant

REFERENCES

1. BROYD T W, KNOWLES N C, et al - "A Director of Computer programs for Assessment of Radioactive Waste Disposal in Geological Formations", CEC, EUR Report 8669 EN 1985.

2. COME B, ed "Computer Modelling of Stresses in Rock", CEC, EUR Report 9355 EN 1985.

3. LOWE M J S, and KNOWLES N C, "The Community Project COSA : Comparison of Geomechanical computer codes for salt". CEC, EUR Report 10760 EN 1986.

4. KNOWLES N C and LOWE M J S, "Project COSA - A Benchmark of Computer Codes for the Geomechanical Behaviour of Rocksalt" in "Reliability Methods of Engineering Analysis, Pineridge Press 1986.

5. KNOWLES N C, LOWE M J S and PIPER D, "An Update on Project COSA" in Transactions of 9th Int. Conf. Structural Mechanics in Reactor Technology, Lausanne 1987.

6. COME B, "Benchmarking rock-mechanics codes : the Community Project COSA" Proc. 6th ISRM Congress on Rock Mechanics, Montreal, August 1987.

7. PRIJ J, et al "Measurements in the 300m deep dry-drilled borehole and feasibility study on the dry-drilling of a 600m deep borehole in the Asse II salt mine", CEC, EUR Report 10737 EN 1986.

8. PUDEWILLS A, KORTHAUS E, "Modellrechnungen zur Bestimmung der Spannungsverteilung im Salzstock Asse unter dem Einfluss von unversetzten Abbaukammern", Kernforschungszentrum Karlsruhe Gmbh, July 1987.

9. VOUILLE G, "Creep tests on Asse Rock Salt" as Addendum to COSA I final report WS Atkins Engineering Sciences, April 1987.

10. LEMAITRE J, "Sur la determination des lois de comportement des materiaux elastovisco-plastiques. Publ. ONERA No. 135, Paris 1970.

11. TIJANI S M, "Thermomechanical behaviour of Rock Salt" - Topical contribution to Project COSA, EMP Dec. 1987.

The authors are grateful to all participants in the COSA project for their unremitting help throughout the exercise and for their selfless generosity in submitting their calculations for scrutiny and comparison.

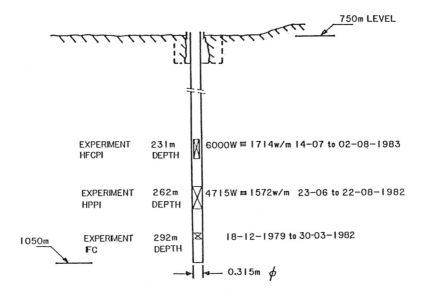

EXPERIMENT HFCPI	231m DEPTH	6000W ≡ 1714w/m 14-07 to 02-08-1983
EXPERIMENT HPPI	262m DEPTH	4715W ≡ 1572w/m 23-06 to 22-08-1982
EXPERIMENT FC	292m DEPTH	18-12-1979 to 30-03-1982

Figure 1   Location of Experiments at Asse for COSA II

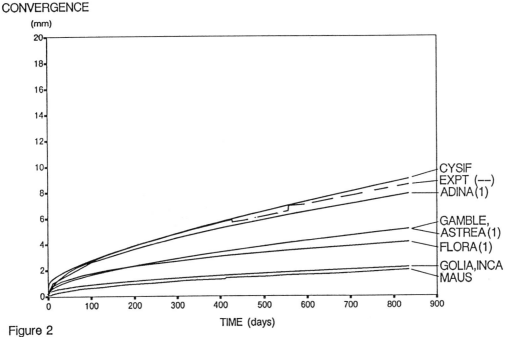

Figure 2

IFC – RADIAL BOREHOLE CONVERGENCE AT 292m DEPTH

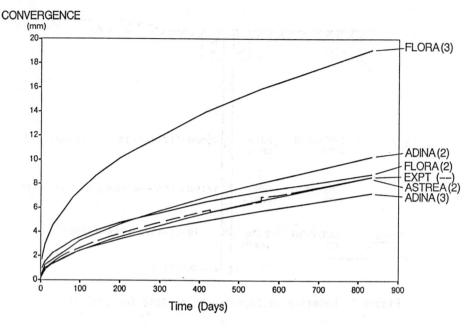

Figure 3   IFC – RADIAL BOREHOLE CONVERGENCE AT 292m DEPTH
– VARIANTS

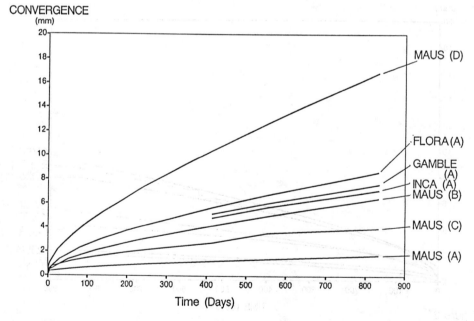

Figure 4   IFC – RADIAL BOREHOLE CONVERGENCE AT 292m DEPTH
– LATER VARIANTS

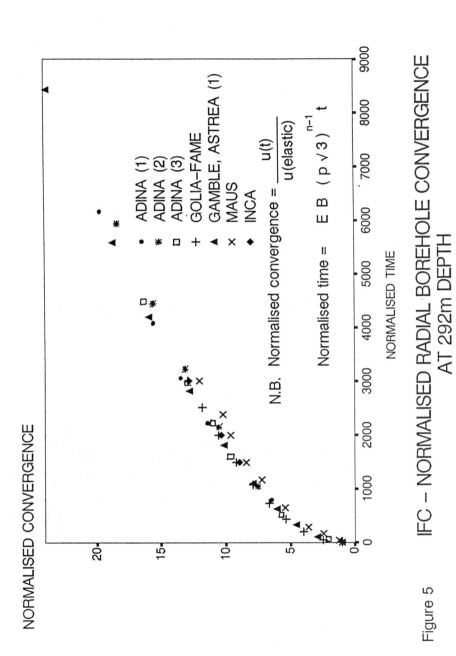

Figure 5   IFC – NORMALISED RADIAL BOREHOLE CONVERGENCE
AT 292m DEPTH

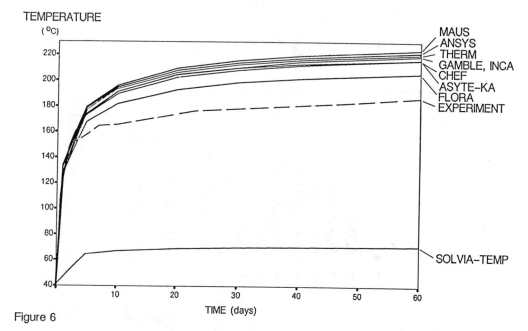

Figure 6

HPPI – TIME HISTORY OF TEMPERATURE AT POINT A

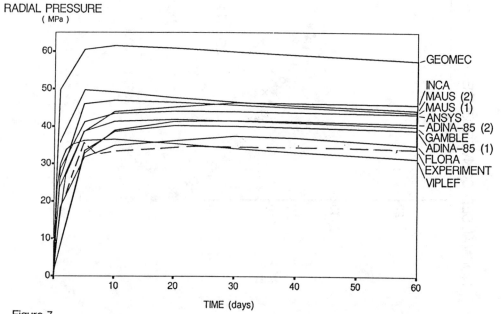

Figure 7

HPPI – TIME HISTORY OF RADIAL PRESSURES AT POINT A

```
* * * * * * * * * * * * * * * * * * * * * * * * * * * * * * * *
```

Session III - Séance III

CHAIRMAN'S SUMMARY - RAPPORT DE SYNTHESE DU PRESIDENT

R.V. MATALUCCI (SNL, United States)

```
* * * * * * * * * * * * * * * * * * * * * * * * * * * * * * *
```

I.   INTRODUCTION

The workshop indicated the importance of understanding excavation effects for nuclear waste repositories in salt. In particular, the workshop highlighted the need for further studies on the disturbed zone, more correlations among current in-situ data, and continued work in model validation and an evaluation of sealing systems (effects and performance assessment).

A systematic study of the disturbed zones is considered necessary. Research in this area should include: a) a history of the zone from its initiation; b) the effects that excavation has on permeability; and c) how the disturbed zone changes in time, particularly in regard to whether the change is considered to be healing.

The workshop also noted that more correlations among in-situ test data are required. Specifically, correlations among permeability and hydrologic monitoring, deformations and geophysical data would be productive.

Studies that were recommended for continuation include: a) fracture model and criteria development; b) validation of models through in-situ data; and c) determining the effects of sealing systems and their performance in-situ. Projects in these areas have been underway for several years, but continued study and evaluation is recommended for a thorough understanding of these areas.

Finally, the topics under discussion at this workshop suggested that an additional workshop on the accomplishments in model validation for repository designs would be especially productive.

A brief review of the papers in the session is provided below.

II.   PAPERS PRESENTED IN SESSION III

The Keynote Paper was presented by M. Langer (FRG), who noted that design and safety requirements for a permanent repository for radioactive waste can be demonstrated only by a site analysis in which the entire system (the geological situation, the repository, and the form of the wastes and the waste containers) and their inter-relationships are taken into consideration. Technology that adequately characterises a repository in salt has developed over the past years, resulting in state-of-the-art tools for site analysis. Various countries have been and are currently conducting intensive basic research in the fields of geotechnics and rock mechanics. This research has produced in principle, but not in every detail, reliable tools (i.e., methodology, scientific principles, technical "know-how" and experience in construction and operations). These tools are available to predict the feasibility of various techniques for the permanent storage of radioactive waste and the consequent safety aspects.

Examples of issues that have been successfully researched or are currently under study include the following:

a) Creep in rock salt can, both theoretically and empirically, be defined sufficiently for a perfectly acceptable engineering assessment of the phenomenon as a basis for a safety analysis.

b) The major objective of creep failure studies is the quantitative determination of the threshold conditions for such failures. The data derived from the tri-axial test rig has proved reliable for calculating the safety of a permanent repository over the long term in terms of fracture mechanics.

c) A calculational method that is particularly suitable for the solution of geomechanical problems is the finite element method (FEM). For the purpose of analysis of waste disposal facilities, codes (e.g., ANSALT) have been developed, with proven capacity.

d) Determining the natural stress in a rock mass is one of the most difficult problems in rock mechanics. Great efforts are being made throughout the world to develop more effective methods of measuring stress. It is currently possible to measure stress and deformation in boreholes 300 m deep and more. Also, newly developed vacuum-measuring equipment can determine permeability in the salt rock in the disturbed zone around a tunnel.

e) Extensive research, including failure analysis, is currently underway in many countries to develop a realistic and testable long-term repository concept.

Central to the calculations for the safety analysis is the construction of a model that simulates as closely as possible the actual conditions of the rock before and after construction of the cavern. The geoscientific approach is continuous improvement of the model, appropriate to the improved knowledge of the input data. At present, the models draw upon the large database for short-term thermo-mechanical effects, with predictive capability being verified through lab and field tests.

In summary, the conclusions and recommendations of the keynote paper are as follows:

- Conclusions:

  . Large database on short-term thermo-mechanical effects
  . Predictive capability being developed via lab/field data
  . FEM codes and models -- tools for safety analysis and design
  . Long-term response and failure criteria under study.

- Suggests:

  . Continuation of testing for model validation
  . Additional studies for fracture development and failure mode.

K. Wieczorek (FRG) presented a paper on deformational rock-mass response during a simulation experiment at the Asse Mine. The Brine Migration Test (BMT), which was performed at the Asse Salt Mine from 1983 through 1985, was a joint US/FRG nuclear waste repository simulation experiment that offered the opportunity to validate deformation/structural models with in-situ data. The BMT consisted of four individual test sites. At two sites, electrical heaters simulated high-level waste (HLW). At the other sites, electrical heaters and cobalt-60-sources were used.

FEM calculations were performed for Test Site 2 (non-radioactive), using a two-dimensional (2D) approach to simplify the problem. The authors compared the calculations to the horizontal and vertical room closures, and found good agreement between the measured and calculated curves. The measured horizontal closure was higher, however, than the calculated closure; the measured vertical closure was lower than calculated. The authors felt that although the calculation results could be improved with a three-dimensional (3D) model, the 2D approach produced good results for the closure calculation at a comparatively low expenditure. Workshop participants discussed the possibility of determining an "acceptable" discrepancy bound (i.e., calculations vs. data). In summary, the conclusions and recommendations are as follows:

- Conclusions:

  . 2D calculations comparable with measured data
  . 3D calculations may improve comparisons.

- Suggests:

  . Deformation structural models being validated with in-situ data
  . What would be an "acceptable" discrepancy bound (calculations vs. data)?

C. Heick (FRG) presented a paper on the investigation of cavity responses by microseismic methods. Because fault planes are generally regarded as possible brine intrusion sites, the potential of microseismic monitoring to determine their location and extent is of great importance. The authors reported on microseismic measurements in rock salt that have been monitored in the Asse mine (FRG) since 1979, and in the abandoned Hope potash mine (FRG) since 1986.

A great advantage of microseismic monitoring is that the zone can be investigated from a remote location. The microseismic arrays in this study consisted of geophone stations arranged around the zone to be investigated. The data from at least four geophones were required to locate an event.

Microseismic monitoring disclosed that seismic activity was occurring in the vicinity of roads and chambers in both mines, and that the fault planes were closely related to the diameters of the neighbouring cavity. Frequencies

of seismic events were a function of rock disturbances. The results of the
study suggest that fractures and fault zones can be located by microseismic
monitoring. Workshop participants suggested that further site evaluations
concurrent with excavation activities are needed. In summary, the conclusions
and recommendations are as follows:

- Conclusions:

  . Microseismic monitoring successful from remote locations
  . Perhaps locate "weak" or "stressed" zones
  . Event frequencies function of rock disturbance.

- Suggests:

  . Fractures/fault zones can be located
  . Need further site evaluations concurrent with excavation
    activities.

An interim report on excavation effect studies at the WIPP site was
presented by D.J. Borns (US). Here, the damaged rock zone (DRZ) has been
characterised using three approaches: a) visual observation; b) geophysical
methods; and c) measurement of hydraulic properties. All three approaches have
defined a DRZ at WIPP that extends laterally throughout the excavation and
varies in depth from 1 to 5 m, according to the size and age of the opening.
An ongoing experimental programme is developing a more detailed 3D definition
of the DRZ.

Visual observations using drillholes indicate that fractures (with
apertures greater than 2 mm and visible without enhancement to the naked eye)
and fluids are common in the wall rock of the underground facility.
Geophysical observations included electromagnetic surveys, resistivity and
seismic technology. According to the authors, the DRZ is characterised by
distinct changes in permeability and water content of the host rock from the
excavation surface out to 1 to 2 m from the excavation. Geophysical studies of
resistivities suggest that the permeability and/or fluid content of the DRZ
also varies laterally.

The DRZ can be envisioned as a series of elliptical structures centred
on the excavation. The magnitude of the structures that develop within the DRZ
appears to be a function of both the size and the age of the opening. Within a
5-year period at WIPP, fractures in the wider openings were observed at all
locations. In the narrower openings at WIPP, within the same 5-year period,
fracturing was not observed at all locations. However, with time the narrow
rooms showed an increasing distribution of fractures (50% of areas after 3
years, and 70% after 5 years).

To further investigate the DRZ, the authors and workshop participants
felt that more fluid-flow measurements were needed, with regard to
time-dependent effects and the healing potential of salt. Geophysical methods
should also be evaluated for the DRZ studies. In summary, the conclusions and
recommendations are as follows:

- Conclusions:

    . Increased gas flow rates within 2 m (DRZ)
    . Function of size and age of excavation
    . Fluid content increases with distance into rock frame
    . Geophysical techniques promising.

- Suggests:

    . Need more fluid-flow measurements
      - Time-dependent effects
      - Salt healing potential
    . Evaluate geophysical methods for DRZ studies.

An evaluation of excavation effects on rock-mass permeability at WIPP was presented by R.L. Beauheim (US). Pulse-injection permeability tests were performed in the Waste-Handling Shaft to evaluate potential changes in permeability caused by stress relief around the shaft. The tests were performed approximately 3 years after excavation, and were designed to provide a "profile" of permeability with increasing distance from the shaft wall.

All lithologies had very low hydraulic conductivities (6×10^{-15} to 1×10^{-13} m/s). No significant changes in permeability as a function of distance from the shaft wall were observed. The testing did reveal the presence of a depressurisation zone around the shaft, apparently caused by drainage of a fluid from the formation into the shaft.

Experimental constraints limited the extent of this test, but future tests are planned for the WIPP Air-Intake Shaft. The Air-Intake Shaft test will determine the nature and extent of the disturbed rock zone (DRZ) around the shaft by: a) testing at different levels and depths; and b) considering time-dependent effects. Results from this test and future shaft permeability tests should also correlate with other measurements, such as geophysical test results, geomechanical response and ultrasonic tests. In summary, the conclusions and recommendations are as follows:

- Conclusions:

    . Low hydraulic conductivity (6×10^{-15} to 1×10^{-13} m/s)
    . No change in permeability with distance from shaft face.

- Suggests:

    . More shaft permeability tests (nature and extent of DRZ)
      - Different levels and depths into rock
      - Time-dependent effects
    . Correlations with other measurements
      - Geophysical tests
      - Geomechanical response (closure)
      - Ultrasonic tests.

A progress report of Project COSA of the CEC was presented by N.C. Knowles (UK). The COSA project was established in 1984 with the broad objective of assessing the current capability to predict the geomechanical response of rock salt to thermal loading. In the latest phase, a series of experimental heat and pressure tests that were performed in the Asse mine (FRG) were modelled. Code predictions from 10 European organisations (including a total of 12 codes) were compared to the thermo-mechanical behaviour. A key feature of the exercise was that the calculations were performed "blind", i.e., without any prior knowledge of the observed behaviour.

The codes were compared to convergence data, temperature histories and radial pressure data. The authors found that human errors were high, and that there is some weakness in the quality assurance procedures. The developers of the code also encountered some difficulty in defining the stress state and material parameters. In general, the differences among the codes resulted from variations in characterising the materials. Results of the project suggest that improved problem definition would help to reduce discrepancies among codes. In summary, the conclusions and recommendations are as follows:

- Conclusions:

    . Human errors high; Some weakness in QA procedures
    . Difficulty in definition of stress state material parameters
    . Differences resulted from variations in material characterisation
    . Compared convergence, temperature histories, radial pressure (12 codes participating).

- Suggests:

    . Improved problem definition; Reduce discrepancies
    . Validation against in-situ data is final step
    . Benchmarking is a training tool for analysts!

SESSION IV

EXCAVATION RESPONSE IN ARGILLACEOUS ROCK

SEANCE IV

EFFETS DE L'EXCAVATION SUR LES FORMATIONS ARGILEUSES

Chairman - Président

R.H. HEREMANS
(NIRAS/ONDRAF, Belgium)

ETUDES SUR LES EFFETS DE L'EXCAVATION
SUR LE SITE NUCLEAIRE DE MOL

EXCAVATION RESPONSE STUDIES AT THE MOL FACILITY

B. Neerdael, D. De Bruyn
CEN/SCK
Mol, Belgium

ABSTRACT

The Hades-project developed at the Mol site (B) is intended to acquire on a site specific basis (Boom clay formation) the necessary data to assess the technico-economical feasibility and the long term safety of a disposal system in this deep tertiary clay formation.

In this context, the realization of the so-called "test drift" had to demonstrate the possibility to mine and to construct galleries at real scale (3.5 m internal diameter) in this plastic clay and to study the long term behaviour of the mined structures.

An overview is given of the most relevant field data recorded in 1987 during the whole construction period. First results of this mine-by-test are discussed.

RESUME

Le projet HADES développé sur le site nucléaire de Mol-Dessel (B) a été conçu pour acquérir les données nécessaires à l'évaluation pour ce site spécifique (argile de Boom), de la faisabilité technico-économique et de la sécurité à long terme d'un dépôt de déchets radioactifs conditionnés dans une formation argileuse profonde d'âge tertiaire.

Dans ce contexte, la réalisation d'une galerie d'essai devait démontrer la possibilité d'excaver cette argile plastique pour y réaliser des galeries à échelle réelle (diamètre intérieur de 3.5 m) dont on étudiera le comportement à long terme.

Les données géotechniques relevées sur terrain pendant cette phase de construction (1987) sont passées en revue. Les premiers résultats de ce test combinant creusement et auscultation sont discutés.

1. INTRODUCTION

In the framework of the Hades project (High Activity Disposal Experimental Site) managed by the Belgian Nuclear Research Establishment (CEN/SCK), the extension of the URL to a demonstration/pilot facility has been launched in 1987. The project is sponsored by the Commission of the European Communities in the frame of part B of the CEC-programme on radioactive waste management and disposal and by ONDRAF/NIRAS, the Belgian Waste Management Authority.

The interest of the tunnel concept for disposing off radioactive waste into the Boom clay has been recognized since the early phase of the research programme but among the different items to be demonstrated, a number of unsatisfactory answered questions related to the behaviour of clay at depth, had to be tackled.
In a first phase, the construction of a test drift had to give a clear indication of the tunnelling capabilities at real scale, applying conventional techniques at an acceptable level of cost without compromising the integrity of the formation (near-field).

Based on prediction by models and according to previous characterizations of Boom clay (e.g. preliminary investigations at smaller scale), a geotechnical investigation programme was developed in the clay surrounding the opening to be excavated and a monitoring programme on the lining was established ; measurements with regard to soil deformation, behaviour of lining made of precast concrete segments and pore water pressure changes were conducted to quantify the Boom clay's response to excavation. The paper gives a description of the overall context in which these measurements are conducted and data recorded in 1987 are discussed.

Further interpretation of the results of this mine-by-test will allow to validate numerical models and point out the main parameters in view to define the most appropriate tunnelling technique and equipment as well as the waste repository concepts thereby improving the performance of the system.
As an example, the development of lining loads with time which is of major importance in deep tunnelling projects has been exploited in an alternative gallery lining concept (sliding steel ribs) developed by Andra (French Agency for Nuclear Waste Management) and experienced at Mol end 1987 in a 12 m long experimental gallery dug in the prolongation of the concrete test drift. Design and results are discussed by a representative of Andra in this workshop /1/.

2. THE MOL FACILITY (fig. 1)

The first investigations related to waste disposal set-up by CEN/SCK date from the seventies /2/. The research programme was rapidly focussed on the Boom clay, the uppermost argillaceous formation underlying the Mol site.

The first decade of research led to the construction of an underground research laboratory (URL) at 223 m depth in the tertiary clay formation to collect data in situ and to confirm laboratory observations (1980-1983).

The ground freezing technique used to dig the access shaft through the water-bearing overburden sands was further applied to build the gallery in the clay at 223 m depth by way of horizontal freezing pipes implanted from the bottom of the shaft /3/. The high cost, the deformations and subsisting related problems caused by the freezing technique led rapidly to the conclusion that the study of the behaviour of "virgin" clay at depth was a first prerequisite before considering any further design study.

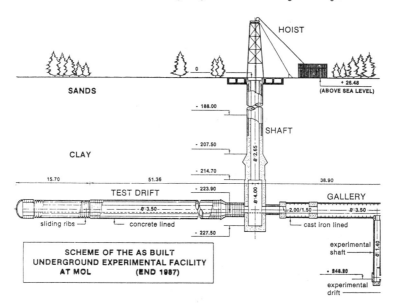

Fig. 1. The Mol facility

The construction of exploratory excavation works in non-frozen clay consisting of a shaft and a gallery, 2 m in diameter, lined with concrete, was started directly, in parallel with the instrumentation of both the clay body and the lining. One of the most important characteristics of clay in this field being its time dependent behaviour, auscultation was developed in order to allow long term measurements /4/. Very encouraging results, resumed in point 4, were obtained.

According to priorities for research and demonstration set in the context of the CEC's programme, the extension of the URL to a demonstration/pilot facility was launched last year, starting with the construction of a test drift. This test drift is a gallery with circular cross section composed of a 5.6 m long access gallery (2.64 m internal diameter), followed by a 42 m long test zone (3.5 m internal diameter). The excavation, about 5 m in diameter, is lined with precast concrete segments, 0.6 m thick, designed to withstand the full lithostatic pressure with a coefficient of earth pressure at rest of 0.7. The segmental liner is equipped with portholes in the sidewall, springing and floor positions.
With the realization of the additional experimental "ANDRA" gallery (sliding steel ribs), and the shotcreting and instrumentation of a concave end front included, the works lasted about 8.5 months.

A work programme is now for the next decade prepared in detail to perform technological tests intended to demonstrate the safety and feasibility of nuclear waste emplacement and consequent tests essentially pertaining to the nuclear technology aspects of waste disposal.

3. BOOM CLAY CHARACTERIZATION

Generic studies have shown that among argillaceous formations, important variances may be found in their composition and thus also in their properties.

Physical and chemical properties of Boom clay were already published /2/ and stratigraphical and sedimentological details were investigated in several studies /5/.
Boom clay pore water, clearly dominated by sodium and carbonate, is weakly alkaline and very reducing.

In geotechnical terms /6/, the Boom clay at a depth of 240 m may be considered as a stiff overconsolidated clay of high plasticity with a moisture content ranging from 20 to 23 %.
An overconsolidation ratio of 2.4 corresponding to a preconsolidation stress of circa 6 MPa was obtained by analysis of one-dimensional consolidation tests. It is however difficult to reconcile the corresponding 600 m thickness of original overburden with the geological evidence.

Bulk and dry density are close to respectively 20 and 16.5 kN/m^3. Swelling pressures (under zero volume change conditions) can reach 0.9 MPa. When considering Atterberg limits, the liquid limit of 64 to 68 % and the plastic limit of 19 to 25 % yield a mean plasticity index of 47 %.

Recent laboratory and field approaches converge to a value for the hydraulic conductivity of 3 to 4 10^{-12} m/s.
The coefficient of consolidation C_v is calculated as 0.65 to 0.80 m²/year for in situ pressure levels.

The strength affects the stability ratio and the plastic zone around the tunnel. The undrained shear strength of 0.6 to 0.8 MPa originally inferred from core samples was determined on less disturbed samples taken at depth yielding 1.1 or 1.2 MPa ; the undrained Young's modulus (secant at 1 % strain) is about 200 MPa under in situ stress conditions.

4. FIRST IN SITU TESTING PROGRAMME

The study of the in situ mechanical behaviour of deep clay was first carried out as mentioned above by reduced scale fully instrumented exploratory works (2 m in diameter). In particular, the 8 m long experimental drift was equipped in a central section with convergence measuring bolts and load cells in the lining as well as with total pressure cells and a multi-points extensometer in the roof /7/.

A set of extremely valuable data was gained as detailed in a previous paper /8/ together with small scale experiments in boreholes dedicated to convergence, tests, stress field measurement, instrumented tube and dilatometer tests :

- the stress relief below the bottom of the 2 m diameter shaft becomes negligible at about 5 m as reflected by extensometric measurements ;
- the 2 m diameter gallery shows ovalization responding to a main vertical stress ; similar indications were given by other experiments ;
- the behaviour of the front left open now since 4 years does not indicate a significant ground displacement problem but the wall convergence of unlined boreholes reaches 30 % within a few days ;
- time dependent effects are clearly demonstrated ; settlements at 1 meter distance in clay above the crown of the gallery increased from 24 ("instantaneous" deformation) to 29 mm during the 3 years measuring campaign following the construction works and are still perceivable ;
- a total convergence (ratio displacement/radius) of 5 % at the excavated wall can be inferred whereas the convergence of the lining is limited to 1.5 % ; the value of the radial pressure deduced from load cells measurements is 1.5 MPa.

5. TEST DRIFT - MINE-BY-TEST

The objective of installing a comprehensive instrumentation system in the clay was threefold :

- to confirm the validity of the design ;
- to monitor the response of the clay to excavation at a real scale ;
- to provide the means for a long-term assessment of the drift's behaviour.

Several weeks prior to construction, instrumentation in the clay mass was installed from the basis of the access shaft ; it consists essentially in an array of settlement gauges of different measuring principles for recording the soil deformation above the drift and an array of piezometers to monitor the pore water pressure changes, a few meters under the tunnel centre line.

Refering to the 3.5 m diameter completed section and the effective working days, the mean progress rate is of about 1.85 m per week. The mining face was always about 1.5 m ahead of the last completed ring at any time.

The lining of the drift consists of unbolted rings each composed of 64 concrete segments (0.6 m thick) separated by wooden plates.
Seven rings along the 42 m long tunnel were instrumented for measurements of diametral convergence and pressure build-up.
The theoretical mined diameter (4.7 m) was slightly increased, particularly at the upper part of a ring where an overbreak of 15 to 30 cm was necessary for placing the wedge into the completed ring and using a temporary support ; this gap was even sometimes higher at the crown due to loose clay blocks. The gaps were backfilled regularly after placing two segmented rings, using a 1 to 14 cement/sand grout with 5 % water.
As no mechanized techniques were used, alignement difficulties were sometimes encountered which can influence some of the recorded measurements in the lining.

During the construction works, a multi-point extensometer was lowered and sealed off in a borehole drilled from ring 35 radially downwards to evaluate long term displacements in the clay close to the drift.

The stability of the end front, shotcreted over a thickness of 10 to 15 cm is monitored by measuring displacements from the concave surface until a few meters behind.

Very recently, self boring pressuremeter tests have been performed ; profiles of the in situ total lateral stress, shear modulus and shear strength will be deduced from tests conducted at a fast expansion rate not allowing any drainage, the membrane being expanded to approximately 10 % strain.

5.1. Soil deformation

The deflectometer or settlingmeter devices placed horizontally above the roof of the gallery are intended to record displacements in clay resulting from the construction of the test drift including the instantaneous deformations occurring at the excavated wall and responsible for more than 3/4 of the total deformation. They are shown in figure 2 and consist of :

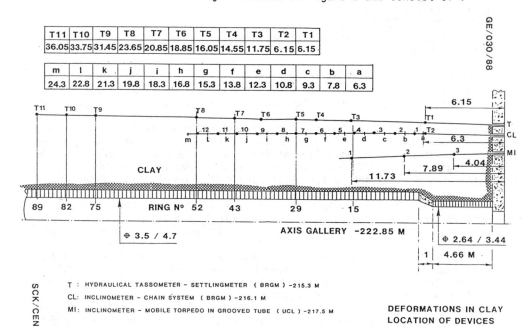

Fig. 2. Test drift : Location of settlement gauges

- an inclinometer using a mobile torpedo in a 12 m long fluted tube at level -217.5 m (MI).
  The displacement induced by the construction is quite significant, reaching a mean value of 90 mm corresponding to a convergence (u/r) of about 2 % at 4.2 m distance of the drift axis.
- one hydraulic tassometer or settlingmeter (T) at level -215.3 m composed of 11 hydraulical sensors distributed along a 36 m long borehole and one 24 m long horizontal inclinometer at level -216.1 m in a chain configuration with one hydraulical sensor, representing the fictive

reference point at 6.3 m into the clay, followed by 12 potentiometric inclinometers (MIDORI, range ± 10°) every 1.5 m (CL).
Fig. 3 illustrates the displacements which occurred between April 1987 and October 1987 in each of the 2 measuring systems which are represented on this diagram as starting from the same point for comparison. The curves dated April, 6 represent the initial slope of each drilled hole. The total displacements recorded in October are ranging from 37 mm for the Tassometer Y to a mean value of 48 mm for the Inclinometer (CL).

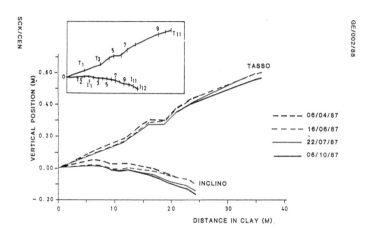

Fig. 3. Test drift - Displacements in clay for different time periods

The diagram of figure 4 has been drawn to illustrate the response of the clay mass by way of the cumulative deformation recorded by each inclinometer successively affected along the 24 m long instrumented borehole as a function of tunneling progress. Although being located more than 4 m in the clay above the lining, the measuring sensors are projected, for scale evidence, at the drift axis level.
The zone of influence of 3 times the radius inferred from this graph is therefore an under-estimated value of about 25 % ; the same extension can be considered ahead or behind the front when taken into account, no delayed effects are taken into account.

This interpretation is corroborated by figure 5 where the evolution of the voltage output of clinometric sensors is plotted versus time.

To investigate more in detail the plastified zone surrounding the test drift, an electrical multi-point extensometer (Distofor) has been placed during the gallery construction in a 20 m long vertical borehole. Seven sensors are distributed along the first 5.7 m of the borehole and the displacements are measured radially to this drift section with regard to a fixed reference materialized in the lower part of the 20 m long casing.
The informations recorded by this multi-point extensometer where "instantaneous" deformations are of course not included indicate up to now maximal deformations of 9 mm at 0.3 m behind the lining and 3 mm some 5.5 m further. According to our present hypotheses, the radius of the plastified zone surrounding the drift could be estimated to 3.6 m.

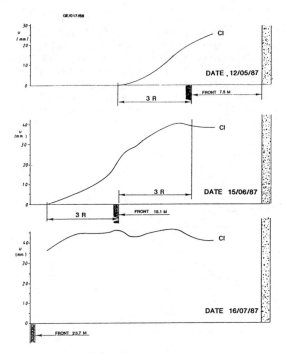

Fig. 4. Displacements in clay above the test drift
Inclinometric measurements (projected at drift level) versus
excavation progress

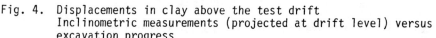
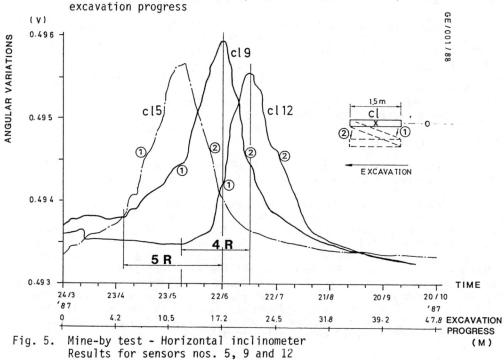

Fig. 5. Mine-by test - Horizontal inclinometer
Results for sensors nos. 5, 9 and 12

5.2. Pore water pressure measurements

The location of the different piezometer screens and hydraulical pressure cells (PPK) to measure the pore water pressure dissipation is given on fig. 6.

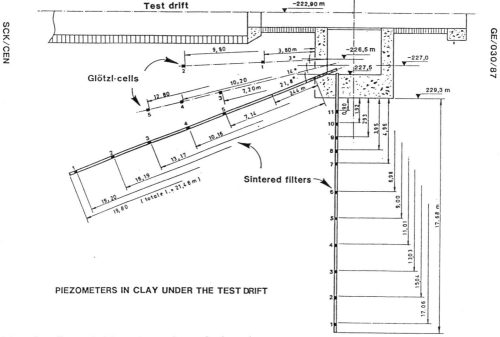

PIEZOMETERS IN CLAY UNDER THE TEST DRIFT

Fig. 6. Test drift - Location of the piezometers array

The variations recorded in 1987 are plotted on fig. 7. After a pseudo-equilibrium state related to its initial positioning with regard to the access shaft, each sensor is affected by the excavation. A similar interpretation as made for the soil deformations indicates again a distance of 4 times the radius for the zone of influence.

For Glötzl cell "PPK1", closest to the underground structures, the pore water pressure, after a preliminary reaction caused in March by pilot boreholes, began to drop rapidly when the concrete lining of the access chamber was pulled down (9 April 1987).

From June onwards, the pressures gradually increased towards a new equilibrium governed by the drift construction. Assuming the tunnel to act as a drain (with length L), the measurements of the piezometer array can be satisfactorily reproduced by a law of the type : $\ln (1 + L/2R)$.

5.3. Convergence measurements on the lining

The invar thread method is used to record the diametral convergence of the lining, along the 7 instrumented sections where convergence studs were embedded into special concrete blocks.

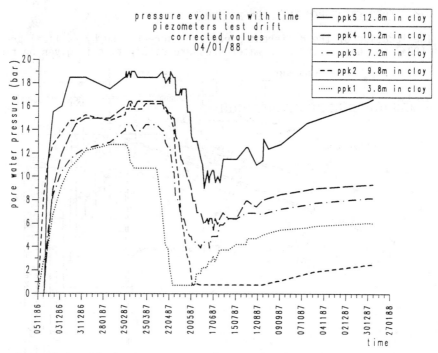

Fig. 7.a. Hydraulical piezometers - Pore water pressure versus time

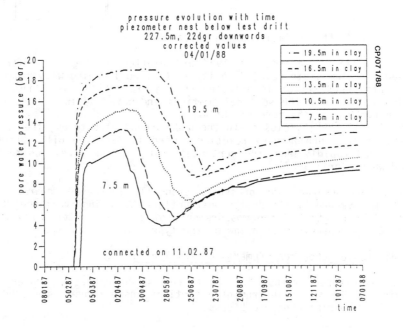

Fig. 7.b. Piezometer screens - Pore water pressure versus time

The evolution of the mean convergence for each section within an equivalent time period is given by figure 8 ; after 90 days, the convergence (u/r) is ranging from 1.2 to 1.5 %. Improvement of the construction works with time resulting from experience seems to be the main explanation to the differences observed although slight modifications in the clay properties and fracturation have also been noticed.

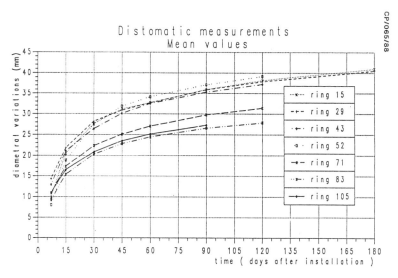

Fig. 8. Diametral convergence of the lining

5.4. Pressure build-up

Different types of sensors are monitoring the pressure acting on the lining :

- load cells between concrete segments (2 cells ⌀ 30 cm) from which equivalent radial pressures have been derived ;
- total pressure cells (15x25 cm) and experimental flat jacks embedded at the external face of especially designed concrete segments.

Measurements indicate that the drift is not subject to the full overburden stress ; the maximum values recorded are ranging between 1.5 and 2.2 MPa.
Load cells show higher values than expected ; the measurements given by the pressure cells have to be considered in such working conditions as under-estimated values for about 10 to 20 %. The variation in the results traduces the influence of the mechanical properties of the grouting material (sand/cement mixture) which is shotcreted between the lining and the clay mass and whose thickness is more important at the upper part.

Pressure readings, given in fig. 9 for ring n° 105, are particularly representative for this situation. Average pressures however are similar for each ring and the evolution with time is in total agreement with the convergence rate recorded.

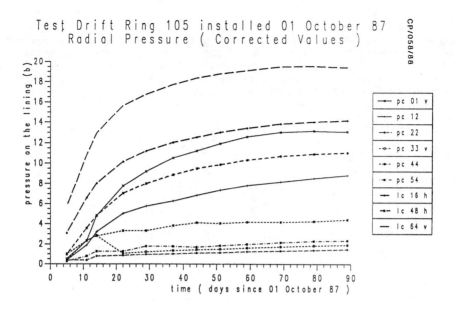

Fig. 9.a. Pressure build-up on ring 105

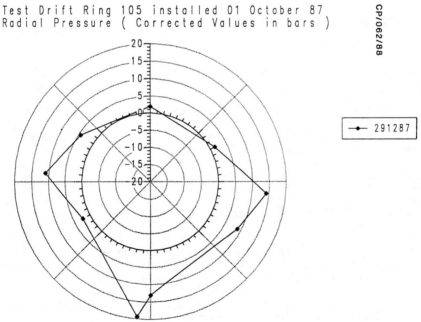

Fig. 9.b. Polar view of pressure distribution on ring 105

5.5. Clay front auscultation

The displacement of fifteen optical reflectors, distributed on the hemispherical surface are measured by way of an electro-optical distance meter.
Four single plastic rod extensometers anchored respectively at depths of 2, 4, 7 and 10 meters behind the front have been mounted on a head plate included in the shotcreting of the front.

Between 22.12.1987 and 01.04.1988, the absolute displacements recorded at 2 and 4 m in clay are respectively of 5.5 and 0.9 mm whereas the mean value of the front displacement remains between 20 and 25 mm.

6. INTERPRETATION AND MODELLING

Several computer codes were used since 1982 to predict or to reproduce the clay response to excavation and the forces acting on the lining.

CREEP/FLORA is a finite element code used by the contractor taking into account creep behaviour of clay /9/.

NONSAP ; a well-known code, was used by our consultant TRACTEBEL for the design of the test drift.

EPLAST, a small analytical code based on the convergence/confinement principle theory with a perfect elastoplastic model, was used e.g. for predicting deformations above the test drift and designing measuring devices. Numerical predictions are plotted on the diagram in figure 10 and compared with in situ measurements recorded in a vertical cross section at 10 m distance from the shaft. This graph clearly indicates that the elasto-plastic calculations considerably underestimate displacements in the formation.

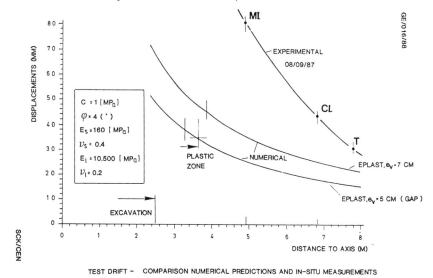

TEST DRIFT - COMPARISON NUMERICAL PREDICTIONS AND IN-SITU MEASUREMENTS

Fig. 10. Test drift - Comparison numerical predictions and in-situ measurements

Due to the mechanical properties of the Boom clay at depth, a rock mechanics approach could be used ; simulations performed for the case of the 2 m diameter exploratory gallery by the finite element code GEOMEC using an elasto-viscoplastic model with strain softening (LMS, Ecole Polytechnique, Paris) yield a satisfactory reproduction of the data from a 3 year measuring campaign.

When considering the time dependent behaviour as a consequence of pore water pressure dissipation, a soil mechanics approach based on the "Cam-clay" model also can reproduce field data. Simulations with the finite element code GEFDYN developed at Ecole Centrale (Paris) have also been performed /10/. It can be shown as detailed later on in this workshop that the Cam clay model generalized by Ismes to thermal loadings fits well the laboratory data obtained from Boom clay specimens.

From all these investigations as well as from other similar studies /11,12,13/, factors of primary importance can be pointed out when reproducing sequences of excavations in such materials characterized by a time dependant behaviour ; they concern the time delay before lining emplacement, the required stiffness of this lining and the influence of the extrados gap size and its backfill.

7.   CONCLUSIONS

Tunnelling capabilities in Boom clay have been demonstrated. Collection of data has been achieved in a satisfactory way and the mine-by-test allows to assess the Boom clay response to full scale excavation ; monitoring is going on to quantify the delayed effects.

Results confirm earlier predictions and cross-checking of information gathered allows to draw some important conclusions. As an example, the coefficient of earth pressure at rest (K_o), taken as 0.7 for design calculations, turns out to be much closer to 1.

By accepting or even encouraging controlled deformation of rock, its strength can be called upon and can contribute by the way to reduce early loads on the lining.
The first numerical simulations allow to point out the essential parameters for a better evaluation. Although simplifying assumptions are considered, the various approaches can reproduce fairly well the experimental situations.

8.   REFERENCES

/1/ André-Jehan, R. and al. : "Caractérisation géomécanique de l'argile du site de Mol (Belgique) - Application au dimensionnement d'ouvrages profonds", OCDE/NEA, Workshop on excavation response in deep radioactive waste repositories, Winnipeg, Canada, April 1988.
/2/ De Beer, E. and al. : "Preliminary studies of an underground facility for nuclear waste burial in a tertiary clay formation. Rockstore 77", Volume 3 : 771-781, 1977.
/3/ Funcken, R. and al. : "Construction of an underground laboratory in a deep clay formation", Eurotunnel Conference, 22/24 June 83, 79-86, 1983.

/4/ Manfroy P. and al. "Experience acquise à l'occasion de la réalisation d'un campagne géotechnqiue dans une argile profonde", CEC/NEA Workshop on the design and instrumentation of in situ experiments in underground laboratories for radioactive waste disposal, Brussels, May 1984.

/5/ Vandenberghe, N. : "Sedimentology of the Boom clay (Rupelian) in Belgium", Verhandeling koninklijke Academie Wetenschappen, Letteren en Schone Kunsten, België - Klasse Wetenschappen, Rept. XL147, 1978.

/6/ Horseman, S.T., and al. : "Geotechnical characterization of Boom clay in relation to the disposal of radioactive waste", EUR Report 10987EN, 1987.

/7/ Rousset, G. and al. : "Mechanical behaviour of galleries in deep clay - Study of an experimental case", Int. Meeting on HLW disposal, Pasco, Sept. 1985.

/8/ Neerdael, B. and al. : "In situ testing programme related to the mechanical behaviour of clay at depth", Second International Symposium on Field Measurements in Geomechanics, Kobe, Japan, Volume 2 proceedings, 763-774, 1987.

/9/ De Bruyn, D. and al. : "Time-dependent behaviour of the Boom clay at great depth - An application to the construction of a waste disposal facility", Computers and Geotechnics, Vol. 3, 3-20, 1987.

/10/ Pellegrini, R. and al. : "Factors affecting the thermo-hydro-mechanical response of clay in the safety assessment of nuclear waste repositories", OCDE/NEA Workshop on excavation response in deep radioactive waste repositories, Winnipeg, Canada, April 1988.

/11/ Berest, P. and al. : "Time-dependent behaviour of lined tunnels in soft rocks", Eurotunnel 83 Conference, Basle, Switzerland, paper 7, 1983.

/12/ Ladanyi, B. : "Direct determination of ground pressure on tunnel lining in a non-linear viscoelastic rock", Underground Rock Engineering, Toronto, Canada, proceedings 13th Canada Rock Mech. Symp., Vol. 2, 126-132, 1980.

/13/ Hanafy, E.A. and al. : "Advancing face simulation of tunnel excavation and lining placement", Underground Rock Engineering, Toronto, Canada, Proceedings 13th Can. Rock. Mech. Symp., Vol. 2, 119-125, 1980.

CARACTERISATION GEOMECANIQUE DE L'ARGILE DU
SITE DE MOL (BELGIQUE) - APPLICATION AU DIMENSIONNEMENT
D'OUVRAGES PROFONDS

R. André-Jehan*, M. Bouilleau**, M. Raynal*
* Agence Nationale pour la Gestion des Déchets Radioactifs
(ANDRA), Paris, France

** Bureau de Recherches Géologiques et Minières
(BRGM), Orléans, France

RESUME

L'ANDRA s'est associée depuis quelques années aux essais in-situ
effectués à Mol dans le cadre d'une collaboration avec le Centre des Etudes
Nucléaires belge (CEN/SCK) sous le couvert des programmes de la Communauté
Européenne. Ces études sont orientées vers la connaissance du comportement
rhéologique de l'argile profonde et son application à l'optimisation du
dimensionnement des ouvrages souterrains. L'expérience principale concernait
un essai en vraie grandeur d'un soutènement à "convergence contrôlée".
L'interprétation des résultats de mesures effectués au cours d'expériences à
petite et grande échelles permettent de conclure entre autres que pour
l'argile de Boom : (1) les contraintes en place sont quasi-isotropes ;
(2) l'effet d'échelle des expériences sur les résultats est modéré ; (3) les
effets différés sont très marqués ; et (4) la résistance à long terme
(plusieurs années) est très bonne.

GEOMECHANICAL CHARACTERISATION OF CLAY AT MOL (BELGIUM)
- APPLICATION TO THE SIZING OF DEEP UNDERGROUND WORKINGS

ABSTRACT

For several years, ANDRA has been conducting in-situ tests at Mol in
collaboration with SCK/CEN, within the research programme of the Commission of
the European Communities. These studies have been oriented towards the
understanding of the rheological behaviour of deep clay, and the application
of this knowledge to optimising the sizing of underground workings. The main
experiment concerned a full-scale test on controlled convergence gallery
support. The interpretation of the measurements made during small- and large-
scale tests lead to the conclusion, among others, that for Boom clay: (1) in-
situ stresses are quasi-isotropic; (2) there is a moderate effect of
experimental scale on the results; (3) delayed effects are very marked; and
(4) long-term resistance (over several years) is very high.

CARACTERISATION GEOMECANIQUE DE L'ARGILE DU SITE DE MOL (BELGIQUE)

APPLICATION AU DIMENSIONNEMENT D'OUVRAGES PROFONDS

~~~~~~~~~~~~~

I - INTRODUCTION

L'ANDRA s'intéresse à l'argile depuis un certain temps au même titre que
d'autres milieux géologiques susceptibles d'accueillir un stockage souterrain de déchets radioactifs à vie longue. L'argile représente un matériau à
faibles caractéristiques mécaniques qui, du point de vue de la faisabilité
du projet, posent un certain nombre de problèmes géotechniques d'autant que
ces installations souterraines resteront opérationnelles plusieurs années
avant leur fermeture définitive.

Dans cette perspective, l'ANDRA s'est associée depuis quelques années aux
essais in-situ effectués à Mol dans le cadre d'une collaboration avec le
Centre des Etudes Nucléaires belge (CEN/SCK) sous le couvert des programmes
de la Communauté Européenne. Ces études sont orientées vers la connaissance
du comportement rhéologique de l'argile profonde et son application à
l'optimisation du dimensionnement des ouvrages souterrains. L'ANDRA a confié
au Laboratoire de Mécanique de Solides de l'Ecole Polytechnique la responsabilité scientifique des études (conception, interprétation) et au Bureau
de Recherches Géologiques et Minières (BRGM) la réalisation technique et la
mise en oeuvre sur le terrain (instrumentation, auscultation).

Les essais sont réalisés dans la couche d'argile de Boom située entre 180 m
et 280 m à l'aplomb du Centre. Ces essais font partie d'un programme plus
global comprenant des essais en laboratoire sur carottes et la réalisation
de modèles.

II - CARACTERISATION AU LABORATOIRE DE L'ARGILE DE BOOM

Les nombreux essais en laboratoire permettent actuellement de cerner assez bien les caractéristiques fondamentales du comportement de cette argile à teneur en eau relativement élevée (W = 22 à 25 %).

La couche d'argile prélevée est très homogène, sans trace de discontinuités de nature géologique.

* Les essais à court terme montrent que :

1) La résistance du matériau à court terme est modérée (cohésion non drainée égale à 1,3 MPa) ;

2) Le matériau est "plastique" avec une déformation importante pour un pic de résistance faible.

3) Le matériau présente une perte de résistance quand la déformation dépasse un certain seuil (radoucissement) sensible même sous fort confinement (figure 1) ;

4) La déformation volumique montre une phase de contractance du matériau suivie d'une phase de dilatance, fonction du confinement.

* Les essais à long terme de fluage en condition non drainée (figure 2) mettent en évidence l'importance des phénomènes différés.

III - CARACTERISATION IN-SITU

L'étude du comportement mécanique in-situ de l'argile est développée par différents essais réalisés dans la couche d'argile de Boom à partir des aménagements souterrains du CEN.

Ces essais peuvent être classés en trois grandes catégories :

1) Un essai à échelle moyenne dans une galerie de petit diamètre revêtue d'un soutènement rigide (2 m),

2) Des essais à petite échelle dans des forages (∅ 100 mm à 150 mm) revêtus ou non,

3) Un essai en vraie grandeur dans une galerie circulaire revêtue d'un soutènement de type industriel (4 m).

III-1 Comportement de la petite galerie expérimentale

En 1984, l'ANDRA a participé à l'instrumentation d'une petite galerie, de 2 mètres de diamètre excavé et de 1,4 m de diamètre utile, revêtue de claveaux de béton (figure 3).

Des mesures sont effectuées périodiquement depuis près de quatre ans. Les résultats, homogènes dans l'ensemble, sont les suivants :

1) les effets différés sont très marqués : juste après la phase de creusement les déplacements mesurés dans le massif augmentent rapidement. Après une période transitoire qui n'excède pas le mois, l'ensemble de ces paramètres continue à évoluer à des vitesses très lentes suivant une courbe asymptotique très atténuée (figure 4),

2) l'ovalisation de la paroi de la galerie est très nette avec un raccourcissement vertical traduisant une contrainte verticale dans le massif légèrement supérieure à la contrainte horizontale (figure 5).

3) la tenue générale de la galerie est excellente : la convergence totale de la paroi du massif n'excède pas 7 % alors que la pression exercée sur le soutènement est inférieure à 1,4 MPa, c'est-à-dire trois fois plus faible que la pression lithostatique qui règne à cette profondeur,

III-2 Essais en forage à petite échelle

Compte-tenu de l'homogénéité du matériau et de ses propriétés mécaniques mises en évidence en laboratoire, et de la quasi-isotropie des contraintes lithostatiques, l'effet d'échelle est certainement très réduit dans le massif argileux à Mol.

Dans ces conditions, on conçoit l'intérêt que présentent les essais en forage à petite échelle : il est possible de réaliser simplement à des coûts raisonnables des études de prédimensionnement d'ouvrages souterrains.

Les trois essais décrits ci-après peuvent être considérés comme des essais sur modèles réduits de galerie circulaire.

III-2-1 Essai "convergence sans soutènement"

La convergence d'un trou de forage à nu est mesurée au cours du temps par un diamétreur à 3 patins (figure 6). Le comportement observé est très différent de celui de la petite galerie. La convergence est très importante : plus de 30 % de convergence en moins d'un mois. Le processus se ralentit après cette phase et la fermeture devient totale quelques mois après la réalisation des forages.

III-2-2 Essai "modèle réduit de soutènement"

Le principe de l'essai consiste à placer dans le forage, en guise de soutènement, un tube métallique mince instrumenté et de mesurer au cours du temps la déformation de ce tube (capteurs de déplacement) ainsi que les efforts de poussée qu'il subit de la part du terrain (jauges de contrainte) (figure 7).

Après une période de mise en place, pendant laquelle le tube s'ovalise légèrement selon un grand axe vertical (figure 8), les efforts de poussée du massif sur le tube augmentent progressivement, en inversant l'évolution de l'ovalisation, pour se stabiliser au bout d'une centaine de jours.

La combinaison des déformations à l'extrados et à l'intrados du tube permet de calculer aisement les efforts globaux : effort normal, moment fléchissant. A partir de là, on en déduit les efforts de poussée du terrain. La pression moyenne sur le soutènement atteint Pm = 2,2 MPa (soit environ la moitié de la pression lithostatique), le déviateur P = 0,2 MPa, pour une convergence totale de la paroi de 6 %.

III-2-3 Essai "fluage in-situ au dilatomètre"

Contrairement au tube de convergence qui est un essai à déformation imposée, l'essai de fluage au dilatomètre permet de travailler soit à déformation bloquée, soit à pression imposée.

Ainsi il est possible de suivre au cours du temps l'évolution de la déformée d'une section de forage soumise en paroi à une pression radiale constante inférieure à la pression lithostatique.

La sonde utilisée est ici munie de quatre capteurs de déplacement radial et d'un capteur de pression ; la sonde étant introduite puis mise en pression aussitôt après la foration , la mesure de la variation de volume de la sonde permet d'apprécier la convergence globale.

L'exemple en figure 9 des courbes expérimentales obtenues sur le dilatomètre horizontal montre par palier de diminution de pression (0,6 MPa) une bonne corrélation des courbes d'évolution du volume et de la convergence.

La convergence totale moyenne est de :

- 1,3 % sous 3,3 MPa au bout de 30 jours,
- 2,1 % sous 2,7 MPa au bout de 150 jours.

Il est commode, pour juger de la représentativité de tous ces essais et pour justifier l'absence d'un effet d'échelle, de reporter les valeurs d'équilibre à long terme (P_i^{oo}, U_i^{oo}) de chaque essai sur le diagramme de convergence de la figure 10.

La courbe tracée sur ce diagramme donne la valeur de la convergence U_i pour une pression intérieure (ou pression de soutènement) donnée.

On remarque notamment la forte non linéarité de cette courbe, due au caractère plastique radoucissant du comportement de l'argile de Boom.

IV - ESSAI EN VRAIE GRANDEUR DE SOUTENEMENT A "CONVERGENCE CONTROLEE"

Comme on vient de le montrer, le comportement des ouvrages de petite dimension est satisfaisant. En particulier, les efforts excercés par le massif sur le soutènement sont largement inférieurs à la pression lithostatique.

Il semble donc naturel de vouloir profiter **des capacités de résistance mécanique à long terme** de l'argile de Boom ainsi mises en évidence, pour optimiser le dimensionnement du soutènement des galeries.

L'"optimisation" concerne ici **le critère économique** (soutènement le plus léger possible, facile à mettre en oeuvre) ainsi que le **critère sûreté** à long terme de l'ouvrage.

IV-1 Objectifs

Deux objectifs essentiels ont été pris en compte :

1) **Le choix d'un revêtement** ne reprenant qu'une fraction de la pression lithostatique règnant à la profondeur de l'ouvrage (massif à l'équilibre avec le soutènement).

   Pour y parvenir, il faut donc un **revêtement permettant la convergence** de la paroi de la galerie.

2) **La préservation de l'intégrité** du massif en évitant le plus possible le développement des ruptures en paroi.

   On y parvient si :

   - le creusement et la mise en oeuvre du revêtement sont rapides,
   - le découvert est minimal,
   - l'efficacité du soutènement est immédiate.

Considérant ces objectifs, le projet d'un galerie expérimentale ANDRA s'est donc porté sur l'utilisation d'un "soutènement à seuil de confinement".

Le concept suppose, dès l'excavation réalisée, que le soutènement entre en contact avec le massif argileux. On réalise pour cela un colmatage de l'interface massif-soutènement. Dès que le confinement est trop important un dispositif spécial, ici les cintres à joints coulissants, permet au soutènement de dissiper les contraintes et autorise la convergence des terrains.

Les deux diagrammes convergence-confinement de la figure 11 avec revêtement rigide ou avec soutènement à seuil de confinement montrent, dans le cas d'un massif supposé homogène et isotrope, l'influence du délai de pose sur l'équilibre à long terme d'une part, et l'intérêt du soutènement à seuil de confinement d'autre part.

L'intersection des 2 courbes, courbe du massif et courbe du soutènement fournit le point d'équilibre à long terme.

IV-2 Réalisation et mise en oeuvre de la galerie cintrée ANDRA

La galerie a été construite du 20 Octobre au 3 Décembre 1987, dans le prolongement d'une galerie CEN/SCK à revêtement en claveaux béton. Elle comprend (figure 12) une zone de transition de 2 m de long, qui a les mêmes caractéristiques que la galerie belge, plus 12 m de galerie à soutènement métallique. Au total 36 anneaux de 4 éléments chacun ont été posés (figure 13). Des plaques de garnissage assurent la continuité du revêtement et permettent un colmatage.

Malgré les conditions d'accès difficiles qui empêchaient la mécanisation des opérations, le déroulement du chantier a été très satisfaisant, dans le respect des règles de tolérance. L'exécution a été régulière, sans arrêt supérieur à 48 heures et avec un découvert maximum de 1,40 m. L'argile a toujours montré une bonne tenue. La durée moyenne de mise en place de 2 cintres est de 22 heures, pour l'ensemble des opérations de terrassement, de pose des cintres puis des tôles de garnissage et de colmatage (avancement moyen de 0,6 m/jour).

IV-3 Instrumentation

Les anneaux sont instrumentés dès leur mise en place, avec un soin particulier pour les 4 sections n° 4, 15, 22 et 32. Six types de mesures sont effectués :

- la convergence des anneaux, par mesures extensométriques de cordes (19 cordes pour les anneaux de mesure, 6 pour les anneaux courants),

- la déformation des cintres, avec 5 plots de mesure pour les anneaux de mesure et 2 pour les anneaux courants, d'où on déduit les efforts dans le soutènement,

- le coulissement au niveau des étriers,

- la contrainte à l'extrados du revêtement, au moyen de cellules placées à l'interface argile-matériau de bourrage ,

- le déplacement au sein du massif, au moyen de 2 extensomètres en forages verticaux à droite des anneaux 15 et 22, en 3 points de mesure pour chacun d'eux,

- le déplacement vertical d'ensemble (nivellement).

IV-4 Premiers résultats

La mise en charge du revêtement commence quasi-immédiatemment et augmente progressivement, assez rapidement pendant les 10 à 20 premiers jours ; l'évolution est ensuite beaucoup plus lente, avec une courbe en dents de scie qui oscille autour d'une valeur à peu près stable de 0,8 MPa (figure 14). L'allure de la courbe est due au relâchement des contraintes après coulissement.

L'évolution du coulissement est représentée en figure 15 ; elle est en phase d'atténuation pour les anneaux n° 4, 15 et 22 avec respectivement une convergence radiale correspondante de 1,61 %, 1,13 % et 1,4 % ; pour l'anneau n° 32, elle atteint la valeur plus faible de 0,66 %, probablement due à la proximité du front de l'excavation.

L'effort normal dans les cintres est d'environ 700 kN au moment du coulissement.

CONCLUSION

Plusieurs essais à petite ou à grande échelle ont permis d'étudier le comportement à long terme d'ouvrages réalisés en profondeur dans une argile (argile de Boom) dont la plasticité est la principale caractéristique.

L'instrumentation mise en oeuvre a permis de connaître avec grande précision l'évolution des paramètres mécaniques essentiels.

Les résultats obtenus se montrent cohérents ; de leur synthèse on retiendra que :

- les contraintes en place sont quasi-isotropes, l'ovalisation des ouvrages étant faible,

- l'effet d'échelle semble très modéré, les comportements étant similaires d'un ouvrage à l'autre,

- les effets différés sont très marqués (il faut attendre plusieurs mois ou années avant d'atteindre la stabilisation des phénomènes).

- l'argile de Boom présente de réelles capacités de **résistance à long terme**, on retiendra qu'une convergence de la paroi de 3 % permet de limiter la poussée du massif à moins de 40 % de la pression lithostatique.

- enfin, l'influence du "chargement" au cours des premières phases (creusement, nature et délai du confinement,...) est importante sur l'équilibre final.

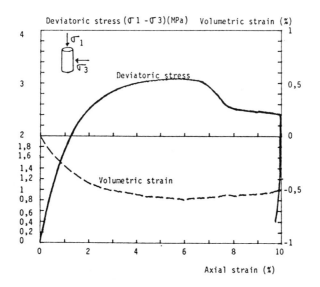

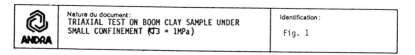

Nature du document: TRIAXIAL TEST ON BOOM CLAY SAMPLE UNDER SMALL CONFINEMENT (σ3 = 1MPa)	Identification: Fig. 1

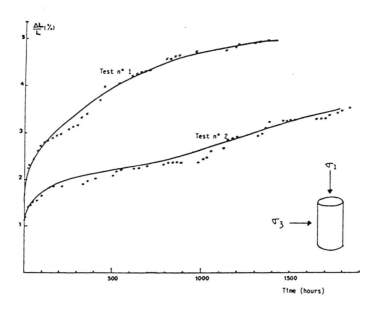

Nature du document: CREEP TRIAXIAL TESTS ON BOOM CLAY SAMPLES WITH σ3 = 5 MPa, σ1 = 7 MPa	Identification: Fig. 2

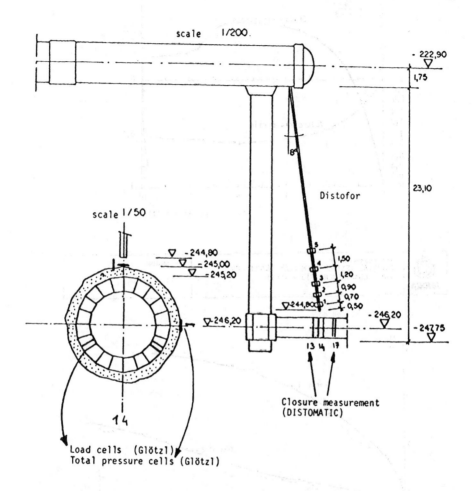

scale   1/200.

- 222,90

1,75

8°

Distofor

23,10

scale 1/50

▽ - 244,80
▽ - 245,00
▽ - 245,20

5
4   1,50
3   1,20
2   0,90
1   0,70
▽244,80   0,50

▽-246,20

- 246,20

- 247,75

13 4 17

Closure measurement
(DISTOMATIC)

1 4

Load cells   (Glötzl)
Total pressure cells (Glötzl)

	Nature du document: MOL (CEN-SCK) – SMALL GALLERY AND INSTRUMENTATION SET UP	Identification: Fig. 3
ANDRA		

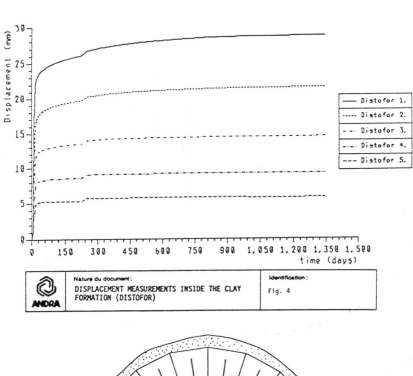

——	Distofor 1.	
·····	Distofor 2.	
– ·· –	Distofor 3.	
–··–	Distofor 4.	
– – –	Distofor 5.	

Nature du document:	Identification:
DISPLACEMENT MEASUREMENTS INSIDE THE CLAY FORMATION (DISTOFOR)	Fig. 4

ANDRA

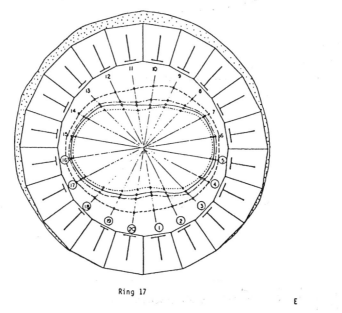

Ring 17

W   E

Scale : 1 cm/12,5 cm   Strain scale : 1 cm = 5 mm

- – – – – – 06/06/84
- – ·· – ·· – 29/01/85
- ———— 29/08/85
- ············ 23/10/86

Nature du document:	Identification:
CLOSURE MEASUREMENTS OF LINING (DISTOMATIC)	Fig. 5

ANDRA

- 491 -

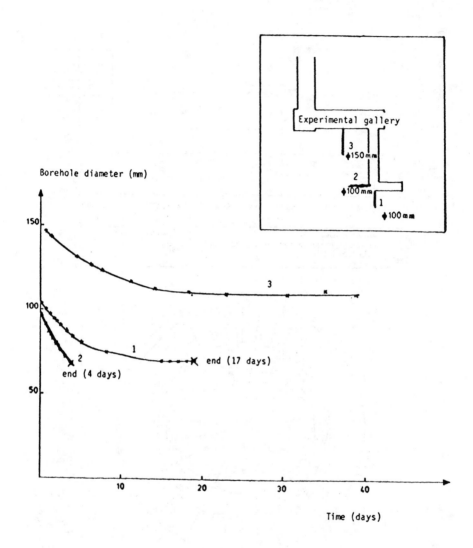

Borehole diameter (mm)

Experimental gallery

3 ♦150 mm

2 ♦100 mm

1 ♦100 mm

end (17 days)

2

end (4 days)

Time (days)

	Nature du document: UNLINED BOREHOLES CLOSURE (CALIPER TEST)	Identification: Fig. 6

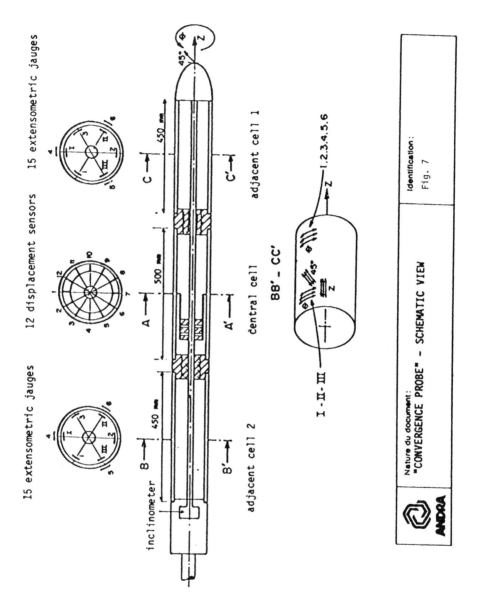

15 extensometric jauges   12 displacement sensors   15 extensometric jauges

15 extensometric jauges

inclinometer

adjacent cell 2

adjacent cell 1

central cell

BB' - CC'

I - II - III

1.2.3.4.5.6

450 mm   500 mm   450 mm

Nature du document :
"CONVERGENCE PROBE" - SCHEMATIC VIEW

Identification :
Fig. 7

ANDRA

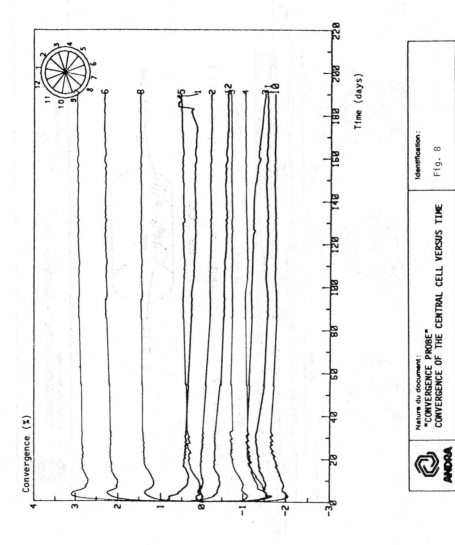

Convergence (%)

Time (days)

Nature du document:
"CONVERGENCE PROBE"
CONVERGENCE OF THE CENTRAL CELL VERSUS TIME

Identification:
Fig. 8

ANDRA

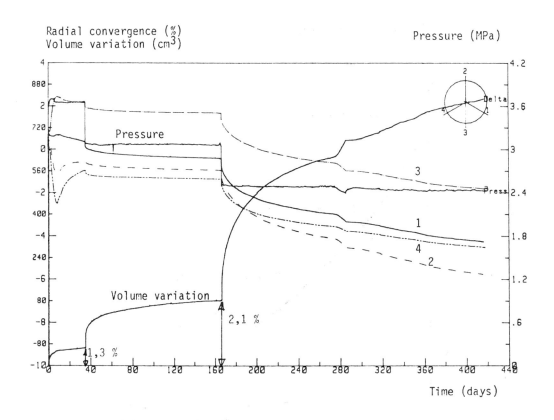

Radial convergence (%)
Volume variation (cm³)

Pressure (MPa)

Nature du document:	Identification:
CREEP DILATOMETRIC TEST **(HORIZONTAL BOREHOLE)**	Fig. 9

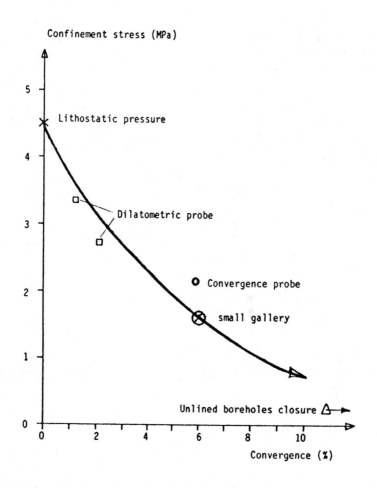

Confinement stress (MPa)

5

Lithostatic pressure

4

Dilatometric probe

3

2 — Convergence probe

small gallery

1

Unlined boreholes closure

0

0   2   4   6   8   10

Convergence (%)

Nature du document : LONG TERM CONVERGENCE CURVE OF CLAY MASS AT MOL	Identification : Fig. 10

ANDRA

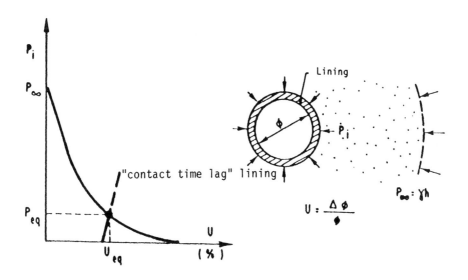

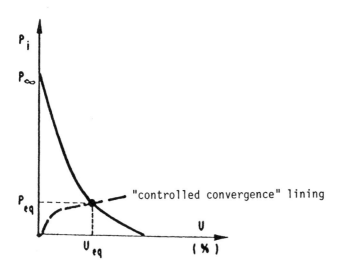

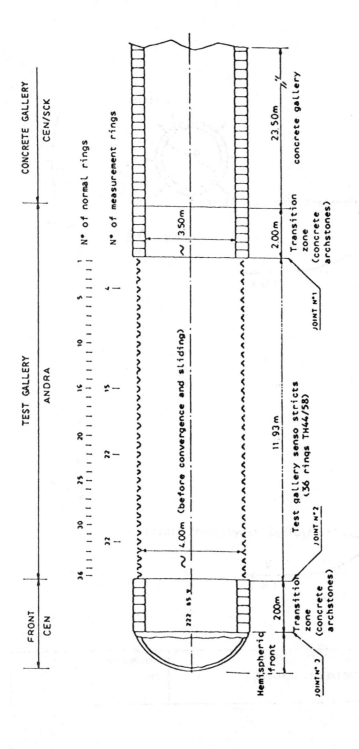

FRONT
CEN

TEST GALLERY
ANDRA

CONCRETE GALLERY
CEN/SCK

N° of normal rings

N° of measurement rings

Nature du document:

ANDRA TEST GALLERY (as built)

Identification:

Fig. 12

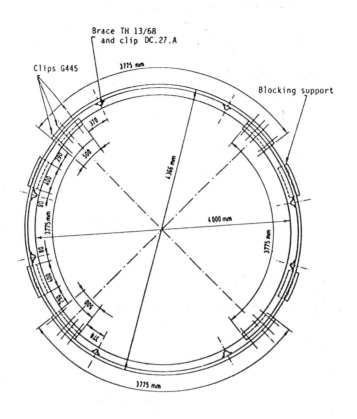

Brace TH 13/68
and clip DC.27.A

Clips G445

Blocking support

3775 mm

370

290

500

4000 mm

4366 mm

3775 mm

80

180

400

500

500

370

3775 mm

3775 mm

3775 mm

	Nature du document :	Identification :
ANDRA	LINING RING	Fig. 13

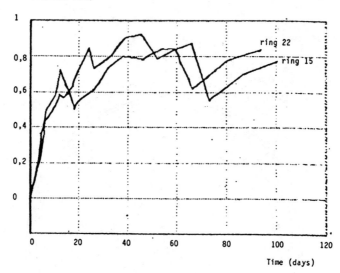

Pressure (MPa)

ring 22

ring 15

Time (days)

	Nature du document:	Identification:
ANDRA	ANDRA TEST GALLERY RADIAL PRESSURE ON THE LINING VERSUS TIME	Fig. 14

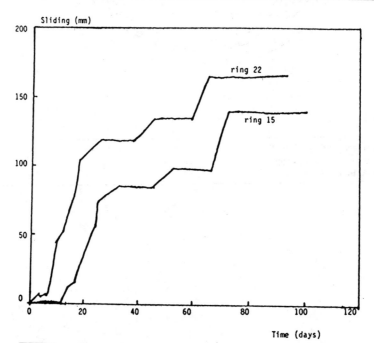

Sliding (mm)

ring 22

ring 15

Time (days)

	Nature du document:	Identification:
ANDRA	ANDRA TEST GALLERY SLIDING VERSUS TIME	Fig. 15

FACTORS AFFECTING THE THERMO-HYDRO-MECHANICAL
RESPONSE OF CLAY IN THE SAFETY ASSESSMENT
OF NUCLEAR WASTE REPOSITORIES

R. Pellegrini, M. Borsetto, A. Peano
ISMES
Italy

E. Tassoni
ENEA
Italy

ABSTRACT
    Decay heat due to waste disposal in a saturated clay medium causes the
development of pore water overpressures.
In this study, the effects of thermal pore pressure increase such as
effective stress reduction, hydraulic gradient increase and possible
yielding have been taken into account.
Numerical computations assess the disturbance on the in situ stress field due
to excavation of an isolated borehole under different assumptions on the
drilling conditions.
Thermo-mechanical response of clay over a zone about 1 metre wide around the
borehole depends significantly on the excavation technique. This zone is
characterized by development of plastic strains; under some drilling
conditions yielding occurs with dilative strains, which are associated to
fracturing of the clay.
Outside this zone, thermo-mechanical response has been found as dependent
mostly on the perturbation of the thermal field.
Significant pore pressure increase and effective stress decrease develop over
a zone 10 metre wide.
This study has been developed for a specific type of clay and it aims to
point out that analogous studies have to be undertaken for sites to be
considered for waste emplacement.

EVALUATION DE LA SURETE DES DEPOTS NUCLEAIRES :
FACTEURS AFFECTANT LES EFFETS THERMO-HYDRO-MECANIQUES DES TRAVAUX
D'EXCAVATION DANS LES FORMATIONS ARGILEUSES

RESUME

La chaleur de décroissance due au stockage de déchets dans un milieu argileux saturé provoque l'apparition d'une surpression de l'eau interstitielle.

Dans cette étude, les auteurs ont pris en compte les effets de l'augmentation de la pression interstitielle due aux phénomènes thermiques notamment la réduction effective des contraintes, l'augmentation du gradient hydraulique et éventuellement les phénomènes de déformation plastique.

Des calculs numériques permettent d'évaluer la perturbation du champ des contraintes in situ imputable à l'excavation d'un trou de forage isolé, compte tenu de diverses hypothèses concernant les conditions de forage.

Dans une zone d'environ 1 m autour du trou de forage, la réponse thermo-mécanique de l'argile dépend largement de la technique d'excavation employée. Cette zone se caractérise par l'apparition de déformations plastiques ; dans d'autres conditions de forage, il se produit un fluage accompagné de déformations par dilatation, elles-mêmes liées à la fissuration de l'argile.

Au-delà de cette zone, on a constaté que la réponse thermo-mécanique dépendait essentiellement de la perturbation du champ thermique. On observe une augmentation significative de la pression interstitielle et une réduction des contraintes effectives dans un rayon de 10 m.

Cette étude a été exécutée pour un type particulier d'argile et vise à mettre en relief la nécessité de procéder à des études analogues pour les sites envisagés en vue de la mise en place de déchets.

1. INTRODUCTION

The host rock formation response to the heat developed by radioactive waste decay is one of the relevant effects to be evaluated when performing the safety assessment of a nuclear waste repository.
Considering clay as the host formation, one of the relevant effects of a temperature increase is the increase in pore pressure and the consequent variation of effective stresses. This effect is highly significant when considering the very low permeability of the clay formations suitable for waste disposal.
Additionally, deformations, with particular reference to the development of irreversible strains, need to be evaluated. They are also important considering the tendency of heated clays to develop significant irreversible strains both in the overconsolidated and normally consolidated states.
These thermo-mechanical characteristics have been verified for all types of clay tested at ISMES [5]. Similar results were achieved by Campanella and Mitchell in their study of oceanic clays [2] and by Heremans et al. [3].
Moreover, in-situ stress and strain conditions are thought to play a considerable role in the development of pore pressure and in the resulting effective stress change under heating due to the stress-strain characteristics of clay response.
The disturbance on the geostatic state due to construction techniques should be therefore evaluated with regard to the consequences on the thermo-mechanical effects induced by decay heat.
In this paper the effects of drilling of an isolated deep borehole have been studied under simplified boundary conditions.

2. SCENARIOS OF CLAY REPOSITORY PERFORMANCE UNDER HEATING

The increase of pore water pressure could in principle lead to two undesirable scenarios.
The first one is the development of a hydraulic field around the emplaced canisters. This effect could be due to the generation of pore pressures as a consequence of the temperature increase. As a result, an increase in the flow of the water coming out of the deposit could be expected.
This effect, if associated with a possible ion migration caused by ill functioning of the insulation system, could contribute to expose human environment to the risk of radionuclides migration.
The second scenario is characterized by the loss of shear resistance of clay resulting from a possible decrease in effective stresses. This is strictly connected with the increase of pore pressure taking into account the principle of effective stress. The decrease in the value of effective stresses could also lead to local material failure.

3. THERMO-HYDRO-MECHANICAL MODELING OF CLAY RESPONSE

When considering the scenarios described in Sect. 2, for safety assessment reasons it is very important to verify that, in a certain situation, they are not realistic. To perform such an analysis the development of a sufficiently comprehensive and accurate model to predict the interaction between the clay skeleton and the fluid deformation and/or flow caused by heating has been considered as necessary. Therefore, a general model has been developed allowing for a fully thermo-mechanical-hydraulic coupling between water and solid [7]; this model has been implemented in a F.E. computer code and the main features are synthetized below with regard to the version adopted in this study.

Under the hypothesis of full saturation, the interaction is described according to the flow continuity condition,

$$\frac{\partial (\underline{v})}{\partial \underline{x}} = \frac{\partial \varepsilon_v}{\partial t} (\delta T, \delta \underline{\sigma}') - (1-n)\frac{\partial \varepsilon_v^*}{\partial t} (\delta T, \delta \underline{\sigma}', \delta u) - \frac{\partial \varepsilon_w}{\partial t} (\delta T, \delta u) \tag{1}$$

stating the equality of the water outflow rate (lhs term) with pore volume variations due to temperature, pore pressure and effective stress dT, du, d$\underline{\sigma}$'in an elementary volume (lhs term).

In (1), $\underline{v}$ is the water flow velocity, ε_v is the skeleton deformation of the soil medium as a whole, ε_v^* and ε_w the deformation of grains and water and n is the porosity.

In order to predict the volume variations in the rhs term of (1), mechanical and thermal compliances of all three constituents must be established.

The most complex task is here the determination of the total deformability of the clay medium. The modeling is performed within the framework of the thermo-plasticity theory. According to this, the well-known modified Cam-clay model has been generalized to thermal effects. As far as deformations are concerned clay can be considered as either thermo-elastic or thermo-plastic depending on whether it is in an over or in a normally consolidated state respectively. The yield function f describes the thermo-elastic domain limit and undergoes modifications due to irreversible volumetric strain or temperature rates.

As far as the overconsolidation domain is concerned, clay is thermo-elastic and the strain vector can be divided into its volumetric and deviatoric strain components.

The dependence of the former on the main stress ($\frac{tr\underline{\sigma}'}{3}$), temperature increase and pore pressure follows the non-linear relationship

$$\varepsilon_v^e = - \frac{K_{\Delta T}}{1+e_o} \, ln \, \frac{tr\underline{\sigma}'}{tr\underline{\sigma}_g'} + \alpha (\Delta T, ln \, \frac{tr\underline{\sigma}'}{tr\underline{\sigma}_g'})\Delta T - C^* u - \varepsilon_g \tag{2}$$

Here $K_{\Delta T}$ is the bulk modulus and is generally non linearly dependent on temperature increase; α is the thermal expansion including the mineral grain expansion: this value too is non linearly dependent on the pressure and temperature increases. Initial void ratio is e_o, C^* the bulk compressibility of the mineral solid and ε_g the strain corresponding to the geostatic stress $\underline{\sigma}'_g$.

The elastic deviatoric strain component is linearly dependent on the deviatoric stress $\underline{S}$, according to the following relationship:

$$\underline{e}^e = \frac{1}{2G} (\underline{S} - \underline{S}_g) \tag{3}$$

where

$$\underline{e}^e = \underline{\varepsilon}^e - \frac{1}{3} \varepsilon_v^e \, \underline{m} \quad ; \quad \underline{S} = \underline{\sigma}' - \frac{1}{3} tr\underline{\sigma}' \, \underline{m}$$

and G is the isothermal shear modulus taken here as constant.

As for the normal consolidation behaviour, a yield function delimits the overconsolidated from the normally consolidated states: it is an ellipses in the triaxial plane. Its analytical expression is:

$$f = \left(\frac{p'}{a(\Delta T, \varepsilon_v^p)} - 1\right)^2 + \left(\frac{q}{Ma(\Delta T, \varepsilon_v^p)}\right)^2 - 1 = 0 \; ; \quad M = const. \tag{4}$$

where $p' = -\frac{1}{3} tr \, \underline{\sigma}'$, $q = \sqrt{\frac{3}{2}}(\underline{S}^T\underline{S})^{\frac{1}{2}}$;

a is the so-called hardening parameter commanding variations in size of the limit surface.
The ellipses expands as a result of plastic volumetric compaction rate and cooling and shrinks in the presence of both plastic dilative volumetric strain rates and temperature increases (Fig. 1).
The irreversible strain increment due to heating and/or mechanical loading is described by means of an associated flow rule

$$d\underline{\varepsilon}^P = d\hat{\lambda} \frac{\partial f}{\partial \underline{\sigma}'} , \tag{5}$$

Following this modelisation, irreversible deviatoric strains can be predicted also in the presence of isotropic loading conditions and/or temperature variation.
The so-called plastic multiplier $d\hat{\lambda}$ depends on the effective stress and temperature increments according to the Prager's consistency equation df = o.

4. MODELING OF THE DRILLING AND EMPLACEMENT PHASES

As mentioned in the Introduction, the case of a deep isolated borehole has been chosen as a case study because it is a modular element of a matrix repository of deep boreholes and, in this sense, it has a strict engineering importance.
This example is also significant in the case of the mined repository disposal option, in order to familiarize with the thermo-mechanical response of the clay and to select some relevant influencing factors.
The borehole selected is drilled into a deep, homogeneous and stiff clay formation. The emplacement zone is 90 m long. The hole radius is 0.25 m. This value has been judged as the minimal suitable to allow for the emplacement of waste canisters (r = 0.15 m) and for convenient backfilling.
The hole has been considered as instantaneously excavated and filled with stabilizing mud. The period of time from the opening of the section until the emplacement of waste canisters is 30 days.
In the modeling of the drilling phase, the situation studied is the one depicted in Fig. 2. The mud has been modeled as a total pressure acting against the hole surface and some assumptions were made on the water exchange between mud and clay at the hole surface.
Four cases were first selected (see again Fig. 2) following the above mentioned criterium.
Refering again to Fig. 2, pressure calculated from the specific weight of 2000 Kg/m3 could be representative of a bentonite type drilling mud, whereas that calculated from 1000 Kg/m3 could model the presence of mud with stabilizing polymers. Two conditions of impervious hole-mud interface have

been considered (case C and D). The other two are an attempt to describe the possibility of water transfer from the mud to the host rock by means of a boundary condition on the water pressure. Case A refers to a pore pressure value equal to the hydrostatic head. In case B a pore pressure equal to that exerted by the mud has been selected as a limiting case.
The initial state of stress is characterized by σ_h = 4.18 MPa and σ_v = 4.4 MPa. The hydrostatic water pressure value has been calculated as 2.2 MPa.
The emplacement of the canisters and the backfilling has been considered istantaneous and modeled with a condition of no further radial displacement. The interface between the backfilling material and the host clay has been considered impervious.
The thermal effect due to the canisters emplacement has been calculated with a power supply variable in time according to the relationship depicted in Fig. 3. Some reduction in power has been allowed for, to account for the presence of alternate Cladding Hull Waste Canisters in the borehole among the High Level Waste ones. The initial power release value is referred to a 30 years delay time, following the indications obtained by thermal optimization studies of DBF repositories, and is equal to 325 W/m.

5. ANALYSIS OF THE RESULTS

5.1 THERMAL ANALYSIS

The temperature distribution resulting from the heat conduction analysis is illustrated in Fig. 4 for the first 17 m after 5, 10 and 50 years from emplacement. Fig. 5 depicts the variation of temperature with time at the interface between clay and backfilling.
As it can be observed, the highest temperatures are located near the hole and reach their maximum value (144°C) 5 years after the emplacement.
Their value start to decrease very slightly 10 years after deposition. After 50 years temperature is however around 88°C.
As far as space distribution is concerned, temperatures distribute in the years uniformly. After 50 years, a 6-metre zone is affected by temperature values higher than 50°C.

5.2 RESPONSE TO DRILLING

The situations considered in Sect. 4 have been studied characterizing the geological host formation with an overconsolidated clay (OCR = 3), whose elastic shear modulus is 100 MPa, and whose logarithmic elastic and elasto-plastic compliance moduli are 0.031 and 0.129, respectively.
The thermal characteristics are in line with the values obtained at ISMES for a stiff plastic clay. Permeability was fixed at 4.3×10^{-7} m/day.
The results have been summarized according to the factors outlined in Sect. 2. Figure 6 depicts the effective stress paths concerning the zone of the clay deposit formation most adjacent to the hole.
As far as excavation response is concerned, in cases A, B and C the effective stress point remains within the yield surface, whereas in case D it reaches the yield limit.
In the A and C cases, behaviour is elastic and almost identical.

The similarity in behaviour between these last two cases is explained by the invariance of pore pressure values. The resulting state of stress is only affected by the slight difference (0.22 MPa) between the horizontal stress present in the clay and the pressure exerted by the mud. Case A was then discontinued.

The excavation induced deviatoric stress decreases with the distance from the borehole and is negligible at a distance of more than 1 m.

Case B is characterized by a relevant decrease in radial stress and an increase in deviatoric stress. They significantly affect a 2-metre clay surface around the hole.

In case D the strong increase in deviatoric stress can be ascribed to the low containment offered by the drilling mud. The state of stress reaches the yield limit.

Subsequently, dilative plastic volumetric strains develop, associated with a certain degree of water suction. This latter disappears at the end of the period studied because of waterflow coming from the outer zones.

Radial stress has considerably decreased in the first few metres of clay. It affects the considerable variations in deviatoric stress that can be noticed. The resulting pore pressure distribution is depicted in Fig. 7.

Appreciable deformations are limited to the first few metres of clay and are characterized by volumetric dilation, except for the C case where no volumetric deformation are present.

The hole becomes wider in the B and C cases and shrinks in the D case. The only appreciable displacement value (about 3 mm) has been noticed in the D case.

6. RESPONSE TO HEATING

Cases B, C and D were studied with respect to the temperature distribution obtained from the thermal conduction analysis described in Sect. 4. .

Referring to Fig. 5 which describes the variation of temperature vs. time at the hole inner side and to Fig. 6 describing the effective stress paths, the following considerations may be drawn.

In case B, response is elastic until temperature is lower than 103°C. In this elapse of time thermal interstial water expansion and thermo-elastic clay skeleton compaction occur. Yield surface decreases with temperature. At 129°C the behaviour is plastic and compactive volumetric strains rates occur which contribute to the enhancement of the pore pressure increase.

The point describing the state of effective stress moves towards lower values of deviatoric stress.

After 5 years (T=143°C) the state of stress is nearly isotropic. Yielding affects approximately 0.5 m of clay.

In any case, the total deformation resulting is dilative (0.2%). Horizontal stress significantly recovers, reaching approximately half the geostatic value.

Excess pore pressure distribution, Fig. 7, drops after 1 year, following temperature distribution and becomes more uniform, in the same way as temperature does. This is also due to seepage effects.

This analysis has been carried through for 50 years, and as a consequence stress has become elastic again due to the higher thermal sensitivity of the material, with respect to the clay seepage characteristics.

It resulted that the state of stress recovers in the deviatoric component and, to a more limited extent, so does the mean stress.
Similar behaviour than in case B is observed for case C, which has been found plastic at 126°C.
After 5 years, a zone 0.5 m wide is characterized by a significant loss of both deviatoric and mean stress.
Once again pore overpressures increase, following temperature distribution. After one year they are very similar to those obtained for case B.
In case D, plastic dilative volumetric deformations continue to develop, with a maximum of 6%, until 113°C temperature is reached.
The state of stress follows the thermal regression of the yield surface.
Consequently, deviatoric and mean stress increase. Pore pressure, however, continues to increase due to thermal water expansion.
At temperatures higher than 113°C, the material starts to undergo plastic compaction. The production of compactive plastic volumetric strains rates reduces the shrinkage speed of the yield surface, until the strain hardening effect exceeds the thermal softening. After that, even if temperature continues to increase the yield surface increases.
The total final deformations remain dilative; their order of magnitude is of some percent. Pore pressure distribution is very similar to that of the C and B cases outside the yielded zone.
The phenomena described concern mainly the first metre of the host clay.

CONCLUSIONS

The thermo-mechanical response of clay surrounding an isolated deep borehole has been studied in plane strain with reference to a certain number of conditions simulating the drilling of the hole into a clay formation and the emplacement of nuclear waste canisters. Calculations were performed for a specific stiff overconsolidated plastic clay under specific in situ conditions: conclusions cannot be extended to clays with different thermo-mechanical properties. Therefore the purpose of this paper is to point out that similar studies have to be performed for any given site.
A zone estimated about one metre radius or probably less is affected by plastic strains. This zone is influenced by the drilling method.
If the drilling method adopted is capable to guarantee a supporting pressure on the borehole comparable with the initial state of stress, the most probable result is that thermally induced yielding causes the production of plastic compactive volumetric strains. Radial stresses around the borehole are almost halved with respect to the initial state in the first decade : subsequently they recover substantially. This type of behaviour should not give negative consequences on the confinement of radionuclides.
If the drilling method implies a supporting pressure low with respect to the initial state of stress, subsequent thermal loads may lead to a significantly different situation. In fact, thermo-mechanical yielding of clay has been observed to occur with plastic dilation, which can be of some percent; fracturing of the material is expected. In this case there is a zone of highly disturbed material which can be a preferred hydraulic path along the borehole.
The above results are based on the assumption of no water inflow from the borehole into the clay formation. As the correct boundary condition during drilling is quite difficult to determine, the extreme case of a large water

inflow modelled by a high water pressure boundary condition has been considered. As a result, material behavior in the region affected by the water intake is much closer to failure. The possibility of failure occurrence is strongly affected by the thermo-mechanical characteristics of the clay considered.

Considering a specified clay type, the drilling method is seen to produce strongly different situations in the clay around the boreholes: therefore thermo-mechanical effects must be taken into account when selecting the drilling procedure.

Significant effects can be observed also in a zone wider (about 10 meter radius) than the one previously discussed.

As illustrated in Figs. 7 and 8, in the first decade pore pressure rise is about 1.0 MPa and it halves in about 50 years. Consequently the isotropic component of stress decreases: about 0.5 MPa after 5 years. Deviatoric stress doubles with respect to the undisturbed state.

This increase contributes, together with the yield limit reduction due to the temperature, to diminishing the elastic domain available for further loading. These effects are substantially independent from drilling techniques, and are definitely affected by the thermal field.

REFERENCES

1.   Baestle L.H. and Mittempergher M., in Radioactive Waste Management and Disposal. Proc. 1st European Comm. Conference, Luxembourg 1980, edited by R. Simon and S. Orlowski (Harwook Academic Publishers, Brussels and Luxembourg 1980) p. 442

2.   Campanella R.G. and Mitchell J.K., K. SMFE PROC. ASCE, 94 NO SM3, 709 (1968)

3.   Heremans R., Barbreau A., Bourke P., Gries H., in Radioactive Waste Management and Disposal, Proc. 1st Eur. Comm. Conference, Luxembourg 1980, edited by R. Simon and S. Orlowski (Harwood Academic Publishers, Brussels and Luxembourg, 1980) p. 468

4.   Davies T.G. and Banerjee P. L., Constitutive relationships for ocean sediments subjected to stress and temperature gradients, AERE Harwell report NO HL 80/2604 (C22), August 1980.

5.   G. Baldi, T. Hueckel, E. Tassoni, presented for Ing., Symp. on Environmental Geotechnology Allentown, PA, 1985 (to be published)

6.   Borsetto M., Cricchi D., Hueckel T., Peano A., in Numerical Methods for Transient and Coupled Problems, edited by R. W. Lewis et al. (Pineridge Press. Swansea U.K., 1984) p. 608

7.   Prager N., Proc. Konickl Nederl. Akad. van Wessenschappen series B, 61, 3, p. 176, 1958

8.   Roscoe K.H. and Burland J.B., in Engineering Plasticity edited by J. Hayman and F.A.A. Leckie, (Cambridge University Press, Cambridge 1968) p. 535.

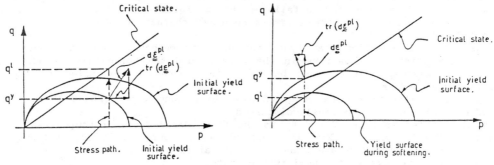

Fig. 1 Yield surface evolution with plastic strain:
hardening (right) and softening (left) range.

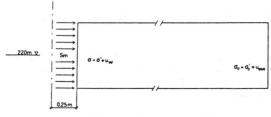

	γ_m (Kg/m³)	S_m (MPa)	U_w (MPa)
A	2000	4.4	2.2
B	2000	4.4	4.4
C	2000	4.4	impervious
D	2000	2.2	impervious

Fig. 2 Modelling of the drilling
mud effect.

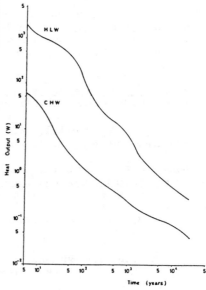

Fig. 3 Heat output of a single canister
of HLW and CHW with time
after production.

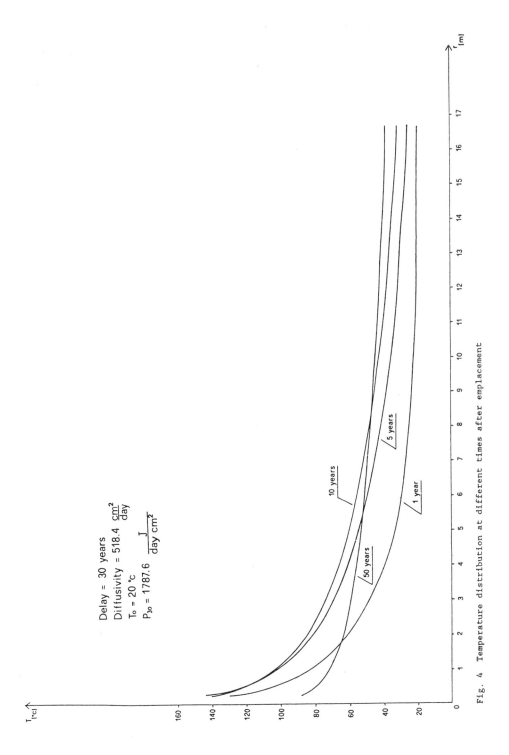

Delay = 30 years
Diffusivity = 518.4 $\frac{cm^2}{day}$
T_o = 20 °c
P_{3o} = 1787.6 $\frac{J}{day\ cm^2}$

Fig. 4 Temperature distribution at different times after emplacement

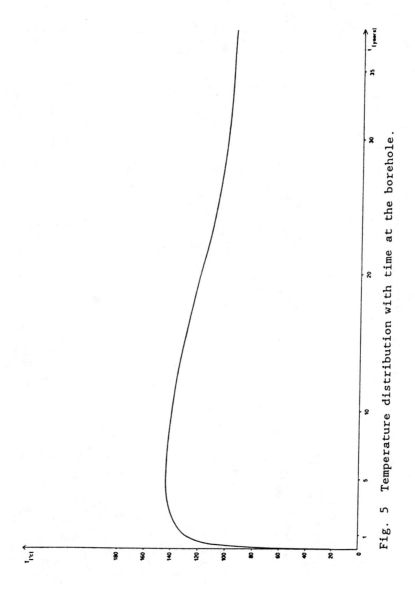

Fig. 5 Temperature distribution with time at the borehole.

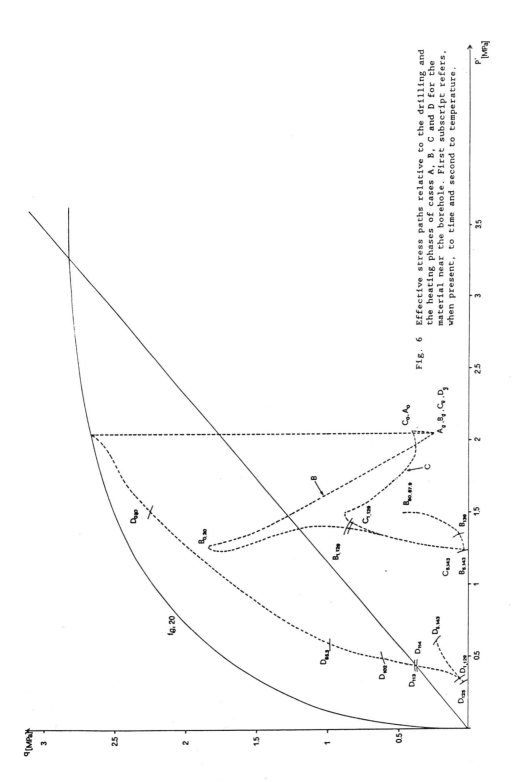

Fig. 6 Effective stress paths relative to the drilling and
the heating phases of cases A, B, C and D for the
material near the borehole. First subscript refers,
when present, to time and second to temperature.

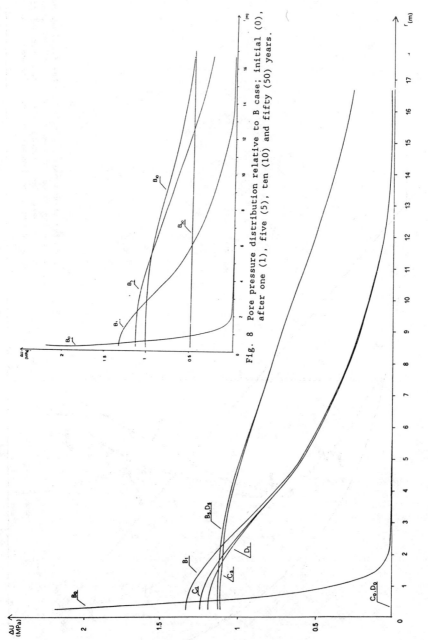

Fig. 8 Pore pressure distribution relative to B case; initial (0), after one (1), five (5), ten (10) and fifty (50) years.

Fig. 7 Pore pressure distribution in B, C and D cases; initial (0), after one year (1) and after five years (5).

EXCAVATION RESPONSE IN RESPECT OF THE REPOSITORY CONCEPT FOR THE KONRAD MINE AS SEEN BY THE MINING AUTHORITY

W. Roehl
Mining Authority of Lower Saxony
Oberbergamt
Clausthal-Zellerfeld
Federal Republic of Germany

ABSTRACT

In connection with the securing technical check-up of the repository concept for the Konrad mine the Mining Authority of Lower Saxony was asked for a comment on the mining concern of the Konrad plan by the Ministry for Environmental Protection of Lower Saxony.

Referring to excavation response the valuation of rock-mechanical data, efficiency of the dams, underground convergences and mining subsidence are important to the realisation of the repository concept.

MODELE DE DEPOT AMENAGE DANS LA MINE KONRAD : INTERPRETATION DES EFFETS DES TRAVAUX D'EXCAVATION PAR LE SERVICE DES MINES

RESUME

En liaison avec la vérification technique de la sûreté du modèle de dépôt aménagé dans la mine Konrad, le Service des mines de Basse-Saxe a été invité à formuler des observations sur l'aspect minier du plan relatif à la mine Konrad mis en cause par le Ministère de la protection de l'environnement de Basse-Saxe.

S'agissant des effets des travaux d'excavation, il importe pour réaliser ce modèle de dépôt de quantifier les propriétés mécaniques des roches, l'efficacité des cloisons étanches, les convergences souterraines et la subsidence liée aux travaux d'excavation.

1. Introduction

Since 1976 the former Konrad iron ore mine in the SE of Lower
Saxony has been investigated for being adapted for an underground
repository for radioactive wastes. The minable thickness of the
mineral deposit, which dips gently by 18 to 22 degrees from East
to West, varies between 4 and 18 m. The deposit is covered by a
solid and volumenous marl clay. At depths of 8oo to 13oo m the
buried iron ore structure appears as a trough of a syncline.

In 1982, on the basis of the results of the first investigations,
the Physical-Technical Federal Institution (PTB) applied for
the performance of license proceedings for the KONRAD mine as an
underground repository.

2. Presentation of the planned repository concept

Before imbedding radioactive wastes the underground structure of
the Konrad mine will be completed and enlarged.

Referring to the location of the deposit panels around the former
focal points of working PTB set up the following principles for
the repository concept (figure 1) [1] :

- Shaft No. 1 remains the downcast ventilating shaft. It will
  also be used for transporting rubbish and all kinds of
  other material and for man haulage. The hauling installation
  will be modernized.
- The waste containers will be transported in shaft No. 2.
  Therefore the complete hoisting plant will be renewed.
- Former extraction chambers will not be used for depositing
  waste containers.
- Broken material, which will yield by drifting new panels,
  will serve as backfilling of abandoned workings and for the
  construction of the dams after filling the chambers.
- Drifting operations will completely be separated from
  disposal operations in all parts of the underground
  repository. There will be no reciprocal influence. Waste
  air out of disposal panels will directly flow to the upcast
  ventilation shaft No. 2 normally without coming into contact
  with the pit crew.
- After been filled by waste containers all excavations of
  the underground repository will be backfilled. It is intended
  to build up impermeable and maintenance-free packers into
  the shafts.

3. Main points of the comment of the Mining Authority on the Konrad plan

On account of the special and professional knowledges the Mining
Authority was asked for a comment on the mining concern of the
Konrad plan by the Ministry for Environmental Protection of
Lower Saxony. The special fields of the report of the Mining
Authority are summarized as follows (figure 2).

In relation to excavation response there are three main points of the comment, which are important to the realization of the repository concept:

1. valuation of rock-mechanical data,
2. valuation of the efficiency and practicability of the dams at the entrances into the deposit chambers and
3. valuation of underground convergences and mining subsidence.

3.1 Valuation of rock-mechanical data

Based on mining experience during the operation period of more than 15 years the Konrad iron ore mine has proved its stability and reliability under former conditions. However, the repository concept differs from the operation period of the iron ore mine by the way as mining experience is not always transferable. A lot of planned excavations had never been comparably performed within the long period of ore mining in Lower Saxony.

The valuation of the submitted rock-mechanical data mainly refers to

1. the stability and reliability of all excavations in the deposit fields
   - by checking up the calculations added to the Konrad plan and
   - by drifting experimentally a part of a deposit panel,
2. the stability and reliability of the new onsetting station at shaft No. 2 and
3. the response of the overlying strata in relation to variations of rock permeability.

3.1.1 Stability of the excavations of a deposit field

The Mining Authority regards the stability and reliability of all excavations of a deposit field and generally of all panels and excavations of the whole underground repository as the central aspect of the comment on the mining concern of the Konrad plan.

Therefore, in the opinion of the Mining Authority, the realization of the Konrad plan must be proved by a suitable indication consisting of two parts: rock-mechanical calculations and the in-situ-test of drifting a panel. It is desirable to compare the results of the calculations with the results of the in-situ-test. At present there are retardments in the projected rock-mechanical measurements, which will be carried out during the testing period.

The following two figures (figures 3 and 4) will show rock-mechanical characteristics of operating the Konrad repository as seen by the Mining Authority.

All chambers are horizontally drifted at room-pillar-ratio 1 : 4 following the strike of the iron ore zone until almost 1ooo m. The cross-sectional area of a chamber is about 4o sqm.

Mining experience has proved that cross-cutting direction of driving is better and less dangerous than excavation following the strike.

About 35 m above the disposal chambers a cross-cutting ventilation drift leads waste air back to shaft No. 2 (figure 4). Cased ventilation boreholes of nearly 1,4 m diameter are connecting each disposal chamber with the ventilation drift.

Stability and reliability of all excavations of a deposit field inclusive boreholes and the ventilation drift are to be guaranteed till the end of the operating period before back-filling and closing the chambers. Up to now excavation responses are mostly unknown in all parts of a deposit field. A reciprocal rock-mechanical influence especially on the ventilation drift or on the boreholes may be possible.

3.1.1.1 Valuation of the rock-mechanical calculations

At the instigation of the Mining Authority a supplementary expert was asked to review the specifications of rock-mechanical calculations of the Konrad plan. The task for the expert is summarized in the following figure 5.

The submitted geological and rock-mechanical data and calculations carried out by the Federal Institute for Geological Science and Natural Recources, Hannover, represent the plastification degree of a pillar between two chambers as being approximately 35 % [2]. The average thickness of the disconsolidated zone at the side walls of a chamber is about 5 m.

More results of the applicant's calculations are shown in figure 6. The greatest vertical displacements come to 2 cm caused by floor lift and to 2,8 cm caused by roof deflection. Both side walls of a chamber converge to 4,7 cm. In opposition to the vertical displacements the horizontal displacements fade quickly away. They disappear in the pillars within a distance of about 2 m.

As all calculations are based on excavation without support the applicant is convinced of the stability of the underground structure. Nevertheless it is intended to set supports in all chambers and drifts.

Having checked up the results of the applicant's calculations the expert of the Mining Authority essentially verifies the implication of the submitted data. In detail the expert preśents the following results:

- The compression zone in the middle of the pillars runs up to 1 cm only.
- The floor lift causes a zone of relaxation penetrating into a depth of about 6 m.
- The disconsolidated zones in the pillars are confined to a rather narrow area at the side walls of the chambers. In connection with the dimensional analysis the pillars have sufficient strength property by calculation.

What does that mean for the comment of the Mining Authority? By using roof strata bolting and mesh wire the planned support of the excavations will protect the mine workers against rock fall during the operation period of a deposit panel.

Strata bolting, even into the side walls of the chambers, is needed to be set as quickly as possible in order to minimize rock disaggregation.

As the trucks and stackers transporting the waste containers in the underground repository are sensitive of lateral slope, the roadways of the transport drifts and the chambers must probably be repaired. This effect is in line with mining experience during the operation period of the Konrad iron ore mine.

Excavated zones of jointing are able to cause caving formation, especially at the disconsolidated zones of the side walls. While setting strata bolting the mine workers have to observe the critical zones very carefully.

As a matter of principle the stability and reliability of the deposit chambers are verified by the Mining Authority.

The rock-mechanical calculations of the applicant and the check up by the expert of the Mining Authority didn't include the stability of the ventilation drift and the ventilation boreholes. Because of the far-reaching influence of the induced vertical stress an unfavourable consequence to the stability may be possible.

3.1.1.2   In-situ-test of drifting a part of a deposit panel

To demonstate the reliability of all excavations of a deposit panel PTB intends to drift experimentally a part of a deposit panel consisting of three chambers with their entrances, unloading stations, ventilation boreholes and the cross-cutting ventilation drift almost 35 m above.

When driving the panel the applicant carries out extensive rock-mechanical measurements proposed by the Mining Authority.

Later the measurements will be explained in detail.

The subaltern Mining Office in Goslar regularly inspects the progress of test working. All exceptional particularities are proved by documentary evidence, especially direction of strike, sequence of strata, inclination of dip, tectonics, dislocations ahd water outlets. Special events like downfalls, floor lifts etc. are also documented. The Mining Office in Goslar has prepared a supplementary documentation by taking pictures of all events and particularities occuring during the test period.

3.1.2 Valuation of the stability and reliability of the new onsetting station at shaft No. 2

The applicant plans to construct a new onsetting station at shaft No. 2 by pervading the shaft with an excavated cross-section area of about 95 sqm. Therefore the stability of the shaft has to be checked up too.

A kind of portal crane in the onsetting station will put the waste containers upon the trucks transporting the containers to the disposal chambers. This portal crane is very sensitive to rock movements, especially to modifications of the lateral slope.

At the instigation of the Mining Authority a supplementary expert was asked to review the submitted data of the new onsetting station also. The task includes

- a valuation of the planned construction,
- a valuation of rock movements during the operation period of the repository,
- a valuation of the additional consumption for safety precautions and
- a check-up of the calculations by using a finite element net for discretizing the new station.

Up to now there are no results on hand.

3.1.3 Response of the overlying strata in relation to variations of rock permeability

In connection with rock movements caused by excavations existing separation planes may be varied as well as new separation planes may come into existence.

Because of the incomplete knowledge of the great number of fractures and their different facing it is impossible to discretize the real joint system by using a finite element net. The Mining Authority is of the opinion that the plürality of assumptions influencing the finite element model would produce such a great dispersion of joint widths, to the interest of a serious revision this possibility of dissolving the problem should not be continued. Therefore the Mining Authority proposes to register and assess all variations of joints in a way that all assumptions are summarized within an integral factor of different influences. Then investigations of rock permeability and numerical calculations as a basis for the safety performance should be carried out by those parameter variations of the integral factor, which represent realistic borderline cases and simulate rock-mechanical disaggregations.

3.2   Valuation of the efficiency and practicability of the dams

3.2.1 Design and projected efficiency of the dams

After backfilling a disposal chamber the applicant intends to construct a dam between the entrance to the filled chamber and the transportation drift (figure 7).

The interior part of a dam consists of stowing material closed by a double partition wall at the entrance to a chamber. Caviations at the roof of a chamber caused by settlements of stowing material are backfilled by injections.

The projected dams have to separate the filled chambers from other parts of the underground structure of the repository in a way that the influence of the radioactive emission by gaseous radioisotopes is minimized during the operating period. Oscillations of the barometric pressure mustn't cause a worth mentioning release of gaseous radioisotopes out of the filled chambers. The dams are planned to absorb gaseous radioisotopes during an air depression period. When the air pressure rises again, the dams have to return the gaseous radioisotopes into the filled chambers.

3.2.2 Performance of an in-situ-test of constructing a dam and proving its efficiency

The requirements on a dam in an underground repository differ from those on a dam usually constructed in a mine. A dam at the entrance to a deposit chamber has to fulfill the following conditions:
- The stowing material is to have a certain, definite quality. Especially its porosity and permeability is to be very constant.
- In every case the permeability of the disaggregation zone within the region of the planned dam must be constant and well known also. The dam is able to exercise the intended function only with the understanding that the permeability of the disaggregation zone around the entrance to a chamber is less than the permeability of the stowing material.
- In spite of convergences and processes of settlement in the stowing material a lasting connection between the roof and the upper part of the dam is necessary.
- The stowing material must be able to absorb gaseous radioisotopes and return them into the chamber.

A dam like this has never been realized before. As measurements of rock permeability at several points of the underground structure proved the existence of disaggregation zones of a length of almost 1o m, the Mining Authority asked for an indication of the efficiency and practicability of the dams. Joint systems caused by excavations render possible a migration or flow of gas. In the opinion of the Mining Authority there are great uncertainties, if the system consisting of dams and disaggregation zones is able to fulfill the mentioned requirements.

Possibly the double partition wall in front of the dam may be destroyed by convergences that the efficiency of the dam may be indured.

While the applicant was engaged in planning details of the in-situ-test, a new concept for closing the deposit chambers was developed. Probably it is now intended to fasten the exposed areas (roof, floor and side walls) by concreting directly after having excavated the entrance to a chamber. By that means the disaggregation of the surrounding rock may be avoided. Beyond this the Mining Authority hasn't got any revised data up to now.

3.3 Valuation of underground convergences, mining subsidence and far-reaching rock responses caused by excavation

During the experimental driving of the part of a repository field rock-mechanical measurements are carried out by the applicant in order to become acquainted with rock responses. These knowledges are important to verify the results of the rock-mechanical calculations. The Mining Authority participated in working out the measuring programme as follows:

1. To find out the amount of underground convergences in all test chambers two measuring stations are installed within a distance of 15 m.
2. Permanent deformations are measured by two extensiometer stations in both pillars between the projected three chambers.
3. The disaggregation of the surrounding rock is measured in the inner chamber and registered also.
4. Roof-bolt dynamometers are installed to measure the stress of a representative number of roof bolts.
5. Beyond this all geologic and tectonic data of the testing field are written down to be proved by documentary evidence.

In situ the Mining Authority puts up an own documentation programme to register all particularities by letter and by photo report.

Having valuated the far-reaching rock responses as a result of excavations the supplementary expert of the Mining Authority confirms the stability of the underground structure of the planned Konrad repository is not recognizably endangered.

REFERENCES:

[1]   Plan Konrad, Sept. 1986, PTB

[2]   N. Diekmann, R. Konieczny, D. Meister, H. Schnier:
      Geotechnical and rockmechanical investigations for the
      design of the Konrad repository;
      paper presented at the "International symposium on the
      siting, design and construction of underground repositories
      for radioactive wastes", Hannover, FRG, 3 - 7 March 1986,
      IAEA - SM 289/47

NOTE: Figures No. 1,3,4,6 and 7 belong to [1]

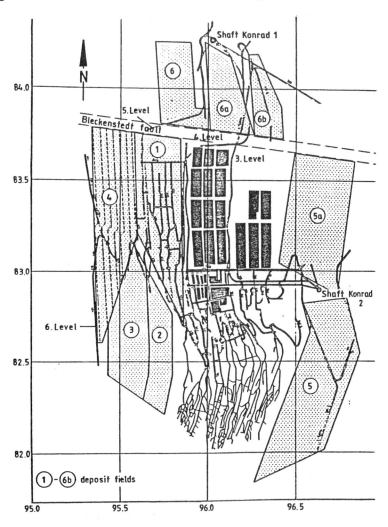

Fig. 1: location of the planned deposit panels in the Konrad iron ore mine

Fig. 2:   Special fields of the report of the Mining Authority

1. Safety of the hoisting plants,

2. Support, maintenance and supervision of excavations;'
   valuation of the stability of the deposit chambers,

3. applicability of the filling method,

4. applicability of the dams

   - constructions
   - design
   - controlling devices,

5. haulage

   - haulage of persons
   - applicability of the vehicles
   - traffic regulations,

6. ventilation

   - equipment
   - organisation of ventilating districts
   - air flow measuring devices
   - climatic conditions,

7. safety precautions against encountering gas and water;
   safety distances,

8. surface and underground fireproofing

   - safety against downcast fumes
   - organisation of fire areas
   - fire - extinguishing installation

9. surface and underground safety precautions against explosion

10. rescue systems

   - mine rescue brigade
   - radiation rescue brigade
   - emergency precautions,

11. stability of excavations

   - mining subsidence
   - stability of the underground structure

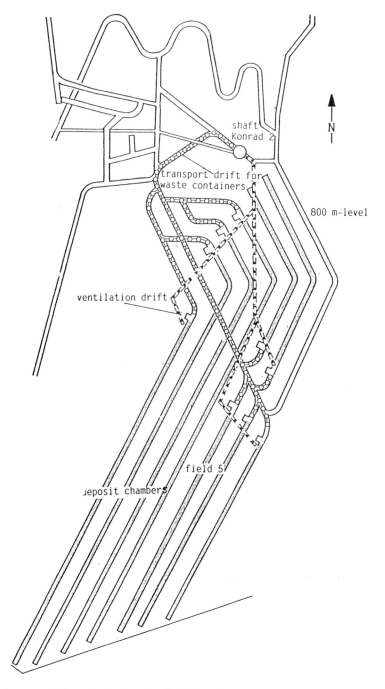

shaft
Konrad 2

N

transport drift for
waste containers

800 m-level

ventilation drift

field 5

deposit chambers

Fig. 3: ground plan of the deposit field No. 5

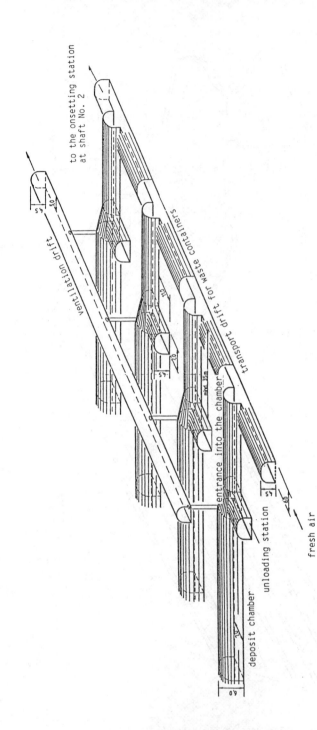

Fig. 4: sketch of a deposit field

Fig. 5:   Assignment for the expert of the Mining Authority

Valuations of   1. cut-out of the basis for calculation,

2. choosen formation characteristics,

3. strata models,

4. completeness and correctness of the
   calculated arrangements,

5. calculations and

6. rock-mechanical results.

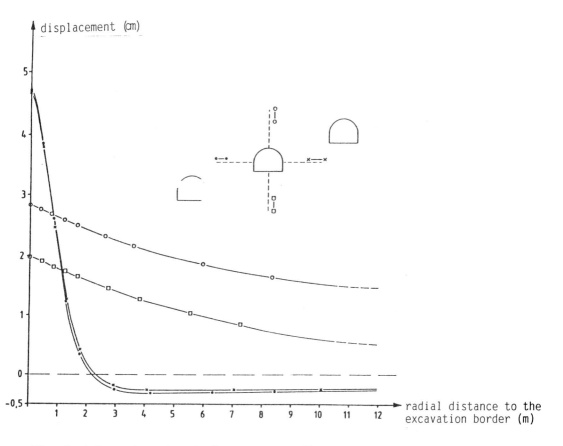

Fig. 6: deformation of a chamber at room-pillar-ratio 1 : 4

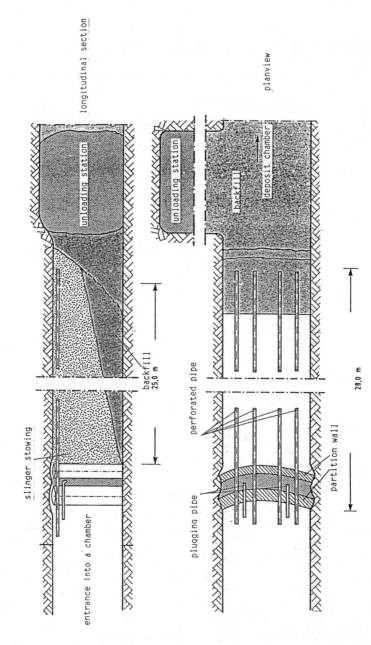

Fig. 7: design of a dam at the entrance into a deposit chamber

longitudinal section

planview

unloading station

unloading station

backfill

deposit chamber

backfill 25,0 m

slinger stowing

entrance into a chamber

perforated pipe

plugging pipe

partition wall

28,0 m

Séance IV – RAPPORT DE SYNTHESE DU PRESIDENT

R.H. HEREMANS (NIRAS/ONDRAF, Bélgique)

L'originalité de la Séance IV de la Réunion de travail réside dans le fait que nous avons pu, au cours de la présentation de M. W.G. Roehl, prendre connaissance du point de vue d'une Autorité officielle concernée par le processus d'autorisation d'exploitation d'un site d'évacuation et de l'importance que cette Autorité accorde entre autres aux expériences et études concernant l'impact potentiel des travaux de creusement sur la conception et la sûreté à long terme d'installations souterraines d'évacuation. Je crois que nous touchons ici à un point important dont, en premier lieu, les responsables et les gestionnaires de programmes de recherches et de développement en la matière doivent être pleinement conscients. Les questions, auxquelles ils auront à répondre un jour, lorsque sera venu le moment d'obtenir les licences d'exploitation, seront celles que leur poseront ces Autorités nationales responsables. Un dialogue régulier entre les chercheurs, d'une part, et les Autorités, d'autre part, me semble donc essentiel et ceci dès l'établissement des programmes de travail en vue d'orienter au mieux ceux-ci vers les objectifs finals. Cela paraît sans doute évident, mais je ne suis pas persuadé que ceci soit toujours et partout le cas dans nos pays respectifs.

Les formations argileuses ont depuis de nombreuses années été considérées comme des formations hôtes susceptibles de convenir pour une évacuation sûre à long terme de déchets radioactifs. Si, comme l'a dit M. A. Bonne au cours de son exposé introductif sur les argiles, il existe une très grande variété de formation de ce type dont les caractéristiques peuvent être aussi diférentes que leur nombre est grand, c'est peut-être là qu'il faut chercher une des raisons pour laquelle les travaux de recherche in-situ n'ont pas connus un développement aussi intensif que ce fut le cas pour les granits ou le sel. Je veux dire par là que dans le cas des argiles, la transposition des résultats obtenus sur un site vers un autre est peut-être moins évidente que pour d'autres types de formations et que des expériences in-situ dans l'argile ne se justifient pleinement que si elles sont faites sur un site qui présente une bonne probabilité d'être retenu comme site final d'évacuation.

Les argiles présentent, tout comme les autres roches considérées, des avantages et des inconvénients. Une des questions fondamentales qui se posait il y a quelques années et plus particulièrement pour les argiles peu indurées, telles que l'argile de Boom en Belgique, était de savoir s'il était possible de créer des cavités ou galeries à une profondeur raisonnable à des prix acceptables. Les techniques de creusement à utiliser, les types de revêtement à prévoir, le comportement des ouvrages à court et moyen terme, etc., étaient autant d'autres questions à considérer. Les travaux effectués ces dix dernières années par le SCK/CEN de Mol ont apporté une première série de réponses à ces questions.

La présentation de M. B. Neerdael décrit les phases successives d'un laboratoire souterrain dans l'argile de Boom à une profondeur entre 223 et 247 m, l'instrumentation mise en place pour l'auscultation du massif et du revêtement des galeries pendant et après les travaux de creusement et de mise en place ainsi que les résultats obtenus. La conclusion la plus importante est sans aucun doute que le creusement à grand diamètre (4 à 5 m) dans l'argile de Boom à Mol est possible avec des techniques minières courantes entraînant une perturbation tout à fait acceptable du massif dans une zone limitée autour de la galerie. D'autre part, il paraît possible de réduire la charge sur le revêtement en faisant participer le terrain au soutènement par le biais d'une déformation contrôlée de ce terrain.

L'exposé de Mme R. André-Jehan se rapporte à une série de travaux effectués par l'ANDRA de France dans ce même laboratoire souterrain dans l'argile à Mol. Ces travaux sont complémentaires à ceux effectués par le SCK/CEN et ont été réalisés dans le cadre d'un accord de coopération entre les deux organismes. L'expérience principale concernait un essai en vraie grandeur d'un soutènement à "convergence contrôlée". L'interprétation des résultats de mesures effectués au cours d'expériences à petite et grande échelles permettent de conclure entre autres que pour l'argile de Boom :

- les contraintes en place sont quasi-isotropes

- l'effet d'échelle des expériences sur les résultats est modéré

- les effets différés sont très marqués

- la résistance à long terme (plusieurs années) est très bonne.

Il reste bien entendu un travail considérable à poursuivre, notamment l'exploitation rationnelle des résultats expérimentaux, dans les modèles qui permettront, d'une part, la conception optimisée d'une installation industrielle (donc avec une zone perturbée réduite au minimum, et, d'autre part, une évaluation correcte du comportement du système à plus long terme. Cette étroite collaboration entre les expérimentateurs, d'une part, et les modéliseurs, d'autre part, est une recommandation qu'il n'est sans doute pas superflu de refaire à l'occasion de la présente réunion.

Une seconde question fondamentale qui se posait il y a quelques années concernait la charge thermique admissible dans une argile contenant de l'eau et ayant une conductibilité thermique réduite. Les caractérisations faites en laboratoire sur échantillons ainsi que quelques expériences avec des sources de chaleur artificielles menées à faible profondeur dans des formations argileuses représentatives ont démontré que le problème était maîtrisable pour autant qu'un temps de refroidissement raisonnable des déchets de haute activité - 50 à 70 ans - soit pris en considération. Le moment est certes venu maintenant d'examiner de plus près et d'estimer à sa juste valeur l'importance de ce problème.

Les résultats de l'étude entreprise par l'ISMES et qui furent présentés par Mlle. R. Pellegrini sont une première approche encourageante de réponse plus précise à cette question. Les premiers résultats de calculs sur les effets thermiques induits dans l'argile utilisant des valeurs paramétriques expérimentales devront être confirmés ou infirmés par d'autres expériences in-situ dès à présent en préparation dans le laboratoire souterrain de Mol.

Les résultats attendus dans les mois à venir des différentes expériences et mesures que nous venons d'évoquer contribueront incontestablement à mieux juger de l'acceptabilité du système d'évacuation des déchets dans une formation argileuse. Je m'en voudrais pour terminer de ne pas évoquer une autre question fondamentale qui nécessitera dans l'avenir une réponse précise et qui n'a été abordée que par un nombre limité d'oraeurs, c'est celle de la possibilité de procéder au remblayage et à l'obturation efficace des installations d'évacuation. Je me permettrais donc de recommander à l'AEN d'organiser, tel qu'elle l'a suggéré elle-même déjà, un réunion de travail sur ce sujet qui, sans nul doute, intéressera tous les responsables et les chercheurs des grands programmes sur l'évacuation des déchets.

* *

Session IV - CHAIRMAN'S SUMMARY

R.H. HEREMANS - (NIRAS/ONDRAF, Belgium)
* *

A highlight of Session IV was W.G. Roehl's paper, which presented the viewpoint of an official authority responsible for issuing licenses to operate disposal sites, also showing the importance attached by the authority to, among other things, experiments and studies about the potential impact of excavation on the design and long-term safety of underground repositories. I believe this is an important point which managers and administrators of research and development programmes on this subject should be fully aware. The questions which they will have to answer when the time comes to obtain operating licences will emanate from the national authorities in question. On-going dialogue between researchers and the authorities is therefore essential, and should begin as soon as programmes of work are established, with a view to directing them towards the final objectives. This no doubt seems self-evident, but I am not at all sure that such is the case in all our respective countries.

For many years, clay formations have been considered as suitable host environments for safe, long-term disposal of radioactive waste. If, as A. Bonne stated during his introductory paper on clay, there is a very wide range of clay formations with equally numerous characteristics, perhaps this explains why in-situ research on clay formations has not advanced as much as for granite and salt. In other words, it is perhaps not as easy to transfer the results obtained from one site to another as with other types of formations, and there can be full justification for in-situ experiments in clay only if they are conducted at a site which is likely to be selected as a final disposal site.

Like the other rocks investigated, clay presents advantages and disadvantages. One of the fundamental questions raised a few years ago, and more specifically for unhardened clay such as that occurring in the Boom clay formation in Belgium, was whether cavities or galleries could be excavated at a reasonable depth and at acceptable cost. The excavation techniques, types of lining and the behaviour of facilities in the short- and medium-term were further questions to be answered. The research conducted in the past ten years by SCK/CEN at Mol have provided a first set of replies to such questions.

B. Neerdael's paper described the successive stages in the establishment of an underground laboratory in Boom clay at a depth of between 223 and 247 metres, the instrumentation installed for investigating the rock mass and lining of the galleries during and after excavation and installation, and the results obtained. The most important conclusion seemed to be that large-diameter excavation (4 to 5 metres) in Boom clay at Mol is possible using routine mining techniques, the ensuing disruption to the rock being fully acceptable in a limited area surrounding the gallery. On the other hand, it seems possible to reduce the load on the lining by using the clay to support the structure through controlled distortion of the clay.

R. André-Jehan's paper reported on a set of studies by ANDRA (France), at the same underground laboratory in clay at Mol. This work complements that being conducted by SCK/CEN, and formed part of a co-operative agreement between the two organisations. The main experiment concerned a full-scale test on controlled convergence gallery support. The interpretation of the measurements made during small- and large-scale tests lead to the conclusion, among others, that for Boom clay:

- in-situ stresses are quasi-isotropic

- there is a moderate effect of experimental scale on the results

- delayed effects are very marked

- long-term resistance (over several years) is very high.

Naturally, considerable work remains to be done, including rational exploitation of the experimental findings using models, which should lead to optimised design of an industrial facility (hence reducing the disrupted area to a minimum) as well as correct evaluation of longer term behaviour of the system. Once more, may we strongly recommend close co-operation between experimenters and modellers of the kind we have just seen.

A second fundamental question raised a few years ago concerned the permissible heat load in a clay containing water and having reduced heat conductibility. Measurements on samples in the laboratory and a few experiments using artificial heat sources at low depth in representative clay formations showed that this problem could be overcome provided that allowance was made for a reasonable cooling time for high-level waste (50 to 70 years). The time has now come to examine this matter more closely and to assess the true status of this problem.

The results of the study undertaken by ISMES, which were presented by R. Pellegrini, provide an encouraging initial attempt to answer the above question more precisely. The first calculations on heat effects induced in clay using experimental parameters should be confirmed or invalidated by further in-situ experiments now being prepared at the Mol underground laboratory.

The results to be obtained in the next few months from the different experiments and measurements described will undoubtedly lead to a clearer assessment of the system for waste disposal in a clay formation. I should not like to conclude without mentioning another fundamental question which will have to be answered accurately in future and which has not been addressed by many speakers, namely the possibility of backfilling and effective sealing of disposal facilities. Consequently, I should like to recommend that the NEA organise a workshop on the subject, as it has already suggested, as this will no doubt be of interest to all managers and researchers of major waste disposal programmes.

SESSION V

DISCUSSION AND CONCLUSIONS

SEANCE V

DISCUSSIONS ET CONCLUSIONS

Chairman – Président

K.W. DORMUTH
(AECL, Canada)

RAPPORTEURS

K.W. DORMUTH   (Canada)
R.A. ROBINSON   (United States)
A. BARBREAU   (France)
R.V. MATALUCCI   (United States)
R.H. HEREMANS   (Belgium)

Based upon the presentations and discussions in Session V, an
"Executive Summary and Conclusions" was prepared by the NEA Secretariat, and
is reported on page 11 of these proceedings.

* *

Sur la base de l'exposé et des discussions de la Séance V, un resumé et
des conclusions ont été établis par le Secrétariat de l'AEN et figurent à la
page 17 de ce compte rendu.

BELGIUM - BELGIQUE

BONNE, A., Centre d'Etudes Nucléaires, Laboratories of the SCK/CEN, Boeretang 200, B-2400 Mol.

HEREMANS, R. H., Organisme National des Déchets Radioactifs et des Matières Fissiles (ONDRAF/NIRAS), Place Madou 1, Boîtes 24/25, B-1030 Bruxelles.

NEERDAEL, B., Centre d'Etudes Nucléaires, Laboratories of the SCK/CEN, Boeretang 200, B-2400 Mol.

CANADA

CHAN, T., Atomic Energy of Canada Limited (AECL), Whiteshell Nuclear Research Establishment, Pinawa, Manitoba ROE 1LO.

DORMUTH, K. W., Atomic Energy of Canada Limited (AECL), Whiteshell Nuclear Research Establishment, Pinawa, Manitoba ROE 1LO.

GRIFFITHS, P. M., Atomic Energy of Canada Limited (AECL), Whiteshell Nuclear Research Establishment, Pinawa, Manitoba ROE 1LO.

HANCOX, W. T., Atomic Energy of Canada Limited (AECL), Whiteshell Nuclear Research Establishment, Pinawa, Manitoba ROE 1LO.

HOLLOWAY, A. L., Atomic Energy of Canada Limited (AECL), Whiteshell Nuclear Research Establishment, Pinawa, Manitoba ROE 1LO.

JAKUBICK, A. T., Ontario Hydro , 800 Kipling Avenue, Toronto, Ontario M8Z 5S4.

KAISER, P. K., Laurentian University, Ramsey Lake Road, Sudbury, Ontario P3E 2C6.

KEIL, L. D., Atomic Energy of Canada Limited (AECL), Whiteshell Nuclear Research Establishment, Pinawa, Manitoba ROE 1LO.

KJARTANSON, B. H., Atomic Energy of Canada Limited (AECL), Whiteshell Nuclear Research Establishment, Pinawa, Manitoba ROE 1LO.

KOOPMANS, R., Ontario Hydro, 800 Kipling Avenue, Toronto, Ontario M8Z 5S4.

KOZAK, E. T., Atomic Energy of Canada Limited (AECL), Whiteshell Nuclear Research Establishment, Pinawa, Manitoba ROE 1LO.

KUZYK, G. W., Atomic Energy of Canada Limited (AECL), Whiteshell Nuclear Research Establishment, Pinawa, Manitoba ROE 1LO.

LADANYI, B., Ecole Polytechnique, Université de Montréal, C.P. 6079, Station A, Montréal, Québec H3C 3A7.

LANG, P. A., Atomic Energy of Canada Limited (AECL), Whiteshell Nuclear Research Establishment, Pinawa, Manitoba ROE 1L0.

MARTIN, C. D., Atomic Energy of Canada Limited (AECL), Whiteshell Nuclear Research Establishment, Pinawa, Manitoba ROE 1L0.

NGUYEN, T. S., Atomic Energy Control Board (AECB), 270 Albert Street, Ottawa, Ontario K1P 5S9.

PETERS, D. A., Atomic Energy of Canada Limited (AECL), Whiteshell Nuclear Research Establishment, Pinawa, Manitoba ROE 1L0.

SARGENT, P., Atomic Energy of Canada Limited (AECL), Whiteshell Nuclear Research Establishment, Pinawa, Manitoba ROE 1L0.

SIMMONS, G. R., Atomic Energy of Canada Limited (AECL), Whiteshell Nuclear Research Establishment, Pinawa, Manitoba ROE 1L0.

THOMPSON, P. M., Atomic Energy of Canada Limited (AECL), Whiteshell Nuclear Research Establishment, Pinawa, Manitoba ROE 1L0.

WHITAKER, S. H., Atomic Energy of Canada Limited (AECL), Whiteshell Nuclear Research Establishment, Pinawa, Manitoba ROE 1L0.

FINLAND - FINLANDE

AIKAS, T. T. J., Industrial Power Company Ltd., Nuclear Waste Office, Fredrikinkatu 51-53 B, SF-00100 Helsinki.

OHBERG, A., Saanio and Riekkola Engineering, Vuorikatu 6 B 20, SF-00100 Helsinki.

VUORELA, P. K., Geological Survey of Finland, Kivimiehentie 1 A, SF-02150 Espoo.

FEDERAL REPUBLIC OF GERMANY - REPUBLIQUE FEDERALE D'ALLEMAGNE

BALTES, B., Gesellschaft für Reaktorsicherheit (GRS), Glockengasse 2, D-5000 Köln 1.

BRASSER, T., Gesellschaft für Strahlen- und Umweltforschung (GSF), Institut für Tieflagerung, Theodor-Heuss-Strasse 4, D-3300 Braunschweig.

BREWITZ, W., Gesellschaft für Strahlen- und Umweltforschung (GSF), Institut für Tieflagerung, Theodor-Heuss-Strasse 4, D-3300 Braunschweig.

HEICK, C., Kavernen Bau- und Betriebs-Gesellschaft mbH, Roscherstr. 7, D-3000 Hannover.

LANGER, M., Federal Institute for Geosciences and Natural Resources, P.O. Box 51 01 53, D-3000 Hannover 51.

LIEDTKE, L., Federal Institute for Geosciences and Natural Resources, P.O. Box 51 01 53, D-3000 Hannover 51.

ROEHL, W. G., Oberbergamt Clausthal-Zellerfeld, Hindenburgplatz 9, Postfach 220, D-3392 Clausthal-Zellerfeld.

WIECZOREK, K., Gesellschaft für Strahlen- und Umweltforschung (GSF), Institut für Tieflagerung, Schachtanlage Asse, D-3346 Remlingen.

FRANCE

ANDRE-JEHAN, R., Agence Nationale pour la Gestion des Déchets Radioactifs (ANDRA), 31-33 rue de la Fédération, F-75752 Paris.

BARBREAU, A., Institut de Protection et de Sûreté Nucléaire - DPT, Commissariat à l'Energie Atomique (CEA), Boîte Postale No. 6, F-92265 Fontenay-aux-Roses Cedex.

HOORELBEKE, J. M., Agence Nationale pour la Gestion des Déchets Radioactifs (ANDRA), 31-33 rue de la Fédération, F-75752 Paris.

ITALY - ITALIE

PELLEGRINI, R., ISMES Spa, Viaie G. Cesare, 29-24100 Bergamo.

JAPAN - JAPON

IWATA, M., Engineering Development Department, Kajima Corporation, 2-1-1, Nishi-Shinjuku, Shinjuku, Tokyo.

NINOMIYA, Y., Waste Isolation Research Section, Chubu Works, Power Reactor and Nuclear Fuel Development Corp. (PNC), 959-31, Sonodo, Jorinji, Izumi-Cho, Toki-shi, Gifu.

SHIMADA, J., Underground Engineering Department, Institute of Technology, Shimizu Corporation, 3-4-17, Koshi-Nakajima, Koto-Ku, Tokyo.

NORWAY - NORVEGE

BARTON, N., Norwegian Geotechnical Institute (NGI), Box 40 Tasen, N-0801 Oslo 8.

SWEDEN - SUEDE

BERGSTROM, A., Swedish Nuclear Fuel and Waste Management Co. (SKB), Box 5864, S-10248 Stockholm.

KAUTSKY, F. E., Swedish Nuclear Power Inspectorate (SKI), Box 27106, S-10252 Stockholm.

PUSCH, R., Clay Technology AB, Ideon Research Center, S-223 70 Lund.

RAMQVIST, G., Stripa Project, Swedish Nuclear Fuel and Waste Management Co. (SKB), Gruvbackevägen, S-71700 Storä.

STILLBORG, B., Swedish Nuclear Fuel and Waste Management Co. (SKB), Box 5864, S-10248 Stockholm.

WINBERG, A., Swedish Geological Company (SGAB), Pusterviksgatan 2, S-413 01, Göteberg.

SWITZERLAND - SUISSE

FRANK, E., Swiss Nuclear Safety Inspectorate (HSK), CH-5303 Würenlingen.

LIEB, R. W., Grimsel Test Site, Nationale Genossenschaft für die Lagerung Radioaktiver Abfälle (NAGRA), Parkstrasse 23, CH-5401 Baden.

UNITED KINGDOM - ROYAUME-UNI

CALDWELL, A. M., British Nuclear Fuels PLC (BNFL), Risley, Warrington, Cheshire WA3 6AS.

COOLING, C., Building Research Establishment, Garston, Watford, Hertfordshire.

KNOWLES, N. C., W.S. Atkins Engineering Sciences, Woodgate Grove, Ashley Road, Epsom, Surrey KT18 5BW.

UNITED STATES - ETATS-UNIS

BAUER, S., Sandia National Laboratories (SNL), Division 6314, P.O. Box 5800, Albuquerque, New Mexico 87185.

BEAUHEIM, R. L., Sandia National Laboratories (SNL), Division 6331, P.O. Box 5800, Albuquerque, New Mexico 87185.

BLEJWAS, T., Sandia National Laboratories (SNL), Division 6313, P.O. Box 5800, Albuquerque, New Mexico 87185.

BORNS, D. J., Sandia National Laboratories (SNL), Division 6331, P.O. Box 5800, Albuquerque, New Mexico 87185.

BRUMLEVE, C. B., Battelle/Office of Waste Technology Development, 7000 S. Adams Street, Willowbrook, Illinois 60521.

FERRIGAN, P. M., United States Department of Energy (US DOE), Repository Technology Programme, 9800 South Cass Avenue, Argonne, Illinois 60439.

LINDNER, E. N., Battelle/Office of Waste Technology Development, 7000 S. Adams Street, Willowbrook, Illinois 60521.

MATALUCCI, R. V., Sandia National Laboratories (SNL), Division 6332, Experimental Programs, P.O. Box 5800, Albuquerque, New Mexico 87185.

PESHEL, J. J., Nuclear Regulatory Commission (NRC), Mail Stop 1 WFN, 4H-3, Washington, D.C. 20555.

ROBINSON, R. A., Battelle/Office of Waste Technology Development, 7000 S. Adams Street, Willowbrook, Illinois 60521.

TANIOUS, N. S., Nuclear Regulatory Commission (NRC), Mail Stop 1 WFN, 4H-3, Washington, D.C. 20555.

UBBES, W. F., Battelle/Office of Waste Technology Development, 7000 S. Adams Street, Willowbrook, Illinois 60521.

VOSS, C., Battelle Memorial Institute, 2030 M Street Northwest, Suite 800, Washington, D.C. 20036.

ZIMMERMAN, R. M., Sandia National Laboratories (SNL), Division 6313, Box 5800, Albuquerque, New Mexico 87185.

NUCLEAR ENERGY AGENCY - AGENCE POUR L'ENERGIE NUCLEAIRE

CHAMNEY, L., Division of Radiation Protection and Waste Management, 38 boulevard Suchet, F-75016 Paris (France).

WHERE TO OBTAIN OECD PUBLICATIONS
OÙ OBTENIR LES PUBLICATIONS DE L'OCDE

ARGENTINA - ARGENTINE
Carlos Hirsch S.R.L.,
Florida 165, 4° Piso,
(Galeria Guemes) 1333 Buenos Aires
Tel. 33.1787.2391 y 30.7122

AUSTRALIA - AUSTRALIE
D.A. Book (Aust.) Pty. Ltd.
11-13 Station Street (P.O. Box 163)
Mitcham, Vic. 3132   Tel. (03) 873 4411

AUSTRIA - AUTRICHE
OECD Publications and Information Centre,
4 Simrockstrasse,
5300 Bonn (Germany)   Tel. (0228) 21.60.45
Gerold & Co., Graben 31, Wien 1 Tel. 52.22.35

BELGIUM - BELGIQUE
Jean de Lannoy,
Avenue du Roi 202
B-1060 Bruxelles   Tel. (02) 538.51.69

CANADA
Renouf Publishing Company Ltd/
Éditions Renouf Ltée,
1294 Algoma Road, Ottawa, Ont. K1B 3W8
Tel: (613) 741-4333
Toll Free/Sans Frais:
Ontario, Quebec, Maritimes:
1-800-267-1805
Western Canada, Newfoundland:
1-800-267-1826
Stores/Magasins:
61 rue Sparks St., Ottawa, Ont. K1P 5A6
Tel: (613) 238-8985
211 rue Yonge St., Toronto, Ont. M5B 1M4
Tel: (416) 363-3171
Federal Publications Inc.,
301-303 King St. W.,
Toronto, Ont. M5V 1J5   Tel. (416)581-1552
Les Éditions la Liberté inc.,
3020 Chemin Sainte-Foy,
Sainte-Foy, P.Q. G1X 3V6, Tel. (418)658-3763

DENMARK - DANEMARK
Munksgaard Export and Subscription Service
35, Nørre Søgade, DK-1370 København K
Tel. +45.1.12.85.70

FINLAND - FINLANDE
Akateeminen Kirjakauppa,
Keskuskatu 1, 00100 Helsinki 10   Tel. 0.12141

FRANCE
OCDE/OECD
Mail Orders/Commandes par correspondance :
2, rue André-Pascal,
75775 Paris Cedex 16   Tel. (1) 45.24.82.00
Bookshop/Librairie : 33, rue Octave-Feuillet
75016 Paris
Tel. (1) 45.24.81.67 ou/ou (1) 45.24.81.81
Librairie de l'Université,
12a, rue Nazareth,
13602 Aix-en-Provence   Tel. 42.26.18.08

GERMANY - ALLEMAGNE
OECD Publications and Information Centre,
4 Simrockstrasse,
5300 Bonn   Tel. (0228) 21.60.45

GREECE - GRÈCE
Librairie Kauffmann,
28, rue du Stade, 105 64 Athens Tel. 322.21.60

HONG KONG
Government Information Services,
Publications (Sales) Office,
Information Services Department
No. 1, Battery Path, Central

ICELAND - ISLANDE
Snæbjörn Jónsson & Co., h.f.,
Hafnarstræti 4 & 9,
P.O.B. 1131 – Reykjavik
Tel. 13133/14281/11936

INDIA - INDE
Oxford Book and Stationery Co.,
Scindia House, New Delhi 110001
Tel. 331.5896/5308
17 Park St., Calcutta 700016   Tel. 240832

INDONESIA - INDONÉSIE
Pdii-Lipi, P.O. Box 3065/JKT.Jakarta
Tel. 583467

IRELAND - IRLANDE
TDC Publishers - Library Suppliers,
12 North Frederick Street, Dublin 1
Tel. 744835-749677

ITALY - ITALIE
Libreria Commissionaria Sansoni,
Via Benedetto Fortini 120/10,
50125 Firenze   Tel. 055/645415
Via Bartolini 29, 20155 Milano   Tel. 365083
La diffusione delle pubblicazioni OCSE viene
assicurata dalle principali librerie ed anche da :
Editrice e Libreria Herder,
Piazza Montecitorio 120, 00186 Roma
Tel. 6794628
Libreria Hœpli,
Via Hœpli 5, 20121 Milano   Tel. 865446
Libreria Scientifica
Dott. Lucio de Biasio "Aeiou"
Via Meravigli 16, 20123 Milano   Tel. 807679

JAPAN - JAPON
OECD Publications and Information Centre,
Landic Akasaka Bldg., 2-3-4 Akasaka,
Minato-ku, Tokyo 107   Tel. 586.2016

KOREA - CORÉE
Kyobo Book Centre Co. Ltd.
P.O.Box: Kwang Hwa Moon 1658,
Seoul   Tel. (REP) 730.78.91

LEBANON - LIBAN
Documenta Scientifica/Redico,
Edison Building, Bliss St.,
P.O.B. 5641, Beirut   Tel. 354429-344425

MALAYSIA/SINGAPORE -
MALAISIE/SINGAPOUR
University of Malaya Co-operative Bookshop
Ltd.,
7 Lrg 51A/227A, Petaling Jaya
Malaysia   Tel. 7565000/7565425
Information Publications Pte Ltd
Pei-Fu Industrial Building,
24 New Industrial Road No. 02-06
Singapore 1953   Tel. 2831786, 2831798

NETHERLANDS - PAYS-BAS
SDU Uitgeverij
Christoffel Plantijnstraat 2
Postbus 20014
2500 EA's-Gravenhage   Tel. 070-789911
Voor bestellingen: Tel. 070-789880

NEW ZEALAND - NOUVELLE-ZÉLANDE
Government Printing Office Bookshops:
Auckland: Retail Bookshop, 25 Rutland Stseet,
Mail Orders, 85 Beach Road
Private Bag C.P.O.
Hamilton: Retail: Ward Street,
Mail Orders, P.O. Box 857
Wellington: Retail, Mulgrave Street, (Head
Office)
Cubacade World Trade Centre,
Mail Orders, Private Bag
Christchurch: Retail, 159 Hereford Street,
Mail Orders, Private Bag
Dunedin: Retail, Princes Street,
Mail Orders, P.O. Box 1104

NORWAY - NORVÈGE
Narvesen Info Center – NIC,
Bertrand Narvesens vei 2,
P.O.B. 6125 Etterstad, 0602 Oslo 6
Tel. (02) 67.83.10, (02) 68.40.20

PAKISTAN
Mirza Book Agency
65 Shahrah Quaid-E-Azam, Lahore 3 Tel. 66839

PHILIPPINES
I.J. Sagun Enterprises, Inc.
P.O. Box 4322 CPO Manila
Tel. 695-1946, 922-9495

PORTUGAL
Livraria Portugal, Rua do Carmo 70-74,
1117 Lisboa Codex   Tel. 360582/3

SINGAPORE/MALAYSIA -
SINGAPOUR/MALAISIE
See "Malaysia/Singapor". Voir
«Malaisie/Singapour»

SPAIN - ESPAGNE
Mundi-Prensa Libros, S.A.,
Castelló 37, Apartado 1223, Madrid-28001
Tel. 431.33.99
Libreria Bosch, Ronda Universidad 11,
Barcelona 7   Tel. 317.53.08/317.53.58

SWEDEN - SUÈDE
AB CE Fritzes Kungl. Hovbokhandel,
Box 16356, S 103 27 STH,
Regeringsgatan 12,
DS Stockholm   Tel. (08) 23.89.00
Subscription Agency/Abonnements:
Wennergren-Williams AB,
Box 30004, S104 25 Stockholm Tel. (08)54.12.00

SWITZERLAND - SUISSE
OECD Publications and Information Centre,
4 Simrockstrasse,
5300 Bonn (Germany)   Tel. (0228) 21.60.45
Librairie Payot,
6 rue Grenus, 1211 Genève 11
Tel. (022) 31.89.50
Maditec S.A.
Ch. des Palettes 4
1020 – Renens/Lausanne   Tel. (021) 35.08.65
United Nations Bookshop/Librairie des Nations-
Unies
Palais des Nations, 1211 – Geneva 10
Tel. 022-34-60-11 (ext. 48 72)

TAIWAN - FORMOSE
Good Faith Worldwide Int'l Co., Ltd.
9th floor, No. 118, Sec.2, Chung Hsiao E. Road
Taipei   Tel. 391.7396/391.7397

THAILAND - THAILANDE
Suksit Siam Co., Ltd., 1715 Rama IV Rd.,
Samyam Bangkok 5   Tel. 2511630
INDEX Book Promotion & Service Ltd.
59/6 Soi Lang Suan, Ploenchit Road
Patjumamwan, Bangkok 10500
Tel. 250-1919, 252-1066

TURKEY - TURQUIE
Kültur Yayinlari Is-Türk Ltd. Sti.
Atatürk Bulvari No: 191/Kat. 21
Kavaklidere/Ankara   Tel. 25.07.60
Dolmabahce Cad. No: 29
Besiktas/Istanbul   Tel. 160.71.88

UNITED KINGDOM - ROYAUME-UNI
H.M. Stationery Office,
Postal orders only: (01)211-5656
P.O.B. 276, London SW8 5DT
Telephone orders: (01) 622.3316, or
Personal callers:
49 High Holborn, London WC1V 6HB
Branches at: Belfast, Birmingham,
Bristol, Edinburgh, Manchester

UNITED STATES - ÉTATS-UNIS
OECD Publications and Information Centre,
2001 L Street, N.W., Suite 700,
Washington, D.C. 20036 - 4095
Tel. (202) 785.6323

VENEZUELA
Libreria del Este,
Avda F. Miranda 52, Aptdo. 60337,
Edificio Galipan, Caracas 106
Tel. 951.17.05/951.23.07/951.12.97

YUGOSLAVIA - YOUGOSLAVIE
Jugoslovenska Knjiga, Knez Mihajlova 2,
P.O.B. 36, Beograd   Tel. 621.992

Orders and inquiries from countries where
Distributors have not yet been appointed should be
sent to:
OECD, Publications Service, 2, rue André-Pascal,
75775 PARIS CEDEX 16.

Les commandes provenant de pays où l'OCDE n'a
pas encore désigné de distributeur doivent être
adressées à :
OCDE, Service des Publications. 2, rue André-
Pascal, 75775 PARIS CEDEX 16.

72233-12-1988

LES ÉDITIONS DE L'OCDE, 2, rue André-Pascal, 75775 PARIS CEDEX 16 - Nº 44614 1989
IMPRIMÉ EN FRANCE
(66 89 02 3)   ISBN 92-64-03148-0

Proceedings of an NEA Workshop on

EXCAVATION RESPONSE IN GEOLOGICAL REPOSITORIES FOR RADIOACTIVE WASTE

Compte rendu d'une réunion de travail de l'AEN sur

L'INCIDENCE DES TRAVAUX D'EXCAVATION SUR LE COMPORTEMENT DES DÉPÔTS DE DÉCHETS RADIOACTIFS EN FORMATIONS GÉOLOGIQUES

WINNIPEG, CANADA
26-28 April/Avril 1988

Organised by the
OECD NUCLEAR ENERGY AGENCY
in co-operation with the
ATOMIC ENERGY OF CANADA LIMITED (AECL)

Organisée par
L'AGENCE DE L'OCDE POUR L'ÉNERGIE NUCLÉAIRE
en coopération avec
L'ÉNERGIE ATOMIQUE DU CANADA, LIMITÉE (EACL)

Pursuant to article 1 of the Convention signed in Paris on 14th December 1960, and which came into force on 30th September 1961, the Organisation for Economic Co-operation and Development (OECD) shall promote policies designed:

- to achieve the highest sustainable economic growth and employment and a rising standard of living in Member countries, while maintaining financial stability, and thus to contribute to the development of the world economy;
- to contribute to sound economic expansion in Member as well as non-member countries in the process of economic development; and
- to contribute to the expansion of world trade on a multilateral, non-discriminatory basis in accordance with international obligations.

The original Member countries of the OECD are Austria, Belgium, Canada, Denmark, France, the Federal Republic of Germany, Greece, Iceland, Ireland, Italy, Luxembourg, the Netherlands, Norway, Portugal, Spain, Sweden, Switzerland, Turkey, the United Kingdom and the United States. The following countries became Members subsequently through accession at the dates indicated hereafter: Japan (28th April 1964), Finland (28th January 1969), Australia (7th June 1971) and New Zealand (29th May 1973).

The Socialist Federal Republic of Yugoslavia takes part in some of the work of the OECD (agreement of 28th October 1961).

The OECD Nuclear Energy Agency (NEA) was established on 1st February 1958 under the name of the OEEC European Nuclear Energy Agency. It received its present designation on 20th April 1972, when Japan became its first non-European full Member. NEA membership today consists of all European Member countries of OECD as well as Australia, Canada, Japan and the United States. The commission of the European Communities takes part in the work of the Agency.

The primary objective of NEA is to promote co-operation among the governments of its participating countries in furthering the development of nuclear power as a safe, environmentally acceptable and economic energy source.

This is achieved by:

- *encouraging harmonisation of national regulatory policies and practices, with particular reference to the safety of nuclear installations, protection of man against ionising radiation and preservation of the environment, radioactive waste management, and nuclear third party liability and insurance;*
- *assessing the contribution of nuclear power to the overall energy supply by keeping under review the technical and economic aspects of nuclear power growth and forecasting demand and supply for the different phases of the nuclear fuel cycle;*
- *developing exchanges of scientific and technical information particularly through participation in common services;*
- *setting up international research and development programmes and joint undertakings.*

In these and related tasks, NEA works in close collaboration with the International Atomic Energy Agency in Vienna, with which it has concluded a Co-operation Agreement, as well as with other international organisations in the nuclear field.

En vertu de l'article 1er de la Convention signée le 14 décembre 1960, à Paris, et entrée en vigueur le 30 septembre 1961, l'Organisation de Coopération et de Développement Économiques (OCDE) a pour objectif de promouvoir des politiques visant :

- à réaliser la plus forte expansion de l'économie et de l'emploi et une progression du niveau de vie dans les pays Membres, tout en maintenant la stabilité financière, et à contribuer ainsi au développement de l'économie mondiale ;
- à contribuer à une saine expansion économique dans les pays Membres, ainsi que les pays non membres, en voie de développement économique ;
- à contribuer à l'expansion du commerce mondial sur une base multilatérale et non discriminatoire conformément aux obligations internationales.

Les pays Membres originaires de l'OCDE sont : la République Fédérale d'Allemagne, l'Autriche, la Belgique, le Canada, le Danemark, l'Espagne, les Etats-Unis, la France, la Grèce, l'Irlande, l'Italie, l'Islande, le Luxembourg, la Norvège, les Pays-Bas, le Portugal, le Royaume-Uni, la Suède, la Suisse et la Turquie. Les pays suivants sont ultérieurement devenus Membres par adhésion aux dates indiquées ci-après : le Japon (28 avril 1964), la Finlande (28 janvier 1969), l'Australie (7 juin 1971) et la Nouvelle-Zélande (29 mai 1973).

La République socialiste fédérative de Yougoslavie prend part à certains travaux de l'OCDE (accord du 28 octobre 1961).

L'Agence de l'OCDE pour l'Energie Nucléaire (AEN) a été créée le 1er février 1958 sous le nom d'Agence Européenne pour l'Énergie Nucléaire de l'OECE. Elle a pris sa dénomination actuelle le 20 avril 1972, lorsque le Japon est devenu son premier pays Membre de plein exercice non européen. L'Agence groupe aujourd'hui tous les pays Membres européens de l'OCDE, ainsi que l'Australie, le Canada, les États-Unis et le Japon. La Commission des Communautés européennes participe à ses travaux.

L'AEN a pour principal objectif de promouvoir la coopération entre les gouvernements de ses pays participants pour le développement de l'énergie nucléaire en tant que source d'énergie sûre, acceptable du point de vue de l'environnement, et économique.

Pour atteindre cet objectif, l'AEN :

- *encourage l'harmonisation des politiques et pratiques réglementaires notamment en ce qui concerne la sûreté des installations nucléaires, la protection de l'homme contre les rayonnements ionisants et la préservation de l'environnement, la gestion des déchets radioactifs, ainsi que la responsabilité civile et l'assurance en matière nucléaire ;*
- *évalue la contribution de l'électronucléaire aux approvisionnements en énergie, en examinant régulièrement les aspects économiques et techniques de la croissance de l'énergie nucléaire et en établissant des prévisions concernant l'offre et la demande de services pour les différentes phases du cycle du combustible nucléaire ;*
- *développe les échanges d'informations scientifiques et techniques notamment par l'intermédiaire de services communs ;*
- *met sur pied des programmes internationaux de recherche et développement, et des entreprises communes.*

Pour ces activités, ainsi que pour d'autres travaux connexes, l'AEN collabore étroitement avec l'Agence Internationale de l'Energie Atomique de Vienne, avec laquelle elle a conclu un Accord de coopération, ainsi qu'avec d'autres organisations internationales opérant dans le domaine nucléaire.

FOREWORD

Currently, one of the most important areas of radioactive waste
management is research and development directed towards assessment of the
performance of potential radioactive waste disposal systems. In particular,
in-situ research and investigations have become an integral and essential part
of national and international programmes for evaluation of geological
repositories, including assessment of the concept, site characterisation and
selection, and repository development. In the current programme of the NEA
Radioactive Waste Management Committee (RWMC), activities related to in-situ
research and investigations for geological repositories play a significant
role. In order to focus work on the most appropriate topics in this field, an
"Advisory Group on In-Situ Research and Investigations for Geological
Disposal" (ISAG) has been established by the RWMC in 1986 to provide relevant
guidance and advice. ISAG is composed of participants with key positions in
national waste management programmes, including the governmental and industry
organisations responsible for the programmes, in-situ research facilities and
regulatory agencies. The objective of the Group is to assist in the
co-ordination of in-situ research, investigation and demonstration activities
in Member countries by providing a forum for exchanging information and
planning joint initiatives at an international level.

One of the first tasks of ISAG was to conduct a review of current needs
in the area of in-situ research and investigations. A primary topic which was
identified concerned excavation response, that is, the potential effects that
may be induced by the construction and development of a geological repository
for radioactive waste. Specifically, it was noted that residual stresses,
potential creep or subsidence, and induced fractures leading to an increased
potential for groundwater flow, are phenomena which require analysis to
determine their influence on the safety performance of a repository, and on
the design of engineered barriers for the repository. All these aspects were
subsequently examined at a workshop organised by the NEA in Canada, with the
co-sponsorship of Atomic Energy of Canada Limited (AECL).

These proceedings reproduce the papers presented at the workshop,
together with a summary and conclusions prepared by the NEA Secretariat based
upon the main findings reported by the session Chairmen and the discussions
held at the workshop. The opinions expressed are those of the authors and do
not commit the Member countries of OECD.